Applied Mathematical Sciences
Volume 93

Applied Mathematical Sciences

(continued following index)

David Colton   Rainer Kress

Inverse Acoustic and Electromagnetic Scattering Theory

Springer-Verlag

Berlin Heidelberg New York London Paris
Tokyo Hong Kong Barcelona Budapest

David Colton
Department of Mathematical Sciences
University of Delaware
Newark, DE 19716, USA

Rainer Kress
Institut für Numerische und Angewandte Mathematik
Universität Göttingen
Lotzestraße 16–18
W-3400 Göttingen, Fed. Rep. of Germany

Editors

F. John	J. E. Marsden	L. Sirovich
Courant Institute of	Department of	Division of
Mathematical Sciences	Mathematics	Applied Mathematics
New York University	University of California	Brown University
New York, NY 10012	Berkeley, CA 94720	Providence, RI 02912
USA	USA	USA

Mathematics Subject Classification (1991):
35J05, 35P25, 35Q60, 35R25, 35R30, 45A05, 65M30, 65R20, 65R30, 76Q05, 78A45

With 6 Illustrations

ISBN 3-540-55518-8 Springer-Verlag Berlin Heidelberg New York
ISBN 0-387-55518-8 Springer-Verlag New York Berlin Heidelberg

Library of Congress Cataloging-in-Publication Data. Colton, David L. Inverse acoustic and electromagnetic scattering theory / David Colton, Rainer Kress. p. cm. – (Applied mathematical sciences; v. 93) Includes bibliographical references and index. ISBN 0-387-55518-8 (alk. paper) 1. Sound-waves-Scattering. 2. Electromagnetic waves-Scattering. 3. Inverse scattering transform. I. Kress, Rainer, 1941-. II. Title. III. Series: Applied mathematical sciences (Springer-Verlag New York Inc.); v. 93. QA1.A647 vol. 93 [QC243.3.S3] 510 s–dc20 [534'.2] 92-16437

© Springer-Verlag Berlin Heidelberg 1992
Printed in the United States of America

Typesetting: Camera ready by authors

41/3140-5 4 3 2 1 0 – Printed on acid-free paper

From Rainer for Marcus

As far as my eyes can see
There are shadows approaching me
And to those I left behind
I wanted you to know
You've always shared my deepest thoughts
I'll miss you when I go -
And oh when I'm old and wise
Bitter words mean little to me
Like autumn winds will blow right through me
And someday in the mist of time
When they asked me if I knew you
I'd smile and say you were a friend of mine
And the sadness would be lifted from my eyes

Alan Parsons

Preface

It has now been almost ten years since our first book on scattering theory appeared [32]. At that time we claimed that "in recent years the development of integral equation methods for the direct scattering problem seems to be nearing completion, whereas the use of such an approach to study the inverse scattering problem has progressed to an extent that a 'state of the art' survey appears highly desirable". Since we wrote these words, the inverse scattering problem for acoustic and electromagnetic waves has grown from being a few theoretical considerations with limited numerical implementations to a well developed mathematical theory with tested numerical algorithms. This maturing of the field of inverse scattering theory has been based on the realization that such problems are in general not only nonlinear but also improperly posed in the sense that the solution does not depend continuously on the measured data. This was emphasized in [32] and treated with the ideas and tools available at that time. Now, almost ten years later, these initial ideas have developed to the extent that a monograph summarizing the mathematical basis of the field seems appropriate. This book is our attempt to write such a monograph.

The inverse scattering problem for acoustic and electromagnetic waves can broadly be divided into two classes, the inverse obstacle problem and the inverse medium problem. In the inverse obstacle problem, the scattering object is a homogeneous obstacle with given boundary data and the inverse problem is to determine the obstacle from a knowledge of the scattered field at infinity, i.e., the far field pattern. The inverse medium problem, in its simplest form, is the situation when the scattering object is an inhomogeneous medium such that the constitutive parameters vary in a continuous manner and the inverse problem is to determine one or more of these parameters from the far field pattern. Only the inverse obstacle problem was considered in [32]. In this book we shall consider both the inverse obstacle and the inverse medium problem using two different methods. In the first method one looks for an obstacle or parameters whose far field pattern best fits the measured data whereas in the second method one looks for an obstacle or parameters whose far field pattern has the same weighted averages as the measured data. The theoretical and numerical development of these two methods for solving the inverse scattering problem for acoustic and electromagnetic waves is the basic subject matter of this book.

We make no claim to cover all the many topics in inverse scattering theory for acoustic and electromagnetic waves. Indeed, with the rapid growth of the field, such a task would be almost impossible in a single volume. In particular, we have emphasized the nonlinear and improperly posed nature of the inverse scattering problem and have paid only passing attention to the various linear methods which are applicable in certain cases. This view of inverse scattering theory has been arrived at through our work in collaboration with a number of mathematicians over the past ten years, in particular Thomas Angell, Peter Hähner, Andreas Kirsch, Ralph Kleinman, Peter Monk, Lassi Päivärinta, Lutz Wienert and Axel Zinn.

As with any book on mathematics, a basic question to answer is where to begin, i.e., what degree of mathematical sophistication is expected of the reader? Since the inverse scattering problem begins with the asymptotic behavior of the solution to the direct scattering problem, it seems reasonable to start with a discussion of the existence and uniqueness of a solution to the direct problem. We have done this for both the Helmholtz and the Maxwell equations. Included in our discussion is a treatment of the numerical solution of the direct problem. In addition to a detailed presentation of the direct scattering problem, we have also included as background material the rudiments of the theory of spherical harmonics, spherical Bessel functions, operator valued analytic functions and ill-posed problems (This last topic has been considerably expanded from the brief discussion given in [32]). As far as more general mathematical background is concerned, we assume that the reader has a basic knowledge of classical and functional analysis.

We have been helped by many peoble in the course of preparing this book. In particular, we would like to thank Wilhelm Grever, Rainer Hartke and Volker Walther for reading parts of the manuscript and Peter Hähner for his many valuable suggestions for improvements. Thanks also go to Ginger Moore for doing part of the typing. We would like to acknowledge the financial support of the Air Force Office of Scientific Research and the Deutsche Forschungsgemeinschaft, both for the long-term support of our research as well as for the funds made available to us for regular visits between Newark and Göttingen to nurture our collaboration. Finally, we want to give special thanks to our friends and colleagues Andreas Kirsch and Peter Monk. Many of the results of this book represent joint work with these two mathematicians and their insights, criticism and support have been an indispensible component of our research efforts.

Newark, Delaware   David Colton
Göttingen, Germany   Rainer Kress
January, 1992

Table of Contents

1. Introduction

The purpose of this chapter is to provide a survey of our book by placing what we have to say in a historical context. We obviously cannot give a complete account of inverse scattering theory in a book of only a few hundred pages, particularly since before discussing the inverse problem we have to give the rudiments of the theory of the direct problem. Hence, instead of attempting the impossible, we have chosen to present inverse scattering theory from the perspective of our own interests and research program. This inevitably means that certain areas of scattering theory are either ignored or given only cursory attention. In view of this fact, and in fairness to the reader, we have therefore decided to provide a few words at the beginning of our book to tell the reader what we are going to do, as well as what we are not going to do, in the forthcoming chapters.

Scattering theory has played a central role in twentieth century mathematical physics. Indeed, from Rayleigh's explanation of why the sky is blue, to Rutherford's discovery of the atomic nucleus, through the modern medical applications of computerized tomography, scattering phenomena have attracted, perplexed and challenged scientists and mathematicians for well over a hundred years. Broadly speaking, scattering theory is concerned with the effect an inhomogeneous medium has on an incident particle or wave. In particular, if the total field is viewed as the sum of an incident field u^i and a scattered field u^s then the *direct scattering problem* is to determine u^s from a knowledge of u^i and the differential equation governing the wave motion. Of possibly even more interest is the *inverse scattering problem* of determining the nature of the inhomogeneity from a knowledge of the asymptotic behavior of u^s, i.e., to reconstruct the differential equation and/or its domain of definition from the behavior of (many of) its solutions. The above oversimplified description obviously covers a huge range of physical concepts and mathematical ideas and for a sample of the many different approaches that have been taken in this area the reader can consult the monographs of Chadan and Sabatier [21], Colton and Kress [32], Jones [79], Lax and Phillips [116], Leis [119], Müller [142], Newton [146], Reed and Simon [160] and Wilcox [195].

1.1 The Direct Scattering Problem

The two basic problems in classical scattering theory (as opposed to quantum scattering theory) are the scattering of time-harmonic acoustic or electromagnetic waves by a penetrable inhomogeneous medium of compact support and by a bounded impenetrable obstacle. Considering first the case of acoustic waves, assume the incident field is given by the time-harmonic acoustic plane wave

$$u^i(x, t) = e^{i(k\,x\cdot d - \omega t)}$$

where $k = \omega/c_0$ is the wave number, ω the frequency, c_0 the speed of sound and d the direction of propagation. Then the simplest scattering problem for the case of an inhomogeneous medium is to find the total field u such that

$$(1.1) \qquad \triangle u + k^2 n(x) u = 0 \quad \text{in } \mathbb{R}^3,$$

$$(1.2) \qquad u(x) = e^{ik\,x\cdot d} + u^s(x),$$

$$(1.3) \qquad \lim_{r \to \infty} r \left(\frac{\partial u^s}{\partial r} - iku^s \right) = 0,$$

where $r = |x|$, $n = c_0^2/c^2$ is the refractive index given by the ratio of the square of the sound speeds, $c = c_0$ in the homogeneous host medium and $c = c(x)$ in the inhomogeneous medium and (1.3) is the *Sommerfeld radiation condition* which guarantees that the scattered wave is outgoing. It is assumed that $1 - n$ has compact support. If the medium is absorbing, n is complex valued and no longer is simply the ratio of the sound speeds. Turning now to the case of scattering by an impenetrable obstacle D, the simplest problem is to find the total field u such that

$$(1.4) \qquad \triangle u + k^2 u = 0 \quad \text{in } \mathbb{R}^3 \setminus \bar{D},$$

$$(1.5) \qquad u(x) = e^{ik\,x\cdot d} + u^s(x),$$

$$(1.6) \qquad u = 0 \quad \text{on } \partial D,$$

$$(1.7) \qquad \lim_{r \to \infty} r \left(\frac{\partial u^s}{\partial r} - iku^s \right) = 0,$$

where the differential equation (1.4) is the *Helmholtz equation* and the boundary condition (1.6) corresponds to a *sound-soft* obstacle. Boundary conditions other than (1.6) can also be considered, for example the Neumann or *sound-hard* boundary condition or the impedance boundary condition

$$\frac{\partial u}{\partial \nu} + i\lambda u = 0 \quad \text{on } \partial D$$

where ν is the unit outward normal to ∂D and λ is a positive constant. Although problems (1.1)–(1.3) and (1.4)–(1.7) are perhaps the simplest examples of physically realistic problems in acoustic scattering theory, they still cannot be considered completely solved, particularly from a numerical point of view, and remain the subject matter of much ongoing research.

Considering now the case of electromagnetic waves, assume the incident field is given by the (normalized) time-harmonic electromagnetic plane wave

$$E^i(x,t) = \frac{i}{k} \operatorname{curl} \operatorname{curl} p\, e^{i(k\,x\cdot d - \omega t)} = ik\,(d \times p) \times d\, e^{i(k\,x\cdot d - \omega t)},$$

$$H^i(x,t) = \operatorname{curl} p\, e^{i(k\,x\cdot d - \omega t)} = ik\, d \times p\, e^{i(k\,x\cdot d - \omega t)},$$

where $k = \omega\sqrt{\varepsilon_0 \mu_0}$ is the wave number, ω the frequency, ε_0 the electric permittivity, μ_0 the magnetic permeability, d the direction of propagation and p the polarization. Then the electromagnetic scattering problem corresponding to (1.1)–(1.3) (assuming variable permittivity but constant permeability) is to find the electric field E and magnetic field H such that

(1.8) $$\operatorname{curl} E - ikH = 0, \quad \operatorname{curl} H + ikn(x)E = 0 \quad \text{in } \mathbb{R}^3,$$

(1.9)
$$E(x) = \frac{i}{k} \operatorname{curl} \operatorname{curl} p\, e^{ik\,x\cdot d} + E^s(x),$$
$$H(x) = \operatorname{curl} p\, e^{ik\,x\cdot d} + H^s(x),$$

(1.10) $$\lim_{r\to\infty} (H^s \times x - rE^s) = 0,$$

where $n = \varepsilon/\varepsilon_0$ is the refractive index given by the ratio of the permittivity $\varepsilon = \varepsilon(x)$ in the inhomogeneous medium and ε_0 in the homogeneous host medium and where (1.10) is the *Silver–Müller radiation condition*. It is again assumed that $1 - n$ has compact support and if the medium is conducting then n is complex valued. Similarly, the electromagnetic analogue of (1.4)–(1.7) is scattering by a perfectly conducting obstacle D which can be mathematically formulated as the problem of finding an electromagnetic field E, H such that

(1.11) $$\operatorname{curl} E - ikH = 0, \quad \operatorname{curl} H + ikE = 0 \quad \text{in } \mathbb{R}^3 \setminus \bar{D},$$

(1.12)
$$E(x) = \frac{i}{k} \operatorname{curl} \operatorname{curl} p\, e^{ik\,x\cdot d} + E^s(x),$$
$$H(x) = \operatorname{curl} p\, e^{ik\,x\cdot d} + H^s(x),$$

(1.13) $$\nu \times E = 0 \quad \text{on } \partial D,$$

(1.14) $$\lim_{r\to\infty} (H^s \times x - rE^s) = 0,$$

where (1.11) are the time-harmonic *Maxwell equations* and ν is again the unit outward normal to ∂D. As in the case of (1.4)–(1.7), more general boundary conditions than (1.13) can also be considered, for example the impedance boundary condition

$$\nu \times \operatorname{curl} E - i\lambda (\nu \times E) \times \nu = 0 \quad \text{on } \partial D$$

where λ is again a positive constant.

This book is primarily concerned with the inverse scattering problems associated with the direct scattering problems formulated above. However, before we can consider the inverse problems, we must say more about the direct problems. The mathematical methods used to investigate the direct scattering problems for acoustic and electromagnetic waves depend heavily on the frequency of the wave motion. In particular, if the wavelength $\lambda = 2\pi/k$ is very small compared with the smallest distance which can be observed with the available apparatus, the scattering obstacle produces a shadow with an apparently sharp edge. Closer examination reveals that the edge of the shadow is not sharply defined but breaks up into fringes. This phenomenon is known as diffraction. At the other end of the scale, obstacles which are small compared with the wavelength disrupt the incident wave without producing an identifiable shadow. Hence, we can distinguish two different frequency regions corresponding to the magnitude of ka where a is a typical dimension of the scattering object. More specifically, the set of values of k such that $ka \gg 1$ is called the *high frequency region* whereas the set of values of k such that ka is less than or comparable to unity is called the *resonance region*. As suggested by the observed physical differences, the mathematical methods used to study scattering phenomena in the resonance region differ sharply from those used in the high frequency region. Because of this reason, as well as our own mathematical preferences, we have decided that in this book we will be primarily concerned with scattering problems in the resonance region.

The first question to ask about the direct scattering problem is that of the uniqueness of a solution. The basic tools used to establish uniqueness are Green's theorems and the unique continuation property of solutions to elliptic equations. Since equations (1.4) and (1.11) have constant coefficients, the uniqueness question for problems (1.4)–(1.7) and (1.11)–(1.14) are the easiest to handle, with the first results being given by Sommerfeld in 1912 for the case of acoustic waves [169]. Sommerfeld's work was subsequently generalized by Rellich [161] and Vekua [179], all under the assumption that $\operatorname{Im} k \geq 0$. The corresponding uniqueness result for problem (1.11)–(1.14) was first established by Müller [136]. The uniqueness of a solution to the scattering problems (1.1)–(1.3) and (1.8)–(1.10) is more difficult since use must now be made of the unique continuation principle for elliptic equations with variable, but non-analytic, coefficients. The first results in this direction were given by Müller [136], again for the case $\operatorname{Im} k \geq 0$. When $\operatorname{Im} k < 0$ then for each of the above problems there can exist values of k for which uniqueness no longer holds. Such values of k are called *resonance states* and are intimately involved with the asymptotic behavior of the time dependent wave equation. Although we shall not treat resonance states in this book, the subject is of considerable interest and we refer the reader to Dolph [52] and Lax and Phillips [116] for further information.

Having established uniqueness, the next question to turn to is the existence and numerical approximation of the solution. The most popular approach to existence has been through the method of integral equations. In particular, for problem (1.1)–(1.3), it is easily verified that for all positive values of k the total field u is the unique solution of the *Lippmann–Schwinger equation*

$$(1.15) \qquad u(x) = e^{ik\,x\cdot d} - k^2 \int_{\mathbb{R}^3} \Phi(x,y)m(y)u(y)\,dy, \quad x \in \mathbb{R}^3,$$

where $m := 1 - n$ and

$$\Phi(x,y) := \frac{1}{4\pi} \frac{e^{ik|x-y|}}{|x-y|}, \quad x \neq y,$$

is the fundamental solution to the Helmholtz equation. The corresponding integral equation for (1.8)–(1.10) is also easily obtained and is given by

$$E(x) = \frac{i}{k} \operatorname{curl} \operatorname{curl} p\, e^{ik\,x\cdot d} - k^2 \int_{\mathbb{R}^3} \Phi(x,y)m(y)E(y)\,dy$$

$$(1.16)$$

$$+ \operatorname{grad} \int_{\mathbb{R}^3} \frac{1}{n(y)} \operatorname{grad} n(y) \cdot E(y)\,\Phi(x,y)\,dy, \quad x \in \mathbb{R}^3,$$

where again $m := 1 - n$ and, if E is the solution of (1.16), we define

$$H(x) := \frac{1}{ik} \operatorname{curl} E(x).$$

The application of integral equation methods to problems (1.4)–(1.7) and (1.11)–(1.14) is more subtle. To see why this is so, suppose, by analogy to Laplace's equation, we look for a solution of problem (1.4)–(1.7) in the form of a double-layer potential

$$(1.17) \qquad u^s(x) = \int_{\partial D} \varphi(y) \frac{\partial \Phi(x,y)}{\partial \nu(y)}\,ds(y), \quad x \in \mathbb{R}^3 \setminus \bar{D},$$

where φ is a continuous density to be determined. Then, letting x tend to the boundary ∂D, it can be shown that φ must be a solution of a boundary integral equation of the second kind in order to obtain a solution of (1.4)–(1.7). Unfortunately, this integral equation is not uniquely solvable if k^2 is a Neumann eigenvalue of the negative Laplacian in D. Similar difficulties occur if we look for a solution of problem (1.11)–(1.14) in the form

$$E^s(x) = \operatorname{curl} \int_{\partial D} a(y)\Phi(x,y)\,ds(y), \quad x \in \mathbb{R}^3 \setminus \bar{D},$$

$$(1.18)$$

$$H^s(x) = \frac{1}{ik} \operatorname{curl} E^s(x), \quad x \in \mathbb{R}^3 \setminus \bar{D},$$

where a is a tangential density to be determined. These difficulties in acoustic and electromagnetic scattering theory were first resolved by Vekua [179], Weyl

[190, 191] and Müller [137, 139, 140]. A more satisfying approach to this problem was initiated by Werner [184, 185] who suggested modifying the representations (1.17) and (1.18) to include further source terms. This idea was further developed by Leis [118], Brakhage and Werner [17], Panich [152], Knauff and Kress [99] and Kress [102, 104] among others. Finally, we note that the numerical solution of boundary integral equations in scattering theory is fraught with difficulties and we refer the reader to Sections 3.5 and 3.6 of this book and to the survey article by Kress [108] for more information on this topic.

Of particular interest in scattering theory is the *far field pattern*, or scattering amplitude, of the scattered acoustic or electromagnetic wave. More specifically, if u^s is the scattered field of (1.1)–(1.3) or (1.4)–(1.7) then u^s has the asymptotic behavior

$$u^s(x) = \frac{e^{ikr}}{r} u_\infty(\hat{x}; d) + O\left(\frac{1}{r^2}\right), \quad r = |x| \to \infty,$$

where $\hat{x} = x/|x|$ and u_∞ is the far field pattern of u^s. Similarly, if E^s, H^s is the scattered field of (1.8)–(1.10) or (1.11)–(1.14), then E^s has the asymptotic behavior

$$E^s(x) = \frac{e^{ikr}}{r} E_\infty(\hat{x}; d, p) + O\left(\frac{1}{r^2}\right), \quad r = |x| \to \infty,$$

where E_∞ is the electric far field pattern of E^s. The interest in far field patterns lies in the fact that the basic inverse problem in scattering theory, and, in fact, the one we shall consider in this book, is to determine either n or D from a knowledge of u_∞ or E_∞ for $\hat{x}$ and d on the unit sphere. Until recently, very little was known concerning the mathematical properties of far field patterns, other than the fact that they are analytic functions of their independent variables. However, in the past few years results have been obtained concerning the completeness properties of far field patterns considered as functions of $\hat{x}$ in $L^2(\Omega)$ where $L^2(\Omega)$ is the space of square integrable functions on the unit sphere Ω. Since these results are of particular relevance to a new method developed by Colton and Monk for solving the inverse scattering problem, we shall now briefly consider the ideas involved in this investigation.

The problem of completeness of far field patterns was first considered by Colton and Kirsch [26] for the case of problem (1.4)–(1.7). In particular, they showed that if $\{d_n : n = 1, 2, \ldots\}$ is a dense set of vectors on the unit sphere Ω then the set $\{u_\infty(\cdot\,; d_n) : n = 1, 2, \ldots\}$ is complete in $L^2(\Omega)$ if and only if k^2 is not an eigenvalue of the interior Dirichlet problem for the negative Laplacian in D or, if k^2 is an eigenvalue, none of the eigenfunctions is a *Herglotz wave function*, i.e., a solution v of the Helmholtz equation in $\mathbb{R}^3$ such that

$$\sup_{R>0} \frac{1}{R} \int_{|x|\leq R} |v(x)|^2 dx < \infty.$$

This result was extended to the case of problem (1.11)–(1.14) by Colton and Kress [33] who also introduced the concept of *electromagnetic Herglotz pairs* which are the analogue for the Maxwell equations of Herglotz wave functions for the Helmholtz equation. Further generalizations were given by Gilbert and Xu [57, 58] who considered sound waves in an ocean of finite depth. The completeness of far field patterns for problem (1.1)–(1.3) is more complicated and was first studied by Kirsch [86, 87], Colton and Monk [39] and Colton, Kirsch and Päivärinta [30]. The result is that the far field patterns are complete provided there does not exist a nontrivial solution of the *interior transmission problem*

$$(1.19) \qquad \triangle\, w + k^2 n(x)w = 0, \quad \triangle v + k^2 v = 0 \quad \text{in } D,$$

$$(1.20) \qquad w = v, \quad \frac{\partial w}{\partial \nu} = \frac{\partial v}{\partial \nu} \quad \text{on } \partial D,$$

such that v is a Herglotz wave function where $D := \{x \in \mathbb{R}^3 : m(x) \neq 0\}$ and again $m := 1 - n$. The generalization of this result to problem (1.8)–(1.10) was given by Colton and Päivärinta [45, 46], whereas the inverse spectral problem associated with (1.19), (1.20) has recently been considered by McLaughlin and Polyakov [128].

1.2 The Inverse Scattering Problem

As indicated above, the direct scattering problem has been thoroughly investigated and a considerable amount of information is available concerning its solution. In contrast, the inverse scattering problem has only recently progressed from a collection of ad hoc techniques with little rigorous mathematical basis to an area of intense activity with at least the beginnings of a solid mathematical foundation. The reason for this is that the inverse scattering problem is inherently nonlinear and, more seriously from the point of view of numerical computations, improperly posed. In particular, small perturbations of the far field pattern in any reasonable norm lead to a function which lies outside the class of far field patterns and, unless regularization methods are used, small variations in the measured data can lead to large errors in the reconstruction of the scatterer. Nevertheless, the inverse scattering problem is basic in areas such as radar, sonar, geophysical exploration, medical imaging and nondestructive testing. Indeed, it is safe to say that the inverse problem is at least of equal interest as the direct problem and, armed with a knowledge of the direct scattering problem, is currently in the foreground of mathematical research in scattering theory.

As with the direct scattering problem, the first question to ask about the inverse scattering problem is uniqueness. The first result in this direction was due to Schiffer (see Lax and Phillips [116]) who showed that for problem (1.4)–(1.7) the far field pattern $u_\infty(\hat{x}; d)$ for all $\hat{x}, d \in \Omega$ and k fixed uniquely determines the scattering obstacle D. The corresponding result for problem (1.1)–(1.3) was recently obtained by Nachman [144], Novikov [148] and Ramm [158]. Uniqueness theorems for the electromagnetic problems (1.8)–(1.10) and (1.11)–(1.14) were obtained while this book was being written (c.f. Section 7.1 of this book and Colton and Päivärinta [47]). Closely related to the uniqueness theorem for the inverse scattering problem is Karp's theorem [82], which states for problem (1.4)–(1.7) that if $u_\infty(\hat{x}; d) = u_\infty(Q\hat{x}; Qd)$ for all rotations Q and all $\hat{x}, d \in \Omega$ then D is a ball centered at the origin. The analogues of this result for problems (1.1)–(1.3) and (1.11)–(1.14) have been given by Colton and Kirsch [27] and Colton and Kress [34] (see also Ramm [159]).

Turning to the question of the existence of a solution to the inverse scattering problem, we first note that this is the wrong question to ask. This is due to the observations made above that the inverse scattering problem is improperly posed, i.e., in any realistic situation the measured data is not exact and hence a solution to the inverse scattering problem does not exist. The proper question to ask is how can the inverse problem be stabilized and approximate solutions found to the stabilized problem. (However, for a different point of view, see Newton [147].) Initial efforts in this direction attempted to linearize the problem by reducing it to the problem of solving a linear integral equation of the first kind. The main techniques used to accomplish this were the Born or Rytov approximation for problems (1.1)–(1.3) and (1.8)–(1.10) and the Kirchhoff, or physical optics, approximation to the solution of problems (1.4)–(1.7) and (1.11)–(1.14). While such linearized models are attractive because of their mathematical simplicity, they have the defect of ignoring the basic nonlinear nature of the inverse scattering problem, e.g., multiple reflections are essentially ignored. For detailed presentations of the linearized approach to inverse scattering theory, including a glimpse at the practical applications of such an approach, we refer the reader to Bleistein [13], Chew [22], Devaney [51] and Langenberg [114].

The earliest attempts to treat the inverse scattering problem without linearizing it were due to Imbriale and Mittra [71] for problem (1.4)–(1.7) and Weston and Boerner [189] for problem (1.11)–(1.14). Their methods were based on analytic continuation with little attention being given to issues of stabilization. Then, beginning in the 1980's, a number of methods were given for solving the inverse scattering problem which explicitly acknowledged the nonlinear and ill-posed nature of the problem. In particular, the *acoustic inverse obstacle problem* of determining ∂D in (1.1)–(1.3) from a knowledge of the far field data u_∞ was considered by Angell, Colton and Kirsch [2, 86], Roger [162], Kristensson and Vogel [112], Tobocman [175], Murch, Tan and Wall [143], Onishi, Ohura and Kobayashi [151] and Wang and Chen [180], among others. Each of these researchers used either integral equations or Green's formulas to reformulate the inverse obstacle problem as a nonlinear optimization problem that required the solution of the direct scattering problem for different domains at each step of the iteration procedure used to arrive at a solution. Two approaches which avoid the problem of solving a direct scattering problem at each iteration step and, furthermore, attempt to separate the inverse obstacle problem into a linear ill-posed part and a nonlinear well-posed part were introduced and theoretically and numerically analysed by Kirsch and Kress in [92, 93, 94] and Colton and Monk in [36, 37, 38]. A method that is closely related to the approach suggested by Kirsch and Kress was also introduced by Angell, Kleinman and Roach [8] and numerically investigated by Angell, Kleinman, Kok and Roach [6, 7] and by Jones and Mao [80, 81]. We shall now briefly outline the main ideas of these last two approaches to the inverse obstacle problem.

In the approach due to Kirsch and Kress, we assume a priori that enough information is known about the unknown scattering obstacle D so we can place a surface Γ inside D such that k^2 is not a Dirichlet eigenvalue of the negative Laplacian for the interior of Γ. For fixed wave number k and fixed incident direction d, we then try to represent the scattered field u^s as a single-layer potential

$$(1.21) \qquad u^s(x) = \int_\Gamma \varphi(y)\Phi(x,y)\,ds(y)$$

with unknown density $\varphi \in L^2(\Gamma)$ and note that in this case the far field pattern u_∞ has the representation

$$(1.22) \qquad u_\infty(\hat{x}; d) = \frac{1}{4\pi} \int_\Gamma e^{-ik\,\hat{x}\cdot y}\varphi(y)\,ds(y), \quad \hat{x} \in \Omega.$$

Given the (measured) far field pattern u_∞ the density φ is now found by solving the ill-posed integral equation of the first kind (1.22). The unknown boundary ∂D is then determined by requiring (1.21) to assume the boundary data (1.5), (1.6) on the surface ∂D. For example, if we assume that ∂D is starlike, i.e., $x(a) = r(a)\,a$ for $x \in \partial D$ and $a \in \Omega$, then this last requirement means solving the nonlinear equation

$$(1.23) \qquad e^{ikr(a)\,a\cdot d} + \int_\Gamma \varphi(y)\Phi(r(a)\,a, y)\,ds(y) = 0, \quad a \in \Omega,$$

for the unknown function r. Numerical examples of the use of this method in three dimensions have recently been provided by Kress and Zinn [109, 110, 111].

The method introduced by Colton and Monk for solving the inverse obstacle problem can be viewed as the dual space analogue of the method of Kirsch and Kress in the sense that we now try to determine a linear functional having prescribed values on the set of far field patterns corresponding to arbitrary directions of the incident field. We assume that the unknown scattering obstacle contains the origin and, given the (measured) far field u_∞ for all $\hat{x}, d \in \Omega$, begin by determining a function $g \in L^2(\Omega)$ such that, for k fixed and p an integer,

$$(1.24) \qquad \int_\Omega u_\infty(\hat{x}; d)g(d)\,ds(d) = \frac{1}{ki^{p+1}} Y_p(\hat{x}), \quad \hat{x} \in \Omega,$$

where Y_p is a spherical harmonic of order p. (By reciprocity, the integration in (1.24) can be replaced by an integration with respect to $\hat{x}$.) If we then define the Herglotz wave function v by

$$(1.25) \qquad v(x) = \int_\Omega e^{ik\,x\cdot d}g(d)\,ds(d), \quad x \in \mathbb{R}^3,$$

then it can be shown that

$$(1.26) \qquad v(x) = -h_p^{(1)}(k|x|) Y_p(\hat{x}), \quad x \in \partial D,$$

where $h_p^{(1)}$ is the spherical Hankel function of the first kind of order p and again $\hat{x} = x/|x|$. The solution of the inverse scattering problem is now obtained by finding ∂D such that (1.26) is satisfied. For example, if ∂D is starlike, i.e., $x(a) = r(a) a$ for $x \in \partial D$ and $a \in \Omega$, then, after solving the ill-posed integral equation (1.24) for g, we must solve the nonlinear equation

$$(1.27) \qquad \int_\Omega e^{ikr(a) a \cdot d} g(d) \, ds(d) = -h_p^{(1)}(kr(a)) Y_p(a), \quad a \in \Omega,$$

for the unknown function r. In addition to Colton and Monk [37, 38], numerical examples of the use of this method have recently been given by Ari and Firth [9] and by Misici and Zirilli [130].

The extension of the above methods to the case of the *electromagnetic inverse obstacle problem* was carried out by Blöhbaum [14] and the first numerical reconstructions were done by Maponi, Misici and Zirilli [124] for axially symmetric obstacles using the method of Colton and Monk. However, at the time of this writing, the numerical solution of the electromagnetic inverse obstacle problem lags well behind the corresponding acoustic case. Hopefully this situation will be rectified in the not too distant future.

The *inverse medium problem* is to determine n in (1.1)–(1.3) or (1.8)–(1.10) from a knowledge of the far field pattern u_∞ or E_∞. As with the inverse obstacle problem, the scalar problem (1.1)–(1.3) has received the most attention. Essentially two methods have been proposed for solving the (nonlinear) acoustic inverse medium problem. The first method is based on noting that from (1.15) we have

$$(1.28) \qquad u_\infty(\hat{x}; d) = -\frac{k^2}{4\pi} \int_{\mathbb{R}^3} e^{-ik \hat{x} \cdot y} m(y) u(y) \, dy, \quad \hat{x} \in \Omega,$$

and then seeking a solution m and u of (1.15) such that the constraint (1.28) is satisfied. This is done by reformulating (1.15) and (1.28) as a nonlinear optimization problem subject to a priori constraints on m such that the optimization problem has a solution that depends continuously on the data. Variations of this approach have been used by Johnson and Tracy [77, 176], Kleinman and van den Berg [98], Tabbara and his co-workers [171], Wang and Chew [181] and Weston [187, 188], among others. There are, of course, important differences in the practical implementation in each of these efforts, for example the far field constraint (1.28) is sometimes replaced by an analogous near field condition and the optimization scheme is numerically solved by different methods, e.g., successive over-relaxation, sinc basis moment methods, steepest descent, etc.

A second method for solving the acoustic inverse medium problem was introduced by Colton and Monk [40, 41] and, as in the case of the Colton and Monk approach for the inverse obstacle problem, can be viewed as a dual space method for approaching the inverse medium problem. The first version of this method [40] begins by determining a function g satisfying (1.24) (where now u_∞ is the far field pattern corresponding to problem (1.1)–(1.3)) and then constructing the Herglotz wave function v defined by (1.25). The solution of the inverse medium problem is then found by looking for a solution m and w of the Lippmann–Schwinger equation

$$(1.29) \qquad w(x) = v(x) - k^2 \int_{\mathbb{R}^3} \Phi(x,y)m(y)w(y)\,dy, \qquad x \in \mathbb{R}^3,$$

such that m and w satisfy the constraint

$$(1.30) \qquad -k^2 \int_{\mathbb{R}^3} \Phi(x,y)m(y)w(y)\,dy = h_p^{(1)}(k|x|)\,Y_p(\hat{x})$$

for $|x| = a$ where a is the radius of a ball B centered at the origin and containing the support of m. We note that if m is real valued, difficulties can occur in the numerical implementation of this method due to the presence of *transmission eigenvalues*, i.e., values of k such that there exists a nontrivial solution of (1.19), (1.20). The second version of the method of Colton and Monk [41, 42] is designed to overcome this problem by replacing the integral equation (1.24) by

$$(1.31) \qquad \int_\Omega [u_\infty(\hat{x}; d) - h_\infty(\hat{x}; d)]\,g(\hat{x})\,ds(\hat{x}) = \frac{i^{p-1}}{k}\,Y_p(d), \quad d \in \Omega,$$

where h_∞ is the (known) far field pattern corresponding to an exterior impedance boundary value problem for the Helmholtz equation in the exterior of the ball B and replacing the constraint (1.30) by

$$(1.32) \qquad \left(\frac{\partial}{\partial\nu} + i\lambda\right)(w(x) - h_p^{(1)}(k|x|)\,Y_p(\hat{x})) = 0, \quad x \in \partial B,$$

where λ is the impedance. A numerical comparison of the two versions of the method due to Colton and Monk can be found in [43].

As previously mentioned, the electromagnetic inverse medium problem is not as thoroughly investigated as the corresponding acoustic case. In particular, although the theoretical basis of the dual space method for electromagnetic waves has been developed by Colton and Päivärinta [45] and Colton and Kress [35], numerical experiments have yet to be done for the three dimensional electromagnetic inverse medium problem using any of the above methods. An important goal of future research in inverse scattering theory is to overcome this defect and hopefully this book will encourage efforts in this direction.

2. The Helmholtz Equation

Studying an inverse problem always requires a solid knowledge of the theory for the corresponding direct problem. Therefore, the following two chapters of our book are devoted to presenting the foundations of obstacle scattering problems for time-harmonic acoustic waves, that is, to exterior boundary value problems for the scalar Helmholtz equation. Our aim is to develop the analysis for the direct problems to an extent which is needed in the subsequent chapters on inverse problems.

In this chapter we begin with a brief discussion of the physical background to scattering problems. We will then derive the basic Green representation theorems for solutions to the Helmholtz equation. Discussing the concept of the Sommerfeld radiation condition will already enable us to introduce the idea of the far field pattern which is of central importance in our book. For a deeper understanding of these ideas, we require sufficient information on spherical wave functions. Therefore, we present in two sections those basic properties of spherical harmonics and spherical Bessel functions that are relevant in scattering theory. We will then be able to derive uniqueness results and expansion theorems for solutions to the Helmholtz equation with respect to spherical wave functions. We also will gain a first insight into the ill-posedness of the inverse problem by examining the smoothness properties of the far field pattern. The study of the boundary value problems will be the subject of the next chapter.

2.1 Acoustic Waves

Consider the propagation of sound waves of small amplitude in a homogeneous isotropic medium in $\mathbb{R}^3$ viewed as an inviscid fluid. Let $v = v(x, t)$ be the velocity field and let $p = p(x, t)$, $\rho = \rho(x, t)$ and $S = S(x, t)$ denote the pressure, density and specific entropy, respectively, of the fluid. The motion is then governed by *Euler's equation*

$$\frac{\partial v}{\partial t} + (v \cdot \text{grad}) v + \frac{1}{\rho} \, \text{grad} \, p = 0,$$

the *equation of continuity*

$$\frac{\partial \rho}{\partial t} + \text{div}(\rho v) = 0,$$

the *state equation*

$$p = f(\rho, S),$$

and the *adiabatic hypothesis*

$$\frac{\partial S}{\partial t} + v \cdot \operatorname{grad} S = 0,$$

where f is a function depending on the nature of the fluid. We assume that v, p, ρ and S are small perturbations of the static state $v_0 = 0$, $p_0 = $ constant, $\rho_0 = $ constant and $S_0 = $ constant and linearize to obtain the linearized Euler equation

$$\frac{\partial v}{\partial t} + \frac{1}{\rho_0} \operatorname{grad} p = 0,$$

the linearized equation of continuity

$$\frac{\partial \rho}{\partial t} + \rho_0 \operatorname{div} v = 0,$$

and the linearized state equation

$$\frac{\partial p}{\partial t} = \frac{\partial f}{\partial \rho}(\rho_0, S_0)\, \frac{\partial \rho}{\partial t}\,.$$

From this we obtain the *wave equation*

$$\frac{1}{c^2}\frac{\partial^2 p}{\partial t^2} = \Delta p$$

where the *speed of sound* c is defined by

$$c^2 = \frac{\partial f}{\partial \rho}(\rho_0, S_0).$$

From the linearized Euler equation, we observe that there exists a velocity potential $U = U(x, t)$ such that

$$v = \frac{1}{\rho_0} \operatorname{grad} U$$

and

$$p = -\frac{\partial U}{\partial t}\,.$$

Clearly, the velocity potential also satisfies the wave equation

$$\frac{1}{c^2}\frac{\partial^2 U}{\partial t^2} = \Delta U.$$

For time-harmonic acoustic waves of the form

$$U(x, t) = \operatorname{Re}\left\{ u(x)\, e^{-i\omega t} \right\}$$

with frequency $\omega > 0$, we deduce that the complex valued space dependent part u satisfies the *reduced wave equation* or *Helmholtz equation*

$$\triangle u + k^2 u = 0$$

where the *wave number* k is given by the positive constant $k = \omega/c$. This equation carries the name of the physicist Hermann Ludwig Ferdinand von Helmholtz (1821 – 1894) for his contributions to mathematical acoustics and electromagnetics.

In the first part of this book we will be concerned with the scattering of time-harmonic waves by obstacles surrounded by a homogeneous medium, that is, with exterior boundary value problems for the Helmholtz equation. However, studying the Helmholtz equation in some detail is also required for the second part of our book where we consider wave scattering from an inhomogeneous medium since we always will assume that the medium is homogeneous outside some sufficiently large sphere.

In obstacle scattering we must distinguish between the two cases of impenetrable and penetrable objects. For a *sound-soft* obstacle the pressure of the total wave vanishes on the boundary. Consider the scattering of a given incoming wave u^i by a sound-soft obstacle D. Then the total wave $u = u^i + u^s$, where u^s denotes the scattered wave, must satisfy the wave equation in the exterior $\mathbb{R}^3 \setminus \bar{D}$ of D and a Dirichlet boundary condition $u = 0$ on ∂D. Similarly, the scattering from *sound-hard* obstacles leads to a Neumann boundary condition $\partial u/\partial \nu = 0$ on ∂D where ν is the unit outward normal to ∂D since here the normal velocity of the acoustic wave vanishes on the boundary. More generally, allowing obstacles for which the normal velocity on the boundary is proportional to the excess pressure on the boundary leads to an *impedance boundary condition* of the form

$$\frac{\partial u}{\partial \nu} + i\lambda u = 0 \quad \text{on } \partial D$$

with a positive constant λ.

The scattering by a penetrable obstacle D with constant density ρ_D and speed of sound c_D differing from the density ρ and speed of sound c in the surrounding medium $\mathbb{R}^3 \setminus \bar{D}$ leads to a transmission problem. Here, in addition to the superposition $u = u^i + u^s$ of the incoming wave u^i and the scattered wave u^s in $\mathbb{R}^3 \setminus \bar{D}$ satisfying the Helmholtz equation with wave number $k = \omega/c$, we also have a transmitted wave v in D satisfying the Helmholtz equation with wave number $k_D = \omega/c_D \neq k$. The continuity of the pressure and of the normal velocity across the interface leads to the *transmission conditions*

$$u = v, \quad \frac{1}{\rho} \frac{\partial u}{\partial \nu} = \frac{1}{\rho_D} \frac{\partial v}{\partial \nu} \quad \text{on } \partial D.$$

Recently, in addition to the transmission conditions, more general *resistive boundary conditions* have been introduced and applied. For their description and treatment we refer to [5].

In order to avoid repeating ourselves by considering all possible types of boundary conditions, we have decided to confine ourselves to working out the basic ideas only for the case of a sound-soft obstacle. On occasion, we will mention modifications and extensions to the other cases.

For the scattered wave u^s, the radiation condition

$$\lim_{r \to \infty} r \left(\frac{\partial u^s}{\partial r} - iku^s \right) = 0, \quad r = |x|,$$

introduced by Sommerfeld [169] in 1912 will ensure uniqueness for the solutions to the scattering problems. From the two possible spherically symmetric solutions

$$\frac{e^{ik|x|}}{|x|} \quad \text{and} \quad \frac{e^{-ik|x|}}{|x|}$$

to the Helmholtz equation, only the first one satisfies the radiation condition. Since via

$$\text{Re} \left\{ \frac{e^{ik|x|-i\omega t}}{|x|} \right\} = \frac{\cos(k|x| - \omega t)}{|x|}$$

this corresponds to an outgoing spherical wave, we observe that physically speaking the *Sommerfeld radiation condition* characterizes outgoing waves. Throughout the book by $|x|$ we denote the Euclidean norm of a point x in $\mathbb{R}^3$.

For more details on the physical background of linear acoustic waves, we refer to the article by Morse and Ingard [135] in the Encyclopedia of Physics and to Jones [79] and Werner [183].

2.2 Green's Theorem and Formula

We begin by giving a brief outline of some basic properties of solutions to the Helmholtz equation $\Delta u + k^2 u = 0$ with positive wave number k. Most of these can be deduced from the *fundamental solution*

$$(2.1) \qquad \Phi(x, y) := \frac{1}{4\pi} \frac{e^{ik|x-y|}}{|x-y|}, \quad x \neq y.$$

Straightforward differentiation shows that for fixed $y \in \mathbb{R}^3$ the fundamental solution satisfies the Helmholtz equation in $\mathbb{R}^3 \setminus \{y\}$.

A domain $D \subset \mathbb{R}^3$, i.e., an open and connected set, is said to be of *class C^k*, $k \in \mathbb{N}$, if for each point z of the boundary ∂D there exists a neighborhood V_z of z with the following properties: the intersection $V_z \cap \bar{D}$ can be mapped bijectively onto the half ball $\{x \in \mathbb{R}^3 : |x| < 1, x_3 \geq 0\}$, this mapping and its inverse are k-times continuously differentiable and the intersection $V_z \cap \partial D$ is mapped onto the disk $\{x \in \mathbb{R}^3 : |x| < 1, x_3 = 0\}$. On occasion, we will express the property of a domain D to be of class C^k also by saying that its boundary ∂D is of class C^k. By $C^k(D)$ we denote the linear space of real or complex valued functions defined on the domain D which are k–times continuously differentiable. By $C^k(\bar{D})$ we denote the subspace of all functions in $C^k(D)$ which together with all their derivatives up to order k can be extended continuously from D into the closure $\bar{D}$.

One of the basic tools in studying the Helmholtz equation is provided by Green's integral theorems. Let D be a bounded domain of class C^1 and let ν denote the unit normal vector to the boundary ∂D directed into the exterior of D. Then, for $u \in C^1(\bar{D})$ and $v \in C^2(\bar{D})$ we have *Green's first theorem*

$$(2.2) \qquad \int_D \{u \, \Delta \, v + \operatorname{grad} u \cdot \operatorname{grad} v\} \, dx = \int_{\partial D} u \frac{\partial v}{\partial \nu} \, ds,$$

and for $u, v \in C^2(\bar{D})$ we have *Green's second theorem*

$$(2.3) \qquad \int_D (u \, \Delta \, v - v \, \Delta \, u) \, dx = \int_{\partial D} \left(u \frac{\partial v}{\partial \nu} - v \frac{\partial u}{\partial \nu} \right) ds.$$

For two vectors $a = (a_1, a_2, a_3)$ and $b = (b_1, b_2, b_3)$ in $\mathbb{R}^3$ or $\mathbb{C}^3$ we will denote by $a \cdot b := a_1 b_1 + a_2 b_2 + a_3 b_3$ the bilinear scalar product and by $|a| := \sqrt{a \cdot \bar{a}}$ the Euclidean norm. For complex numbers or vectors the bar indicates the complex conjugate. Note that our regularity assumptions on D are sufficient conditions for the validity of Green's theorems and can be weakened (see Kellogg [84]).

Theorem 2.1 *Let D be a bounded domain of class C^2 and let ν denote the unit normal vector to the boundary ∂D directed into the exterior of D. Let $u \in C^2(D) \cap C(\bar{D})$ be a function which possesses a normal derivative on the boundary in the sense that the limit*

$$\frac{\partial u}{\partial \nu}(x) = \lim_{h \to +0} \nu(x) \cdot \operatorname{grad} u(x - h\nu(x)), \quad x \in \partial D,$$

exists uniformly on ∂D. Then we have Green's formula

$$
\begin{aligned}
(2.4) \qquad u(x) = & \int_{\partial D} \left\{ \frac{\partial u}{\partial \nu}(y) \, \Phi(x, y) - u(y) \frac{\partial \Phi(x, y)}{\partial \nu(y)} \right\} ds(y) \\
& - \int_D \{\Delta u(y) + k^2 u(y)\} \, \Phi(x, y) \, dy, \quad x \in D,
\end{aligned}
$$

where the volume integral exists as improper integral. In particular, if u is a solution to the Helmholtz equation

$$\Delta u + k^2 u = 0 \quad \text{in } D,$$

then

$$(2.5) \qquad u(x) = \int_{\partial D} \left\{ \frac{\partial u}{\partial \nu}(y) \, \Phi(x, y) - u(y) \frac{\partial \Phi(x, y)}{\partial \nu(y)} \right\} ds(y), \quad x \in D.$$

Proof. First, we assume that $u \in C^2(\bar{D})$. We circumscribe the arbitrary fixed point $x \in D$ with a sphere $\Omega(x; \rho) := \{y \in \mathbb{R}^3 : |x - y| = \rho\}$ contained in D and direct the unit normal ν to $\Omega(x; \rho)$ into the interior of $\Omega(x; \rho)$. We now apply Green's theorem (2.3) to the functions u and $\Phi(x, \cdot)$ in the domain $D_\rho := \{y \in D : |x - y| > \rho\}$ to obtain

(2.6)
$$\int_{\partial D \cup \Omega(x;\rho)} \left\{ \frac{\partial u}{\partial \nu}(y) \, \Phi(x, y) - u(y) \frac{\partial \Phi(x, y)}{\partial \nu(y)} \right\} ds(y)$$
$$= \int_{D_\rho} \{\Delta u(y) + k^2 u(y)\} \Phi(x, y) \, dy.$$

Since on $\Omega(x; \rho)$ we have

$$\Phi(x, y) = \frac{e^{ik\rho}}{4\pi\rho} \quad \text{and} \quad \operatorname{grad}_y \Phi(x, y) = \left(\frac{1}{\rho} - ik \right) \frac{e^{ik\rho}}{4\pi\rho} \, \nu(y),$$

a straightforward calculation, using the mean value theorem, shows that

$$\lim_{\rho \to 0} \int_{\Omega(x;\rho)} \left\{ u(y) \frac{\partial \Phi(x, y)}{\partial \nu(y)} - \frac{\partial u}{\partial \nu}(y) \, \Phi(x, y) \right\} ds(y) = u(x),$$

whence (2.4) follows by passing to the limit $\rho \to 0$ in (2.6). The existence of the volume integral as an improper integral is a consequence of the fact that its integrand is weakly singular.

The case where u belongs only to $C^2(D) \cap C(\bar{D})$ and has a normal derivative in the sense of uniform convergence is treated by first integrating over parallel surfaces to the boundary of D and then passing to the limit ∂D. For the concept of parallel surfaces, we refer to [32], [106] and [126]. We note that the parallel surfaces for $\partial D \in C^2$ belong to C^1.   □

In the literature, Green's formula (2.5) is also known as the *Helmholtz representation*. Obviously, Theorem 2.1 remains valid for complex values of k.

Theorem 2.2 *If u is a two times continuously differentiable solution to the Helmholtz equation in a domain D, then u is analytic.*

Proof. Let $x \in D$ and choose a closed ball contained in D with center x. Then Theorem 2.1 can be applied in this ball and the statement follows from the analyticity of the fundamental solution for $x \neq y$.   □

As a consequence of Theorem 2.2, a solution to the Helmholtz equation that vanishes in an open subset of its domain of definition must vanish everywhere.

In the sequel, by saying u is a solution to the Helmholtz equation we always tacitly imply that u is twice continuously differentiable, and hence analytic, in the interior of its domain of definition.

Definition 2.3 *A solution u to the Helmholtz equation whose domain of definition contains the exterior of some sphere is called* radiating *if it satisfies the* Sommerfeld radiation condition

$$(2.7) \qquad \lim_{r\to\infty} r\left(\frac{\partial u}{\partial r} - iku\right) = 0$$

where $r = |x|$ and the limit is assumed to hold uniformly in all directions $x/|x|$.

Theorem 2.4 *Assume the bounded set D is the open complement of an unbounded domain of class C^2 and let ν denote the unit normal vector to the boundary ∂D directed into the exterior of D. Let $u \in C^2(\mathbb{R}^3 \setminus \bar{D}) \cap C(\mathbb{R}^3 \setminus D)$ be a radiating solution to the Helmholtz equation*

$$\Delta u + k^2 u = 0 \quad \text{in } \mathbb{R}^3 \setminus \bar{D}$$

which possesses a normal derivative on the boundary in the sense that the limit

$$\frac{\partial u}{\partial \nu}(x) = \lim_{h\to+0} \nu(x) \cdot \operatorname{grad} u(x + h\nu(x)), \quad x \in \partial D,$$

exists uniformly on ∂D. Then we have Green's formula

$$(2.8) \qquad u(x) = \int_{\partial D} \left\{ u(y)\frac{\partial \Phi(x,y)}{\partial \nu(y)} - \frac{\partial u}{\partial \nu}(y)\,\Phi(x,y) \right\} ds(y), \quad x \in \mathbb{R}^3 \setminus \bar{D}.$$

Proof. We first show that

$$(2.9) \qquad \int_{\Omega_r} |u|^2 ds = O(1), \quad r \to \infty,$$

where Ω_r denotes the sphere of radius r and center at the origin. To accomplish this, we observe that from the radiation condition (2.7) it follows that

$$\int_{\Omega_r} \left\{ \left|\frac{\partial u}{\partial \nu}\right|^2 + k^2|u|^2 + 2k\,\mathrm{Im}\left(u\,\frac{\partial \bar{u}}{\partial \nu}\right) \right\} ds = \int_{\Omega_r} \left|\frac{\partial u}{\partial \nu} - iku\right|^2 ds \to 0, \quad r \to \infty,$$

where ν is the unit outward normal to Ω_r. We take r large enough such that D is contained in Ω_r and apply Green's theorem (2.2) in the domain $D_r := \{y \in \mathbb{R}^3 \setminus \bar{D} : |y| < r\}$ to obtain

$$\int_{\Omega_r} u\,\frac{\partial \bar{u}}{\partial \nu}\,ds = \int_{\partial D} u\,\frac{\partial \bar{u}}{\partial \nu}\,ds - k^2 \int_{D_r} |u|^2 dy + \int_{D_r} |\operatorname{grad} u|^2 dy.$$

We now insert the imaginary part of the last equation into the previous equation and find that

$$(2.10) \qquad \lim_{r\to\infty} \int_{\Omega_r} \left\{ \left|\frac{\partial u}{\partial \nu}\right|^2 + k^2|u|^2 \right\} ds = -2k\,\mathrm{Im} \int_{\partial D} u\,\frac{\partial \bar{u}}{\partial \nu}\,ds.$$

Both terms on the left hand side of (2.10) are nonnegative. Hence, they must be individually bounded as $r \to \infty$ since their sum tends to a finite limit. Therefore, (2.9) is proven.

Now from (2.9) and the radiation condition

$$\frac{\partial \Phi(x,y)}{\partial \nu(y)} - ik\Phi(x,y) = O\left(\frac{1}{r^2}\right), \quad r \to \infty,$$

which is valid uniformly for $y \in \Omega_r$, by the Schwarz inequality we see that

$$I_1 := \int_{\Omega_r} u(y) \left\{ \frac{\partial \Phi(x,y)}{\partial \nu(y)} - ik\Phi(x,y) \right\} ds(y) \to 0, \quad r \to \infty,$$

and the radiation condition (2.7) for u and $\Phi(x,y) = O(1/r)$ for $y \in \Omega_r$ yield

$$I_2 := \int_{\Omega_r} \Phi(x,y) \left\{ \frac{\partial u}{\partial \nu}(y) - iku(y) \right\} ds(y) \to 0, \quad r \to \infty.$$

Hence,

$$\int_{\Omega_r} \left\{ u(y) \frac{\partial \Phi(x,y)}{\partial \nu(y)} - \frac{\partial u}{\partial \nu}(y) \Phi(x,y) \right\} ds(y) = I_1 - I_2 \to 0, \quad r \to \infty.$$

The proof is now completed by applying Theorem 2.1 in the bounded domain D_r and passing to the limit $r \to \infty$. $\qquad \Box$

From Theorem 2.4 we deduce that radiating solutions u to the Helmholtz equation automatically satisfy Sommerfeld's finiteness condition

$$(2.11) \qquad\qquad u(x) = O\left(\frac{1}{|x|}\right), \quad |x| \to \infty,$$

uniformly for all directions and that the validity of the Sommerfeld radiation condition (2.7) is invariant under translations of the origin. Wilcox [193] first established that the representation formula (2.8) can be derived without the additional condition (2.11) of finiteness. Our proof of Theorem 2.4 has followed Wilcox's proof. It also shows that (2.7) can be replaced by the weaker formulation

$$\int_{\Omega_r} \left| \frac{\partial u}{\partial r} - iku \right|^2 ds \to 0, \quad r \to \infty,$$

with (2.8) still being valid. Of course, (2.8) then implies that (2.7) also holds.

Solutions to the Helmholtz equation which are defined in all of $\mathbb{R}^3$ are called *entire solutions*. An entire solution to the Helmholtz equation satisfying the radiation condition must vanish identically. This follows immediately from combining Green's formula (2.8) and Green's theorem (2.3).

We are now in a position to introduce the definition of the *far field pattern* or the *scattering amplitude* which plays a central role in this book.

Theorem 2.5 *Every radiating solution u to the Helmholtz equation has the asymptotic behavior of an outgoing spherical wave*

$$(2.12) \qquad u(x) = \frac{e^{ik|x|}}{|x|} \left\{ u_\infty(\hat{x}) + O\left(\frac{1}{|x|}\right) \right\}, \quad |x| \to \infty,$$

uniformly in all directions $\hat{x} = x/|x|$ *where the function* u_∞ *defined on the unit sphere* Ω *is known as the* far field pattern *of u. Under the assumptions of Theorem 2.4 we have*

$$(2.13) \quad u_\infty(\hat{x}) = \frac{1}{4\pi} \int_{\partial D} \left\{ u(y) \frac{\partial e^{-ik\hat{x}\cdot y}}{\partial \nu(y)} - \frac{\partial u}{\partial \nu}(y) e^{-ik\hat{x}\cdot y} \right\} ds(y), \quad \hat{x} \in \Omega.$$

Proof. From

$$|x - y| = \sqrt{|x|^2 - 2\,x\cdot y + |y|^2} = |x| - \hat{x}\cdot y + O\left(\frac{1}{|x|}\right),$$

we derive

$$(2.14) \qquad \frac{e^{ik|x-y|}}{|x-y|} = \frac{e^{ik|x|}}{|x|} \left\{ e^{-ik\hat{x}\cdot y} + O\left(\frac{1}{|x|}\right) \right\},$$

$$(2.15) \qquad \frac{\partial}{\partial \nu(y)} \frac{e^{ik|x-y|}}{|x-y|} = \frac{e^{ik|x|}}{|x|} \left\{ \frac{\partial e^{-ik\hat{x}\cdot y}}{\partial \nu(y)} + O\left(\frac{1}{|x|}\right) \right\}$$

uniformly for all $y \in \partial D$. Inserting this into (2.8), the theorem follows. $\square$

One of the main themes of our book will be to recover radiating solutions of the Helmholtz equation from a knowledge of their far field patterns. In terms of the mapping $F : u \mapsto u_\infty$ transfering the radiating solution u into its far field pattern u_∞, we want to solve the equation $Fu = u_\infty$ for a given u_∞. In order to establish uniqueness for determining u from its far field pattern u_∞ and to understand the strong ill-posedness of the equation $Fu = u_\infty$, we need to develop some facts on spherical wave functions. This will be the subject of the next two sections. We already can point out that the mapping F is extremely smoothing since from (2.13) we see that the far field pattern is an analytic function on the unit sphere.

2.3 Spherical Harmonics

For convenience and to introduce notations, we summarize some of the basic properties of spherical harmonics which are relevant in scattering theory and briefly indicate their proofs. For a more detailed study we refer to Lebedev [117].

The restriction of a homogeneous harmonic polynomial of degree n to the unit sphere Ω is called a *spherical harmonic* of order n.

Theorem 2.6 *There exist exactly $2n + 1$ linearly independent spherical harmonics of order n.*

Proof. By the maximum-minimum principle for harmonic functions it suffices to show that there exist exactly $2n+1$ linearly independent homogeneous harmonic polynomials H_n of degree n. We can write

$$H_n(x_1, x_2, x_3) = \sum_{k=0}^{n} a_{n-k}(x_1, x_2)\, x_3^k$$

where the a_k are homogeneous polynomials of degree k in the two variables x_1 and x_2. Then, straightforward calculations show that H_n is harmonic if and only if the coefficients satisfy

$$a_{n-k} = -\frac{\triangle a_{n-k+2}}{k(k-1)}, \quad k = 2, \ldots, n.$$

Therefore, choosing the two coefficients a_n and a_{n-1} uniquely determines H_n, and by setting

$$a_n(x_1, x_2) = x_1^{n-j} x_2^j, \quad a_{n-1}(x_1, x_2) = 0, \quad j = 0, \ldots, n,$$

$$a_n(x_1, x_2) = 0, \quad a_{n-1}(x_1, x_2) = x_1^{n-1-j} x_2^j, \quad j = 0, \ldots, n-1,$$

we get $2n + 1$ linearly independent homogeneous harmonic polynomials of degree n.   □

In principle, the proof of the preceding theorem allows a construction of all spherical harmonics. However, it is more convenient and appropriate to use polar coordinates for the representation of spherical harmonics. In polar coordinates (r, θ, φ), homogeneous polynomials clearly are of the form

$$H_n = r^n Y_n(\theta, \varphi),$$

and $\triangle H_n = 0$ is readily seen to be satisfied if

(2.16) $$\frac{1}{\sin\theta} \frac{\partial}{\partial\theta} \sin\theta \frac{\partial Y_n}{\partial\theta} + \frac{1}{\sin^2\theta} \frac{\partial^2 Y_n}{\partial\varphi^2} + n(n+1)Y_n = 0.$$

From Green's theorem (2.3), applied to two homogeneous harmonic polynomials H_n and $H_{n'}$, we have

$$0 = \int_\Omega \left\{ \overline{H}_{n'} \frac{\partial H_n}{\partial r} - H_n \frac{\partial \overline{H}_{n'}}{\partial r} \right\} ds = (n - n') \int_\Omega Y_n \overline{Y}_{n'}\, ds.$$

Therefore spherical harmonics satisfy the orthogonality relation

(2.17) $$\int_\Omega Y_n \overline{Y}_{n'}\, ds = 0, \quad n \neq n'.$$

We first construct spherical harmonics which only depend on the polar angle θ. Choose points x and y with $r = |x| < |y| = 1$, denote the angle between x and y by θ and set $t = \cos\theta$. Consider the function

$$(2.18) \qquad \frac{1}{|x-y|} = \frac{1}{\sqrt{1-2tr+r^2}}$$

which for fixed y is a solution to Laplace's equation with respect to x. Since for fixed t with $-1 \leq t \leq 1$ the right hand side is an analytic function in r, we have the Taylor series

$$(2.19) \qquad \frac{1}{\sqrt{1-2tr+r^2}} = \sum_{n=0}^{\infty} P_n(t)r^n.$$

The coefficients P_n in this expansion are called *Legendre polynomials* and the function on the left hand side consequently is known as the *generating function* for the Legendre polynomials. For each $0 < r_0 < 1$ the Taylor series

$$(2.20) \qquad \frac{1}{\sqrt{1-r\exp(\pm i\theta)}} = 1 + \sum_{n=1}^{\infty} \frac{1\cdot 3\cdots(2n-1)}{2\cdot 4\cdots 2n} r^n e^{\pm in\theta}$$

and all its term by term derivatives with respect to r and θ are absolutely and uniformly convergent for all $0 \leq r \leq r_0$ and all $0 \leq \theta \leq \pi$. Hence, by multiplying the equations (2.20) for the plus and the minus sign, we note that the series (2.19) and all its term by term derivatives with respect to r and θ are absolutely and uniformly convergent for all $0 \leq r \leq r_0$ and all $-1 \leq t = \cos\theta \leq 1$. Setting $\theta = 0$ in (2.20) obviously provides a majorant for the series for all θ. Therefore, the geometric series is a majorant for the series in (2.19) whence the inequality

$$(2.21) \qquad |P_n(t)| \leq 1, \quad -1 \leq t \leq 1, \quad n = 0,1,2,\ldots,$$

follows.

Differentiating (2.19) with respect to r, multiplying by $1-2tr+r^2$, inserting (2.19) on the left hand side and then equating powers of r shows that the P_n satisfy the recursion formula

$$(2.22) \quad (n+1)P_{n+1}(t) - (2n+1)tP_n(t) + nP_{n-1}(t) = 0, \quad n = 1,2,\ldots.$$

Since, as easily seen from (2.19), we have $P_0(t) = 1$ and $P_1(t) = t$, the recursion formula shows that P_n indeed is a polynomial of degree n and that P_n is an even function if n is even and an odd function if n is odd.

Since for fixed y the function (2.18) is harmonic, differentiating (2.19) term by term, we obtain that

$$\sum_{n=0}^{\infty} \left\{ \frac{1}{\sin\theta} \frac{d}{d\theta} \sin\theta \frac{dP_n(\cos\theta)}{d\theta} + n(n+1)P_n(\cos\theta) \right\} r^{n-2} = 0.$$

Equating powers of r shows that the Legendre polynomials satisfy the *Legendre differential equation*

$$(2.23) \quad (1-t^2)P_n''(t) - 2tP_n'(t) + n(n+1)P_n(t) = 0, \quad n = 0,1,2,\ldots,$$

and that the homogeneous polynomial $r^n P_n(\cos\theta)$ of degree n is harmonic. Therefore, $P_n(\cos\theta)$ represents a spherical harmonic of order n. The orthogonality (2.17) implies that

$$\int_{-1}^{1} P_n(t)P_{n'}(t)\,dt = 0, \quad n \neq n'.$$

Since we have uniform convergence, we may integrate the square of the generating function (2.19) term by term and use the preceding orthogonality to arrive at

$$\int_{-1}^{1}\frac{dt}{1-2tr+r^2} = \sum_{n=0}^{\infty}\int_{-1}^{1}[P_n(t)]^2\,dt\;r^{2n}.$$

On the other hand, we have

$$\int_{-1}^{1}\frac{dt}{1-2tr+r^2} = \frac{1}{r}\ln\frac{1+r}{1-r} = \sum_{n=0}^{\infty}\frac{2}{2n+1}r^{2n}.$$

Thus, we have proven the orthonormality relation

(2.24) $$\int_{-1}^{1}P_n(t)P_m(t)\,dt = \frac{2}{2n+1}\delta_{nm}, \quad n,m = 0,1,2,\ldots,$$

with the usual meaning for the Kronecker symbol δ_{nm}. Since $\mathrm{span}\{P_0,\ldots,P_n\} = \mathrm{span}\{1,\ldots,t^n\}$ the Legendre polynomials P_n, $n = 0,1,\ldots$, form a complete orthogonal system in $L^2[-1,1]$.

We now look for spherical harmonics of the form

$$Y_n^m(\theta,\varphi) = f(\cos\theta)\,e^{im\varphi}.$$

Then (2.16) is satisfied provided f is a solution of the *associated Legendre differential equation*

(2.25) $$(1-t^2)f''(t) - 2tf'(t) + \left\{n(n+1) - \frac{m^2}{1-t^2}\right\}f(t) = 0.$$

Differentiating the Legendre differential equation (2.23) m–times shows that $g = P_n^{(m)}$ satisfies

$$(1-t^2)g''(t) - 2(m+1)tg'(t) + (n-m)(n+m+1)g(t) = 0.$$

From this it can be deduced that the *associated Legendre functions*

(2.26) $$P_n^m(t) := (1-t^2)^{m/2}\frac{d^m P_n(t)}{dt^m}, \quad m = 0,1,\ldots,n,$$

solve the associated Legendre equation for $n = 0,1,2,\ldots$. In order to make sure that the functions $Y_n^m(\theta,\varphi) = P_n^m(\cos\theta)\,e^{im\varphi}$ are spherical harmonics, we have to prove that the harmonic functions $r^n Y_n^m(\theta,\varphi) = r^n P_n^m(\cos\theta)\,e^{im\varphi}$ are homogeneous polynomials of degree n. From the recursion formula (2.22) for the P_n and the definition (2.26) for the P_n^m, we first observe that

$$P_n^m(\cos\theta) = \sin^m\theta\, u_n^m(\cos\theta)$$

where u_n^m is a polynomial of degree $n - m$ which is even if $n - m$ is even and odd if $n - m$ is odd. Since in polar coordinates we have

$$r^m\sin^m\theta\, e^{im\varphi} = (x_1 + ix_2)^m,$$

it follows that

$$r^n\, Y_n^m(\theta,\varphi) = (x_1 + ix_2)^m\, r^{n-m}\, u_n^m(\cos\theta).$$

For $n - m$ even we can write

$$r^{n-m}\, u_n^m(\cos\theta) = r^{n-m} \sum_{k=0}^{\frac{1}{2}(n-m)} a_k \cos^{2k}\theta = \sum_{k=0}^{\frac{1}{2}(n-m)} a_k\, x_3^{2k}\,(x_1^2 + x_2^2 + x_3^2)^{\frac{1}{2}(n-m)-k}$$

which is a homogeneous polynomial of degree $n-m$ and this is also true for $n-m$ odd. Putting everything together, we see that the $r^n\, Y_n^m(\theta,\varphi)$ are homogeneous polynomials of degree n.

Theorem 2.7 *The spherical harmonics*

$$(2.27)\qquad Y_n^m(\theta,\varphi) := \sqrt{\frac{2n+1}{4\pi}\frac{(n-|m|)!}{(n+|m|)!}}\; P_n^{|m|}(\cos\theta)\, e^{im\varphi}$$

for $m = -n,\ldots,n$, $n = 0,1,2,\ldots$, *form a complete orthonormal system in* $L^2(\Omega)$.

Proof. Because of (2.17) and the orthogonality of the $e^{im\varphi}$, the Y_n^m given by (2.27) are orthogonal. For $m > 0$ we evaluate

$$A_n^m := \int_0^\pi [P_n^m(\cos\theta)]^2 \sin\theta\, d\theta$$

by m partial integrations to get

$$A_n^m \doteq \int_{-1}^1 (1-t^2)^m \left[\frac{d^m P_n(t)}{dt^m}\right]^2 dt = \int_{-1}^1 P_n(t)\,\frac{d^m}{dt^m}\, g_n^m(t)\, dt$$

where

$$g_n^m(t) = (t^2-1)^m\,\frac{d^m P_n(t)}{dt^m}.$$

Hence

$$\frac{d^m}{dt^m}\, g_n^m(t) = \frac{(n+m)!}{(n-m)!}\, a_n t^n + \cdots$$

is a polynomial of degree n with a_n the leading coefficient in $P_n(t) = a_n t^n + \cdots$
Therefore, by the orthogonality (2.24) of the Legendre polynomials we derive

$$\frac{(n-m)!}{(n+m)!}\, A_n^m = \int_{-1}^{1} a_n t^n\, P_n(t)\, dt = \int_{-1}^{1} [P_n(t)]^2\, dt = \frac{2}{2n+1}\,,$$

and the proof of the orthonormality of the Y_n^m is finished.

In particular, for fixed m the associated Legendre functions P_n^m for $n = m, m+1, \ldots$ are orthogonal and they are complete in $L^2[-1,1]$ since we have

$$\mathrm{span}\left\{P_m^m, \ldots, P_{m+n}^m\right\} = (1-t^2)^{m/2}\,\mathrm{span}\,\{1, \ldots, t^n\}\,.$$

Writing $Y := \mathrm{span}\,\{Y_n^m : m = -n, \ldots, n,\ n = 0, 1, 2, \ldots\}$, it remains to show that Y is dense in $L^2(\Omega)$. Let $g \in C(\Omega)$. For fixed θ we then have Parseval's equality

(2.28)
$$2\pi \sum_{m=-\infty}^{\infty} |g_m(\theta)|^2 = \int_{0}^{2\pi} |g(\theta, \varphi)|^2\, d\varphi$$

for the Fourier coefficients

$$g_m(\theta) = \frac{1}{2\pi} \int_{0}^{2\pi} g(\theta, \varphi) e^{-im\varphi}\, d\varphi$$

with respect to φ. Since the g_m and the right hand side of (2.28) are continuous in θ, by Dini's theorem the convergence in (2.28) is uniform with respect to θ. Therefore, given $\varepsilon > 0$ there exists $M = M(\varepsilon) \in \mathbb{N}$ such that

$$\int_{0}^{2\pi} \left| g(\theta, \varphi) - \sum_{m=-M}^{M} g_m(\theta) e^{im\varphi} \right|^2 d\varphi = \int_{0}^{2\pi} |g(\theta, \varphi)|^2\, d\varphi - 2\pi \sum_{m=-M}^{M} |g_m(\theta)|^2 < \frac{\varepsilon}{4\pi}$$

for all $0 \le \theta \le \pi$. The finite number of functions g_m, $m = -M, \ldots, M$, can now be simultaneously approximated by the associated Legendre functions, i.e., there exist $N = N(\varepsilon)$ and coefficients a_n^m such that

$$\int_{0}^{\pi} \left| g_m(\theta) - \sum_{n=|m|}^{N} a_n^m P_n^{|m|}(\cos\theta) \right|^2 \sin\theta\, d\theta < \frac{\varepsilon}{8\pi(2M+1)^2}$$

for all $m = -M, \ldots, M$. Then, combining the last two inequalities with the help of the Schwarz inequality, we find

$$\int_{0}^{\pi} \int_{0}^{2\pi} \left| g(\theta, \varphi) - \sum_{m=-M}^{M} \sum_{n=|m|}^{N} a_n^m P_n^{|m|}(\cos\theta)\, e^{im\varphi} \right|^2 \sin\theta\, d\varphi d\theta < \varepsilon.$$

Therefore, Y is dense in $C(\Omega)$ with respect to the L^2 norm and this completes the proof since $C(\Omega)$ is dense in $L^2(\Omega)$. $\square$

We conclude our brief survey of spherical harmonics by proving the important *addition theorem*.

Theorem 2.8 *Let Y_n^m, $m = -n, \ldots, n$, be any system of $2n + 1$ orthonormal spherical harmonics of order n. Then for all $\hat{x}, \hat{y} \in \Omega$ we have*

$$(2.29) \qquad \sum_{m=-n}^{n} Y_n^m(\hat{x}) \overline{Y_n^m(\hat{y})} = \frac{2n + 1}{4\pi} P_n(\cos\theta)$$

where θ denotes the angle between $\hat{x}$ and $\hat{y}$.

Proof. We abbreviate the left hand side of (2.29) by $Y(\hat{x}, \hat{y})$ and first show that Y only depends on the angle θ. Each orthogonal matrix Q in $\mathbb{R}^3$ transforms homogeneous harmonic polynomials of degree n again into homogeneous harmonic polynomials of degree n. Hence, we can write

$$Y_n^m(Q\hat{x}) = \sum_{k=-n}^{n} a_{mk} Y_n^k(\hat{x}), \quad m = -n, \ldots, n.$$

Since Q is orthogonal and the Y_n^m are orthonormal, we have

$$\int_\Omega Y_n^m(Q\hat{x}) \overline{Y_n^{m'}(Q\hat{x})} \, ds = \int_\Omega Y_n^m(\hat{x}) \overline{Y_n^{m'}(\hat{x})} \, ds = \delta_{mm'}.$$

From this it can be seen that the matrix $A = (a_{mk})$ also is orthogonal and we obtain

$$Y(Q\hat{x}, Q\hat{y}) = \sum_{m=-n}^{n} \sum_{k=-n}^{n} a_{mk} Y_n^k(\hat{x}) \sum_{l=-n}^{n} \overline{a_{ml} Y_n^l(\hat{y})} = \sum_{k=-n}^{n} Y_n^k(\hat{x}) \overline{Y_n^k(\hat{y})} = Y(\hat{x}, \hat{y})$$

whence $Y(\hat{x}, \hat{y}) = f(\cos\theta)$ follows. Since for fixed $\hat{y}$ the function Y is a spherical harmonic, by introducing polar coordinates with the polar axis given by $\hat{y}$ we see that $f = a_n P_n$ with some constant a_n. Hence, we have

$$\sum_{m=-n}^{n} Y_n^m(\hat{x}) \overline{Y_n^m(\hat{y})} = a_n P_n(\cos\theta).$$

Setting $\hat{y} = \hat{x}$ and using $P_n(1) = 1$ (this follows from the generating function (2.19)) we obtain

$$a_n = \sum_{m=-n}^{n} |Y_n^m(\hat{x})|^2.$$

Since the Y_n^m are normalized, integrating the last equation over Ω we finally arrive at $4\pi a_n = 2n + 1$ and the proof is complete. $\qquad\square$

2.4 Spherical Bessel Functions

We continue our study of spherical wave functions by introducing the basic properties of spherical Bessel functions. For a more detailed analysis we again refer to Lebedev [117].

We look for solutions to the Helmholtz equation of the form

$$u(x) = f(k|x|)\, Y_n(\hat{x})$$

where Y_n is a spherical harmonic of order n. From the differential equation (2.16) for the spherical harmonics, it follows that u solves the Helmholtz equation provided f is a solution of the *spherical Bessel differential equation*

$$(2.30) \qquad t^2 f''(t) + 2t f'(t) + [t^2 - n(n+1)] f(t) = 0.$$

We note that for any solution f to the spherical Bessel differential equation (2.30) the function $g(t) := \sqrt{t}\, f(t)$ solves the Bessel differential equation with half integer order $n + 1/2$ and vice versa. By direct calculations, we see that for $n = 0, 1, \ldots$ the functions

$$(2.31) \qquad j_n(t) := \sum_{p=0}^{\infty} \frac{(-1)^p t^{n+2p}}{2^p p!\, 1 \cdot 3 \cdots (2n + 2p + 1)}$$

and

$$(2.32) \qquad y_n(t) := -\frac{(2n)!}{2^n n!} \sum_{p=0}^{\infty} \frac{(-1)^p t^{2p-n-1}}{2^p p!(-2n+1)(-2n+3)\cdots(-2n+2p-1)}$$

represent solutions to the spherical Bessel differential equation (the first coefficient in the series (2.32) has to be set equal to one). By the ratio test, the function j_n is seen to be analytic for all $t \in \mathbb{R}$ whereas y_n is analytic for all $t \in (0, \infty)$. The functions j_n and y_n are called *spherical Bessel functions* and *spherical Neumann functions* of order n, respectively, and the linear combinations

$$h_n^{(1,2)} := j_n \pm i y_n$$

are known as *spherical Hankel functions* of the first and second kind of order n.

From the series representation (2.31) and (2.32), by equating powers of t, it is readily verified that both $f_n = j_n$ and $f_n = y_n$ satisfy the recurrence relation

$$(2.33) \qquad f_{n+1}(t) + f_{n-1}(t) = \frac{2n+1}{t}\, f_n(t), \quad n = 1, 2, \ldots.$$

Straightforward differentiation of the series (2.31) and (2.32) shows that both $f_n = j_n$ and $f_n = y_n$ satisfy the differentiation formulas

$$(2.34) \qquad f_{n+1}(t) = -t^n \frac{d}{dt}\left\{ t^{-n} f_n(t) \right\}, \quad n = 0, 1, 2, \ldots,$$

and

(2.35) $$t^{n+1} f_{n-1}(t) = \frac{d}{dt} \left\{ t^{n+1} f_n(t) \right\}, \quad n = 1, 2, \ldots .$$

Finally, from (2.30), the Wronskian

$$W(j_n(t), y_n(t)) := j_n(t) y_n'(t) - y_n(t) j_n'(t)$$

is readily seen to satisfy

$$W' + \frac{2}{t} W = 0,$$

whence $W(j_n(t), y_n(t)) = C/t^2$ for some constant C. This constant can be evaluated by passing to the limit $t \to 0$ with the result

(2.36) $$j_n(t) y_n'(t) - j_n'(t) y_n(t) = \frac{1}{t^2} .$$

From the series representation of the spherical Bessel and Neumann functions, it is obvious that

(2.37) $$j_n(t) = \frac{t^n}{1 \cdot 3 \cdots (2n+1)} \left(1 + O\left(\frac{1}{n} \right) \right), \quad n \to \infty,$$

uniformly on compact subsets of $\mathbb{R}$ and

(2.38) $$h_n^{(1)}(t) = \frac{1 \cdot 3 \cdots (2n-1)}{i t^{n+1}} \left(1 + O\left(\frac{1}{n} \right) \right), \quad n \to \infty,$$

uniformly on compact subsets of $(0, \infty)$. With the aid of Stirling's formula $n! = \sqrt{2\pi n} \, (n/e)^n \, (1 + o(1))$, $n \to \infty$, which implies that

$$\frac{(2n)!}{n!} = 2^{2n+\frac{1}{2}} \left(\frac{n}{e} \right)^n (1 + o(1)), \quad n \to \infty,$$

from (2.38) we obtain

(2.39) $$h_n^{(1)}(t) = O\left(\frac{2n}{et} \right)^n, \quad n \to \infty,$$

uniformly on compact subsets of $(0, \infty)$.

The spherical Bessel and Neumann functions can be expressed in terms of trigonometric functions. Setting $n = 0$ in the series (2.31) and (2.32) we have that

$$j_0(t) = \frac{\sin t}{t}, \quad y_0(t) = -\frac{\cos t}{t}$$

and consequently

(2.40) $$h_0^{(1,2)}(t) = \frac{e^{\pm it}}{\pm it} .$$

Hence, by induction, from (2.40) and (2.34) it follows that the spherical Hankel functions are of the form

$$h_n^{(1)}(t) = (-i)^n \frac{e^{it}}{it} \left\{ 1 + \sum_{p=1}^{n} \frac{a_{pn}}{t^p} \right\}, \quad h_n^{(2)}(t) = i^n \frac{e^{-it}}{-it} \left\{ 1 + \sum_{p=1}^{n} \frac{\bar{a}_{pn}}{t^p} \right\}$$

with complex coefficients $a_{1n}, \ldots, a_{nn}$. From this we readily obtain the following asymptotic behavior of the spherical Hankel functions for large argument

$$h_n^{(1,2)}(t) = \frac{1}{t} e^{\pm i \left(t - \frac{n\pi}{2} - \frac{\pi}{2} \right)} \left\{ 1 + O\left(\frac{1}{t} \right) \right\}, \quad t \to \infty,$$

(2.41)

$$h_n^{(1,2)\prime}(t) = \frac{1}{t} e^{\pm i \left(t - \frac{n\pi}{2} \right)} \left\{ 1 + O\left(\frac{1}{t} \right) \right\}, \quad t \to \infty.$$

Taking the real and the imaginary part of (2.41) we also have asymptotic formulas for the spherical Bessel and Neumann functions.

For solutions to the Helmholtz equation in polar coordinates, we can now state the following theorem on *spherical wave functions*.

Theorem 2.9 *Let Y_n be a spherical harmonic of order n. Then*

$$u_n(x) = j_n(k|x|)\, Y_n(\hat{x})$$

is an entire solution to the Helmholtz equation and

$$v_n(x) = h_n^{(1)}(k|x|)\, Y_n(\hat{x})$$

is a radiating solution to the Helmholtz equation in $\mathbb{R}^3 \setminus \{0\}$.

Proof. Since we can write $j_n(kr) = k^n r^n w_n(r^2)$ with an analytic function $w_n : \mathbb{R} \to \mathbb{R}$ and since $r^n Y_n(\hat{x})$ is a homogeneous polynomial in x_1, x_2, x_3, the product $j_n(kr)\, Y_n(\hat{x})$ is regular at $x = 0$, i.e., u_n also satisfies the Helmholtz equation at the origin. That the radiation condition is satisfied for v_n follows from the asymptotic behavior (2.41) of the spherical Hankel functions of the first kind. $\square$

We conclude our brief discussion of spherical wave functions by the following *addition theorem* for the fundamental solution.

Theorem 2.10 *Let Y_n^m, $m = -n, \ldots, n$, $n = 0, 1, \ldots$, be a set of orthonormal spherical harmonics. Then for $|x| > |y|$ we have*

(2.42)
$$\frac{e^{ik|x-y|}}{4\pi|x-y|} = ik \sum_{n=0}^{\infty} \sum_{m=-n}^{n} h_n^{(1)}(k|x|)\, Y_n^m(\hat{x})\, j_n(k|y|)\, \overline{Y_n^m(\hat{y})}$$

where $\hat{x} = x/|x|$ and $\hat{y} = y/|y|$. The series and its term by term first derivatives with respect to $|x|$ and $|y|$ are absolutely and uniformly convergent on compact subsets of $|x| > |y|$.

Proof. From Green's theorem (2.3) applied to $u_n^m(z) = j_n(k|z|) Y_n^m(\hat{z})$ with $\hat{z} = z/|z|$ and $\Phi(x, z)$, we have

$$\int_{|z|=r} \left\{ u_n^m(z) \frac{\partial \Phi(x, z)}{\partial \nu(z)} - \frac{\partial u_n^m}{\partial \nu}(z) \, \Phi(x, z) \right\} ds(z) = 0, \quad |x| > r,$$

and from Green's formula (2.8), applied to the radiating solution $v_n^m(z) = h_n^{(1)}(k|z|) Y_n^m(\hat{z})$, we have

$$\int_{|z|=r} \left\{ v_n^m(z) \frac{\partial \Phi(x, z)}{\partial \nu(z)} - \frac{\partial v_n^m}{\partial \nu}(z) \, \Phi(x, z) \right\} ds(z) = v_n^m(x), \quad |x| > r.$$

From the last two equations, noting that on $|z| = r$ we have

$$u_n^m(z) = j_n(kr) Y_n^m(\hat{z}), \quad \frac{\partial u_n^m}{\partial \nu}(z) = k j_n'(kr) Y_n^m(\hat{z})$$

and

$$v_n^m(z) = h_n^{(1)}(kr) Y_n^m(\hat{z}), \quad \frac{\partial v_n^m}{\partial \nu}(z) = k h_n^{(1)\prime}(kr) Y_n^m(\hat{z})$$

and using the Wronskian (2.36), we see that

$$\frac{1}{ikr^2} \int_{|z|=r} Y_n^m(\hat{z}) \, \Phi(x, z) \, ds(z) = j_n(kr) h_n^{(1)}(k|x|) Y_n^m(\hat{x}), \quad |x| > r,$$

and by transforming the integral into one over the unit sphere we get

$$(2.43) \qquad \int_\Omega Y_n^m(\hat{z}) \Phi(x, r\hat{z}) \, ds(\hat{z}) = ik \, j_n(kr) h_n^{(1)}(k|x|) Y_n^m(\hat{x}), \quad |x| > r.$$

We can now apply Theorem 2.7 to obtain from the orthogonal expansion

$$\Phi(x, y) = \sum_{n=0}^{\infty} \sum_{m=-n}^{n} \int_\Omega Y_n^m(\hat{z}) \Phi(x, r\hat{z}) \, ds(\hat{z}) \, \overline{Y_n^m(\hat{y})}$$

and (2.43) that the series (2.42) is valid for fixed x with $|x| > r$ and with respect to y in the L^2 sense on the sphere $|y| = r$ for arbitrary r. With the aid of the Schwarz inequality, the Addition Theorem 2.8 for the spherical harmonics and the inequalities (2.21), (2.37) and (2.38) we can estimate

$$\sum_{m=-n}^{n} \left| h_n^{(1)}(k|x|) Y_n^m(\hat{x}) j_n(k|y|) \overline{Y_n^m(\hat{y})} \right|$$

$$\leq \frac{2n+1}{4\pi} |h_n^{(1)}(k|x|) j_n(k|y|)| = O\left(\frac{|y|^n}{|x|^n}\right), \quad n \to \infty,$$

uniformly on compact subsets of $|x| > |y|$. Hence, we have a majorant implying absolute and uniform convergence of the series (2.42). The absolute and uniform convergence of the derivatives with respect to $|x|$ and $|y|$ can be established analogously with the help of estimates for the derivatives j_n' and $h_n^{(1)'}$ corresponding to (2.37) and (2.38) which follow readily from (2.34).   □

Passing to the limit $|x| \to \infty$ in (2.43) with the aid of (2.14) and (2.41), we arrive at the *Funk–Hecke formula*

$$\int_\Omega e^{-ikr\,\hat{x}\cdot\hat{z}}\, Y_n(\hat{z})\, ds(\hat{z}) = \frac{4\pi}{i^n}\, j_n(kr)\, Y_n(\hat{x}), \quad \hat{x} \in \Omega, \quad r > 0,$$

for spherical harmonics Y_n of order n. Obviously, this may be rewritten in the form

(2.44) $$\int_\Omega e^{-ik\,x\cdot\hat{z}}\, Y_n(\hat{z})\, ds(\hat{z}) = \frac{4\pi}{i^n}\, j_n(k|x|)\, Y_n(\hat{x}), \quad x \in \mathbb{R}^3.$$

Proceeding as in the proof of the previous theorem, from (2.44) and Theorem 2.8 we can derive the *Jacobi–Anger expansion*

(2.45) $$e^{ik\,x\cdot d} = \sum_{n=0}^\infty i^n(2n+1)\, j_n(k|x|)\, P_n(\cos\theta), \quad x \in \mathbb{R}^3,$$

where d is a unit vector, θ denotes the angle between x and d and the convergence is uniform on compact subsets of $\mathbb{R}^3$.

2.5 The Far Field Mapping

In this section we first establish the one-to-one correspondence between radiating solutions to the Helmholtz equation and their far field patterns.

Lemma 2.11 (Rellich) *Assume the bounded set D is the open complement of an unbounded domain and let $u \in C^2(\mathbb{R}^3 \setminus \bar{D})$ be a solution to the Helmholtz equation satisfying*

(2.46) $$\lim_{r\to\infty} \int_{|x|=r} |u(x)|^2 ds = 0.$$

Then $u = 0$ in $\mathbb{R}^3 \setminus \bar{D}$.

Proof. For sufficiently large $|x|$, by Theorem 2.7 we have a Fourier expansion

$$u(x) = \sum_{n=0}^\infty \sum_{m=-n}^n a_n^m(|x|)\, Y_n^m(\hat{x})$$

with respect to spherical harmonics where $\hat{x} = x/|x|$. The coefficients are given by

$$a_n^m(r) = \int_\Omega u(r\hat{x})\overline{Y_n^m(\hat{x})}\, ds(\hat{x})$$

and satisfy Parseval's equality

$$\int_{|x|=r} |u(x)|^2 ds = r^2 \sum_{n=0}^{\infty} \sum_{m=-n}^{n} |a_n^m(r)|^2 .$$

Our assumption (2.46) implies that

(2.47) $$\lim_{r \to \infty} r^2 |a_n^m(r)|^2 = 0$$

for all n and m.

Since $u \in C^2(\mathbb{R}^3 \setminus \bar{D})$, we can differentiate under the integral and integrate by parts using $\Delta u + k^2 u = 0$ and the differential equation (2.16) to conclude that the a_n^m are solutions to the spherical Bessel equation

$$\frac{d^2 a_n^m}{dr^2} + \frac{2}{r} \frac{d a_n^m}{dr} + \left(k^2 - \frac{n(n+1)}{r^2} \right) a_n^m = 0,$$

that is,

$$a_n^m(r) = \alpha_n^m h_n^{(1)}(kr) + \beta_n^m h_n^{(2)}(kr)$$

where α_n^m and β_n^m are constants. Substituting this into (2.47) and using the asymptotic behavior (2.41) of the spherical Hankel functions yields $\alpha_n^m = \beta_n^m = 0$ for all n and m. Therefore, $u = 0$ outside a sufficiently large sphere and hence $u = 0$ in $\mathbb{R}^3 \setminus \bar{D}$ by analyticity (Theorem 2.2). $\qquad \square$

Rellich's lemma ensures uniqueness for solutions to exterior boundary value problems through the following theorem.

Theorem 2.12 *Let D be as in Lemma 2.11, let ∂D be of class C^2 with unit normal ν directed into the exterior of D and assume $u \in C^2(\mathbb{R}^3 \setminus \bar{D}) \cap C(\mathbb{R}^3 \setminus D)$ is a radiating solution to the Helmholtz equation with wave number $k > 0$ which has a normal derivative in the sense of uniform convergence and for which*

$$\operatorname{Im} \int_{\partial D} u \frac{\partial \bar{u}}{\partial \nu} \, ds \geq 0.$$

Then $u = 0$ in $\mathbb{R}^3 \setminus \bar{D}$.

Proof. From the identity (2.10) and the assumption of the theorem, we conclude that (2.46) is satisfied. Hence, the theorem follows from Rellich's Lemma 2.11. $\square$

Rellich's lemma also establishes the one-to-one correspondence between radiating waves and their far field patterns.

Theorem 2.13 *Let D be as in Lemma 2.11 and let $u \in C^2(\mathbb{R}^3 \setminus \bar{D})$ be a radiating solution to the Helmholtz equation for which the far field pattern vanishes identically. Then $u = 0$ in $\mathbb{R}^3 \setminus \bar{D}$.*

Proof. Since from (2.12) we deduce

$$\int_{|x|=r} |u(x)|^2 ds = \int_{\Omega} |u_\infty(\hat{x})|^2 ds + O\left(\frac{1}{r}\right), \quad r \to \infty,$$

the assumption $u_\infty = 0$ on Ω implies that (2.46) is satisfied. Hence, the theorem follows from Rellich's Lemma 2.11. $\qquad\square$

Theorem 2.14 *Let u be a radiating solution to the Helmholtz equation in the exterior $|x| > R > 0$ of a sphere. Then u has an expansion with respect to spherical wave functions of the form*

$$(2.48) \qquad u(x) = \sum_{n=0}^{\infty} \sum_{m=-n}^{n} a_n^m\, h_n^{(1)}(k|x|)\, Y_n^m(\hat{x})$$

that converges absolutely and uniformly on compact subsets of $|x| > R$. Conversely, if the series (2.48) converges in the mean square sense on the sphere $|x| = R$ then it also converges absolutely and uniformly on compact subsets of $|x| > R$ and u represents a radiating solution to the Helmholtz equation for $|x| > R$.

Proof. For a radiating solution u to the Helmholtz equation, we insert the addition theorem (2.42) into Green's formula (2.8), applied to the boundary surface $|y| = \tilde{R}$ with $R < \tilde{R} < |x|$, and integrate term by term to obtain the expansion (2.48).

Conversely, L^2 convergence of the series (2.48) on the sphere $|x| = R$, implies by Parseval's equality that

$$\sum_{n=0}^{\infty} \sum_{m=-n}^{n} \left|h_n^{(1)}(kR)\right|^2 |a_n^m|^2 < \infty.$$

Using the Schwarz inequality, the asymptotic behavior (2.38) and the addition theorem (2.29) for $R < R_1 \le |x| \le R_2$ and for $N \in \mathbb{N}$ we can estimate

$$\left[\sum_{n=0}^{N} \sum_{m=-n}^{n} \left|h_n^{(1)}(k|x|)\, a_n^m Y_n^m(\hat{x})\right|\right]^2$$

$$\le \sum_{n=0}^{N} \left|\frac{h_n^{(1)}(k|x|)}{h_n^{(1)}(kR)}\right|^2 \sum_{m=-n}^{n} |Y_n^m(\hat{x})|^2 \sum_{n=0}^{N} \sum_{m=-n}^{n} \left|h_n^{(1)}(kR)\right|^2 |a_n^m|^2$$

$$\le C \sum_{n=0}^{N} (2n+1) \left(\frac{R}{|x|}\right)^{2n}$$

for some constant C depending on R, R_1 and R_2. From this we conclude absolute and uniform convergence of the series (2.48) on compact subsets of $|x| > R$.

Similarly, it can be seen that the term by term first derivatives with respect to $|x|$ are absolutely and uniformly convergent on compact subsets of $|x| > R$. To establish that u solves the Helmholtz equation and satisfies the Sommerfeld radiation condition, we show that Green's formula is valid for u. Using the addition Theorem 2.10, the orthonormality of the Y_n^m and the Wronskian (2.36), we indeed find that

$$\int_{|y|=\tilde{R}} \left\{ u(y) \frac{\partial \Phi(x,y)}{\partial \nu(y)} - \frac{\partial u}{\partial \nu}(y) \Phi(x,y) \right\} ds(y)$$

$$= ik\tilde{R}^2 \sum_{n=0}^{\infty} \sum_{m=-n}^{n} a_n^m h_n^{(1)}(k|x|) Y_n^m(\hat{x}) k W(h_n^{(1)}(k\tilde{R}), j_n(k\tilde{R}))$$

$$= \sum_{n=0}^{\infty} \sum_{m=-n}^{n} a_n^m h_n^{(1)}(k|x|) Y_n^m(\hat{x}) = u(x)$$

for $|x| > \tilde{R} > R$. From this it is now obvious that u represents a radiating solution to the Helmholtz equation.   □

Let R be the radius of the smallest closed ball with center at the origin containing the bounded domain D. Then, by the preceding theorem, each radiating solution $u \in C^2(\mathbb{R}^3 \setminus \bar{D})$ to the Helmholtz equation has an expansion with respect to spherical wave functions of the form (2.48) that converges absolutely and uniformly on compact subsets of $|x| > R$. Conversely, the expansion (2.48) is valid in all of $\mathbb{R}^3 \setminus \bar{D}$ if the origin is contained in D and if u can be extended as a solution to the Helmholtz equation in the exterior of the largest closed ball with center at the origin contained in $\bar{D}$.

Theorem 2.15 *The far field pattern of the radiating solution to the Helmholtz equation with the expansion (2.48) is given by the uniformly convergent series*

$$(2.49) \qquad u_\infty = \frac{1}{k} \sum_{n=0}^{\infty} \frac{1}{i^{n+1}} \sum_{m=-n}^{n} a_n^m Y_n^m.$$

The coefficients in this expansion satisfy the growth condition

$$(2.50) \qquad \sum_{n=0}^{\infty} \left(\frac{2n}{ker} \right)^{2n} \sum_{m=-n}^{n} |a_n^m|^2 < \infty$$

for all $r > R$.

Proof. We cannot pass to the limit $|x| \to \infty$ in (2.48) by using the asymptotic behavior (2.41) because the latter does not hold uniformly in n. Since by Theorem 2.5 the far field pattern u_∞ is analytic, we have an expansion

$$u_\infty = \sum_{n=0}^{\infty} \sum_{m=-n}^{n} b_n^m Y_n^m$$

with coefficients

$$b_n^m = \int_\Omega u_\infty(\hat{x}) \overline{Y_n^m(\hat{x})} \, ds(\hat{x}).$$

On the other hand, the coefficients a_n^m in the expansion (2.48) clearly are given by

$$a_n^m h_n^{(1)}(kr) = \int_\Omega u(r\hat{x}) \overline{Y_n^m(\hat{x})} \, ds(\hat{x}).$$

Therefore, with the aid of (2.41) we find that

$$b_n^m = \int_\Omega \lim_{r\to\infty} r \, e^{-ikr} u(r\hat{x}) \overline{Y_n^m(\hat{x})} \, ds(\hat{x})$$

$$= \lim_{r\to\infty} r \, e^{-ikr} \int_\Omega u(r\hat{x}) \overline{Y_n^m(\hat{x})} \, ds(\hat{x}) = \frac{a_n^m}{k \, i^{n+1}},$$

and the expansion (2.49) is valid in the L^2 sense.

Parseval's equation for the expansion (2.48) reads

$$r^2 \sum_{n=0}^\infty \sum_{m=-n}^n |a_n^m|^2 \left| h_n^{(1)}(kr) \right|^2 = \int_{|x|=r} |u(x)|^2 ds(x).$$

From this, using the asymptotic behavior (2.39) of the Hankel functions for large order n, the condition (2.50) follows. In particular, by the Schwarz inequality, we can now conclude that (2.49) is uniformly valid on Ω.   □

Theorem 2.16 *Let the Fourier coefficients b_n^m of $u_\infty \in L^2(\Omega)$ with respect to the spherical harmonics satisfy the growth condition*

(2.51)
$$\sum_{n=0}^\infty \left(\frac{2n}{keR} \right)^{2n} \sum_{m=-n}^n |b_n^m|^2 < \infty$$

with some $R > 0$. Then

(2.52)
$$u(x) = k \sum_{n=0}^\infty i^{n+1} \sum_{m=-n}^n b_n^m h_n^{(1)}(k|x|) Y_n^m(\hat{x}), \quad |x| > R,$$

is a radiating solution of the Helmholtz equation with far field pattern u_∞.

Proof. By the asymptotic behavior (2.39), the assumption (2.51) implies that the series (2.52) converges in the mean square sense on the sphere $|x| = R$. Hence, by Theorem 2.14, u is a radiating solution to the Helmholtz equation. The fact that the far field pattern coincides with the given function u_∞ follows from Theorem 2.15.   □

The last two theorems indicate that the equation

(2.53)
$$Fu = u_\infty$$

with the linear operator F mapping a radiating solution u to the Helmholtz equation onto its far field u_∞ is ill-posed. Following Hadamard [63], a problem is called *properly posed* or *well-posed* if a solution exists, if the solution is unique and if the solution continuously depends on the data. Otherwise, the problem is called *improperly posed* or *ill-posed*. Here, for equation (2.53), by Theorem 2.13 we have uniqueness of the solution. However, since by Theorem 2.15 the existence of a solution requires the growth condition (2.50) to be satisfied, for a given function u_∞ in $L^2(\Omega)$ a solution of equation (2.53) will, in general, not exist. Furthermore, if a solution u does exist it will not continuously depend on u_∞ in any reasonable norm. This is illustrated by the fact that for the radiating solutions

$$u_n(x) = \frac{1}{n} \, h_n^{(1)}(k|x|) \, Y_n(\hat{x})$$

where Y_n is a normalized spherical harmonic of degree n the far field patterns are given by

$$u_{n,\infty} = \frac{1}{ki^{n+1}n} \, Y_n.$$

Hence, we have convergence $u_{n,\infty} \to 0$, $n \to \infty$, in the L^2 norm on Ω whereas, as a consequence of the asymptotic behavior (2.39) of the Hankel functions for large order n, the u_n will not converge in any suitable norm. Later in this book we will study the ill-posedness of the reconstruction of a radiating solution of the Helmholtz equation from its far field pattern more closely. In particular, we will describe stable methods for approximately solving improperly posed problems such as this one.

3. Direct Acoustic Obstacle Scattering

This chapter is devoted to the solution of the direct obstacle scattering problem for acoustic waves. As in [32], we choose the method of integral equations for solving the boundary value problems. However, we decided to leave out some of the details in the analysis. In particular, we assume that the reader is familiar with the Riesz–Fredholm theory for operator equations of the second kind in dual systems as described in [32] and [106]. We also do not repeat the technical proofs for the jump relations and regularity properties for single- and double-layer potentials. Leaving aside these two restrictions, however, we will present a rather complete analysis of the forward scattering problem. For the reader interested in a more comprehensive treatment of the direct problem, we suggest consulting our previous book [32] on this subject.

We begin by listing the jump and regularity properties of surface potentials in the classical setting of continuous and Hölder continuous functions and briefly mention extensions to the case of Sobolev spaces. We then proceed to establish the existence of the solution to the exterior Dirichlet problem via boundary integral equations and also describe some results on the regularity of the solution. In particular, we will establish the well-posedness of the Dirichlet to Neumann map in the Hölder space setting. Coming back to the far field pattern, we prove the important reciprocity relation which we then use to derive some completeness results on the set of far field patterns corresponding to the scattering of incident plane waves propagating in different directions. For this we need to introduce the notion of Herglotz wave functions.

Our presentation is in $\mathbb{R}^3$. For the sake of completeness, we include a section where we list the necessary modifications for the two-dimensional theory. We also add a section advertising a Nyström method for the numerical solution of the boundary integral equations in two dimensions. This very simple and effective method has been around for some time already but is not used as much in practice as we think it should be. We present the method both for the case of an analytic boundary and also briefly describe more recent modifications for domains with corners. Finally, we present the main ideas of a Nyström type method for the numerical solution of the boundary integral equations in three dimensions developed by Wienert. Based on approximations via spherical harmonics, it is exponentially convergent for analytic boundaries.

3.1 Single- and Double-Layer Potentials

In this chapter, if not stated otherwise, we always will assume that D is the open complement of an unbounded domain of class C^2, that is, we include scattering from more than one obstacle in our analysis.

We first briefly review the basic jump relations and regularity properties of acoustic single- and double-layer potentials. Given an integrable function φ, the integrals

$$u(x) := \int_{\partial D} \varphi(y)\Phi(x,y)\,ds(y), \quad x \in \mathbb{R}^3 \setminus \partial D,$$

and

$$v(x) := \int_{\partial D} \varphi(y) \frac{\partial \Phi(x,y)}{\partial \nu(y)}\,ds(y), \quad x \in \mathbb{R}^3 \setminus \partial D,$$

are called, respectively, *acoustic single-layer* and *acoustic double-layer potentials* with density φ. They are solutions to the Helmholtz equation in D and in $\mathbb{R}^3 \setminus \bar{D}$ and satisfy the Sommerfeld radiation condition. Green's formulas (2.5) and (2.8) show that any solution to the Helmholtz equation can be represented as a combination of single- and double-layer potentials. For continuous densities, the behavior of the surface potentials at the boundary is described by the following *jump relations*. By $\|\cdot\|_\infty = \|\cdot\|_{\infty,G}$ we denote the usual supremum norm of real or complex valued functions defined on a set $G \subset \mathbb{R}^3$.

Theorem 3.1 *Let ∂D be of class C^2 and let φ be continuous. Then the single-layer potential u with density φ is continuous throughout $\mathbb{R}^3$ and*

$$\|u\|_{\infty,\mathbb{R}^3} \le C\|\varphi\|_{\infty,\partial D}$$

for some constant C depending on ∂D. On the boundary we have

$$(3.1) \qquad u(x) = \int_{\partial D} \varphi(y)\Phi(x,y)\,ds(y), \quad x \in \partial D,$$

$$(3.2) \qquad \frac{\partial u_\pm}{\partial \nu}(x) = \int_{\partial D} \varphi(y) \frac{\partial \Phi(x,y)}{\partial \nu(x)}\,ds(y) \mp \frac{1}{2}\varphi(x), \quad x \in \partial D,$$

where

$$\frac{\partial u_\pm}{\partial \nu}(x) := \lim_{h \to +0} \nu(x) \cdot \operatorname{grad} u(x \pm h\nu(x))$$

is to be understood in the sense of uniform convergence on ∂D and where the integrals exist as improper integrals. The double-layer potential v with density φ can be continuously extended from D to $\bar{D}$ and from $\mathbb{R}^3 \setminus \bar{D}$ to $\mathbb{R}^3 \setminus D$ with limiting values

$$(3.3) \qquad v_\pm(x) = \int_{\partial D} \varphi(y) \frac{\partial \Phi(x,y)}{\partial \nu(y)}\,ds(y) \pm \frac{1}{2}\varphi(x), \quad x \in \partial D,$$

where

$$v_\pm(x) := \lim_{h \to +0} v(x \pm h\nu(x))$$

and where the integral exists as an improper integral. Furthermore,

$$\|v\|_{\infty,\overline{D}} \le C\|\varphi\|_{\infty,\partial D}, \quad \|v\|_{\infty,\mathbb{R}^3\setminus D} \le C\|\varphi\|_{\infty,\partial D}$$

for some constant C depending on ∂D and

$$(3.4) \qquad \lim_{h\to+0} \left\{ \frac{\partial v}{\partial \nu}(x + h\nu(x)) - \frac{\partial v}{\partial \nu}(x - h\nu(x)) \right\} = 0, \quad x \in \partial D,$$

uniformly on ∂D.

Proof. For a proof, we refer to Theorems 2.12, 2.13, 2.19 and 2.21 in [32]. The estimates for the double-layer potential can be established with the aid of the uniform boundedness principle. $\qquad\square$

An appropriate framework for formulating additional regularity properties of these surface potentials is provided by the concept of Hölder spaces. A real or complex valued function φ defined on a set $G \subset \mathbb{R}^3$ is called *uniformly Hölder continuous* with *Hölder exponent* $0 < \alpha \le 1$ if there is a constant C such that

$$(3.5) \qquad |\varphi(x) - \varphi(y)| \le C|x - y|^\alpha$$

for all $x, y \in G$. We define the *Hölder space* $C^{0,\alpha}(G)$ to be the linear space of all functions defined on G which are bounded and uniformly Hölder continuous with exponent α. It is a Banach space with the norm

$$(3.6) \qquad \|\varphi\|_\alpha := \|\varphi\|_{\alpha,G} := \sup_{x\in G}|\varphi(x)| + \sup_{\substack{x,y\in G\\x\ne y}} \frac{|\varphi(x) - \varphi(y)|}{|x - y|^\alpha}.$$

Clearly, for $\alpha < \beta$ each function $\varphi \in C^{0,\beta}(G)$ is also contained in $C^{0,\alpha}(G)$. For this imbedding, from the Arzelà–Ascoli theorem, we have the following compactness property (for a proof we refer to [32], p. 38 or [106], p. 84).

Theorem 3.2 *Let $0 < \alpha < \beta \le 1$ and let G be compact. Then the imbedding operators*

$$I^\beta : C^{0,\beta}(G) \to C(G), \quad I^{\alpha,\beta} : C^{0,\beta}(G) \to C^{0,\alpha}(G)$$

are compact.

For a vector field, Hölder continuity and the Hölder norm are defined analogously by replacing absolute values in (3.5) and (3.6) by Euclidean norms. We can then introduce the Hölder space $C^{1,\alpha}(G)$ of uniformly Hölder continuously differentiable functions as the space of differentiable functions φ for which grad φ (or the surface gradient Grad φ in the case $G = \partial D$) belongs to $C^{0,\alpha}(G)$. With the norm

$$\|\varphi\|_{1,\alpha} := \|\varphi\|_{1,\alpha,G} := \|\varphi\|_\infty + \|\operatorname{grad}\varphi\|_{0,\alpha}$$

$C^{1,\alpha}(G)$ is again a Banach space and we also have an imbedding theorem corresponding to Theorem 3.2.

Extending Theorem 3.1, we can now formulate the following regularity properties of single- and double-layer potentials in terms of Hölder continuity.

Theorem 3.3 *Let ∂D be of class C^2 and let $0 < \alpha < 1$. Then the single-layer potential u with density $\varphi \in C(\partial D)$ is uniformly Hölder continuous throughout $\mathbb{R}^3$ and*

$$\|u\|_{\alpha,\mathbb{R}^3} \leq C_\alpha \|\varphi\|_{\infty,\partial D}.$$

The first derivatives of the single-layer potential u with density $\varphi \in C^{0,\alpha}(\partial D)$ can be uniformly Hölder continuously extended from D to $\bar{D}$ and from $\mathbb{R}^3 \setminus \bar{D}$ to $\mathbb{R}^3 \setminus D$ with boundary values

$$(3.7) \quad \operatorname{grad} u_\pm(x) = \int_{\partial D} \varphi(y) \operatorname{grad}_x \Phi(x,y)\, ds(y) \mp \frac{1}{2}\, \varphi(x)\nu(x), \quad x \in \partial D,$$

where

$$\operatorname{grad} u_\pm(x) := \lim_{h \to +0} \operatorname{grad} u(x \pm h\nu(x))$$

and we have

$$\|\operatorname{grad} u\|_{\alpha,\bar{D}} \leq C_\alpha \|\varphi\|_{\alpha,\partial D}, \quad \|\operatorname{grad} u\|_{\alpha,\mathbb{R}^3 \setminus D} \leq C_\alpha \|\varphi\|_{\alpha,\partial D}.$$

The double-layer potential v with density $\varphi \in C^{0,\alpha}(\partial D)$ can be uniformly Hölder continuously extended from D to $\bar{D}$ and from $\mathbb{R}^3 \setminus \bar{D}$ to $\mathbb{R}^3 \setminus D$ such that

$$\|v\|_{\alpha,\bar{D}} \leq C_\alpha \|\varphi\|_{\alpha,\partial D}, \quad \|v\|_{\alpha,\mathbb{R}^3 \setminus D} \leq C_\alpha \|\varphi\|_{\alpha,\partial D}.$$

The first derivatives of the double-layer potential v with density $\varphi \in C^{1,\alpha}(\partial D)$ can be uniformly Hölder continuously extended from D to $\bar{D}$ and from $\mathbb{R}^3 \setminus \bar{D}$ to $\mathbb{R}^3 \setminus D$ such that

$$\|\operatorname{grad} v\|_{\alpha,\bar{D}} \leq C_\alpha \|\varphi\|_{1,\alpha,\partial D}, \quad \|\operatorname{grad} v\|_{\alpha,\mathbb{R}^3 \setminus D} \leq C_\alpha \|\varphi\|_{1,\alpha,\partial D}.$$

In all inequalities, C_α denotes some constant depending on ∂D and α.

Proof. For a proof, we refer to the Theorems 2.12, 2.16, 2.17 and 2.23 in [32]. □

For the direct values of the single- and double-layer potentials on the boundary ∂D, we have more regularity. This can be conveniently expressed in terms of the mapping properties of the single- and double-layer operators S and K, given by

$$(3.8) \qquad (S\varphi)(x) := 2 \int_{\partial D} \Phi(x,y)\varphi(y)\, ds(y), \quad x \in \partial D,$$

$$(3.9) \qquad (K\varphi)(x) := 2 \int_{\partial D} \frac{\partial \Phi(x,y)}{\partial \nu(y)}\, \varphi(y)\, ds(y), \quad x \in \partial D,$$

and the normal derivative operators K' and T, given by

$$(3.10) \qquad (K'\varphi)(x) := 2 \int_{\partial D} \frac{\partial \Phi(x,y)}{\partial \nu(x)} \varphi(y) \, ds(y), \quad x \in \partial D,$$

$$(3.11) \qquad (T\varphi)(x) := 2 \frac{\partial}{\partial \nu(x)} \int_{\partial D} \frac{\partial \Phi(x,y)}{\partial \nu(y)} \varphi(y) \, ds(y), \quad x \in \partial D.$$

Theorem 3.4 *Let ∂D be of class C^2. Then the operators S, K and K' are bounded operators from $C(\partial D)$ into $C^{0,\alpha}(\partial D)$, the operators S and K are also bounded from $C^{0,\alpha}(\partial D)$ into $C^{1,\alpha}(\partial D)$, and the operator T is bounded from $C^{1,\alpha}(\partial D)$ into $C^{0,\alpha}(\partial D)$.*

Proof. The statements on S and T are contained in the preceding theorem and proofs for the operators K and K' can be found in Theorems 2.15, 2.22, and 2.30 of [32].   □

We wish to point out that all these jump and regularity properties essentially are deduced from the corresponding results for the classical single- and double-layer potentials for the Laplace equation by smoothness arguments on the difference between the fundamental solutions for the Helmholtz and the Laplace equation.

Clearly, by interchanging the order of integration, we see that S is self adjoint and K and K' are adjoint with respect to the bilinear form

$$\langle \varphi, \psi \rangle := \int_{\partial D} \varphi \psi \, ds,$$

that is,

$$\langle S\varphi, \psi \rangle = \langle \varphi, S\psi \rangle$$

and

$$\langle K\varphi, \psi \rangle = \langle \varphi, K'\psi \rangle$$

for all $\varphi, \psi \in C(\partial D)$. To derive further properties of the boundary integral operators, let u and v denote the double-layer potentials with densities φ and ψ in $C^{1,\alpha}(\partial D)$, respectively. Then by the jump relations of Theorem 3.1, Green's theorem (2.3) and the radiation condition we find that

$$\int_{\partial D} T\varphi\psi \, ds = 2 \int_{\partial D} \frac{\partial u}{\partial \nu} (v_+ - v_-) \, ds = 2 \int_{\partial D} (u_+ - u_-) \frac{\partial v}{\partial \nu} \, ds = \int_{\partial D} \varphi T\psi \, ds,$$

that is, T also is self adjoint. Now, in addition, let w denote the single-layer potential with density $\varphi \in C(\partial D)$. Then

$$\int_{\partial D} S\varphi T\psi \, ds = 4 \int_{\partial D} w \frac{\partial v}{\partial \nu} \, ds = 4 \int_{\partial D} v_- \frac{\partial w_-}{\partial \nu} \, ds = \int_{\partial D} (K - I)\psi (K' + I)\varphi \, ds,$$

whence

$$\int_{\partial D} \varphi ST\psi \, ds = \int_{\partial D} \varphi(K^2 - I)\psi \, ds$$

follows for all $\varphi \in C(\partial D)$ and $\psi \in C^{1,\alpha}(\partial D)$. Thus, we have proven the relation

$$(3.12) \qquad\qquad ST = K^2 - I$$

and similarly it can be shown that the adjoint relation

$$(3.13) \qquad\qquad TS = K'^2 - I$$

is also valid. Throughout the book I stands for the identity operator.

Looking at the regularity and mapping properties of surface potentials, we think it is natural to start with the classical Hölder space case. As worked out in detail by Kirsch [86, 89], the corresponding results in the Sobolev space setting can be deduced from these classical results through the use of a functional analytic tool provided by Lax [115], that is, the classical results are stronger. Since we shall be referring to Lax's theorem several times in the sequel, we prove it here.

Theorem 3.5 *Let X and Y be normed spaces both of which have a scalar product $(\cdot,\cdot)$ and assume that there exists a positive constant c such that*

$$(3.14) \qquad\qquad |(\varphi,\psi)| \leq c\|\varphi\|\,\|\psi\|$$

for all $\varphi, \psi \in X$. Let $U \subset X$ be a subspace and let $A : U \to Y$ and $B : Y \to X$ be bounded linear operators satisfying

$$(3.15) \qquad\qquad (A\varphi, \psi) = (\varphi, B\psi)$$

for all $\varphi \in U$ and $\psi \in Y$. Then $A : U \to Y$ is bounded with respect to the norms induced by the scalar products.

Proof. We denote the norms induced by the scalar products by $\|\cdot\|_s$. Consider the bounded operator $M : U \to X$ given by $M := BA$ with $\|M\| \leq \|B\|\,\|A\|$. Then, as a consequence of (3.15), M is self adjoint, that is, $(M\varphi, \psi) = (\varphi, M\psi)$ for all $\varphi, \psi \in U$. Therefore, using Schwarz's inequality, we obtain

$$\|M^n\varphi\|_s^2 = (M^n\varphi, M^n\varphi) = (\varphi, M^{2n}\varphi) \leq \|M^{2n}\varphi\|_s$$

for all $\varphi \in U$ with $\|\varphi\|_s \leq 1$ and all $n \in \mathbb{N}$. From this, by induction, it follows that

$$\|M\varphi\|_s \leq \|M^{2^n}\varphi\|_s^{2^{-n}}.$$

By (3.14) we have $\|\varphi\|_s \leq \sqrt{c}\,\|\varphi\|$ for all $\varphi \in X$. Hence,

$$\|M\varphi\|_s \leq \{\sqrt{c}\,\|M^{2^n}\varphi\|\}^{2^{-n}} \leq \{\sqrt{c}\,\|\varphi\|\,\|M\|^{2^n}\}^{2^{-n}} = \{\sqrt{c}\,\|\varphi\|\}^{2^{-n}}\|M\|.$$

Passing to the limit $n \to \infty$ now yields

$$\|M\varphi\|_s \leq \|M\|$$

for all $\varphi \in U$ with $\|\varphi\|_s \leq 1$. Finally, for all $\varphi \in U$ with $\|\varphi\|_s \leq 1$, we again have from Schwarz's inequality that

$$\|A\varphi\|_s^2 = (A\varphi, A\varphi) = (\varphi, M\varphi) \leq \|M\varphi\|_s \leq \|M\|.$$

From this the statement follows. $\square$

We now use Lax's Theorem 3.5 to prove the mapping properties of surface potentials in Sobolev spaces. (For an introduction into Sobolev spaces, we refer to Adams [1].)

Theorem 3.6 *Let ∂D be of class C^2 and let $H^1(\partial D)$ denote the usual Sobolev space. Then the operator S is bounded from $L^2(\partial D)$ into $H^1(\partial D)$. Assume further that ∂D belongs to $C^{2,\alpha}$. Then the operators K and K' are bounded from $L^2(\partial D)$ into $H^1(\partial D)$ and the operator T is bounded from $H^1(\partial D)$ into $L^2(\partial D)$.*

Proof. We prove the boundedness of $S : L^2(\partial D) \to H^1(\partial D)$. Let $X = C^{0,\alpha}(\partial D)$ and $Y = C^{1,\alpha}(\partial D)$ be equipped with the usual Hölder norms and introduce scalar products on X by the L^2 scalar product and on Y by the H^1 scalar product

$$(u, v)_{H^1(\partial D)} := \int_{\partial D} \{\varphi \bar{\psi} + \operatorname{Grad} \varphi \cdot \operatorname{Grad} \bar{\psi}\} ds.$$

By interchanging the order of integration, we have

$$(3.16) \qquad \int_{\partial D} S\varphi \, \psi \, ds = \int_{\partial D} \varphi \, S\psi \, ds$$

for all $\varphi, \psi \in C(\partial D)$. For $\varphi \in C^{0,\alpha}(\partial D)$ and $\psi \in C^2(\partial D)$, by Gauss' surface divergence theorem and (3.16) we have

$$(3.17) \qquad \int_{\partial D} \operatorname{Grad} S\varphi \cdot \operatorname{Grad} \psi \, ds = - \int_{\partial D} \varphi \, S(\operatorname{Div} \operatorname{Grad} \psi) \, ds.$$

(For the reader who is not familiar with vector analysis on surfaces, we refer to Section 6.3.) Using again Gauss' surface divergence theorem and the relation $\operatorname{grad}_x \Phi(x, y) = -\operatorname{grad}_y \Phi(x, y)$, we find that

$$\int_{\partial D} \Phi(x, y) \operatorname{Div} \operatorname{Grad} \psi(y) \, ds(y) = \operatorname{div} \int_{\partial D} \Phi(x, y) \operatorname{Grad} \psi(y) \, ds(y), \quad x \notin \partial D.$$

Hence, with the aid of the jump relations of Theorem 3.1 and 3.3 (see also Theorem 6.12), for $\psi \in C^2(\partial D)$ we obtain

$$S(\operatorname{Div} \operatorname{Grad} \psi) = \tilde{S}(\operatorname{Grad} \psi)$$

where the bounded operator $\tilde{S} : C^{0,\alpha}(\partial D) \to C^{0,\alpha}(\partial D)$ is given by

$$(\tilde{S}a)(x) := 2 \operatorname{div} \int_{\partial D} \Phi(x, y) a(y) \, ds(y), \quad x \in \partial D,$$

for Hölder continuous tangential fields a on ∂D. Therefore, from (3.17) we have

$$(3.18) \qquad \int_{\partial D} \operatorname{Grad} S\varphi \cdot \operatorname{Grad} \psi \, ds = -\int_{\partial D} \varphi \, \tilde{S}(\operatorname{Grad} \psi) \, ds$$

for all $\varphi \in C^{0,\alpha}(\partial D)$ and $\psi \in C^2(\partial D)$. Since, for fixed φ, both sides of (3.18) represent bounded linear functionals on $C^{1,\alpha}(\partial D)$, (3.18) is also true for all $\varphi \in C^{0,\alpha}(\partial D)$ and $\psi \in C^{1,\alpha}(\partial D)$. Hence, from (3.16) and (3.18) we have that the operators $S : C^{0,\alpha}(\partial D) \to C^{1,\alpha}(\partial D)$ and $S^* : C^{1,\alpha}(\partial D) \to C^{0,\alpha}(\partial D)$ given by

$$S^*\psi := \overline{S\bar{\psi}} - \overline{\tilde{S} \operatorname{Grad} \bar{\psi}}$$

are adjoint, i.e.,

$$(S\varphi, \psi)_{H^1(\partial D)} = (\varphi, S^*\psi)_{L^2(\partial D)}$$

for all $\varphi \in C^{0,\alpha}(\partial D)$ and $\psi \in C^{1,\alpha}(\partial D)$. By Theorem 3.3, both S and S^* are bounded with respect to the Hölder norms. Hence, from Lax's Theorem 3.5 we see that there exists a positive constant C such that

$$\|S\varphi\|_{H^1(\partial D)} \leq C\|\varphi\|_{L^2(\partial D)}$$

for all $\varphi \in C^{0,\alpha}(\partial D)$. The proof of the boundedness of $S : L^2(\partial D) \to H^1(\partial D)$ is now finished by observing that $C^{0,\alpha}(\partial D)$ is dense in $L^2(\partial D)$.

The proofs of the assertions on K, K' and T are similar in structure and for details we refer the reader to [86, 89]. □

The jump relations of Theorem 3.1 can also be extended through the use of Lax's theorem from the case of continuous densities to L^2 densities. This was done by Kersten [85]. In the L^2 setting, the jump relations (3.1)–(3.4) have to be replaced by

$$(3.19) \qquad \lim_{h \to +0} \int_{\partial D} |2u(x \pm h\nu(x)) - (S\varphi)(x)|^2 ds(x) = 0,$$

$$(3.20) \qquad \lim_{h \to +0} \int_{\partial D} \left| 2\frac{\partial u}{\partial \nu}(x \pm h\nu(x)) - (K'\varphi)(x) \pm \varphi(x) \right|^2 ds(x) = 0$$

for the single-layer potential u with density $\varphi \in L^2(\partial D)$ and

$$(3.21) \qquad \lim_{h \to +0} \int_{\partial D} |2v(x \pm h\nu(x)) - (K\varphi)(x) \mp \varphi(x)|^2 ds(x) = 0,$$

$$(3.22) \qquad \lim_{h \to +0} \int_{\partial D} \left| \frac{\partial v}{\partial \nu}(x + h\nu(x)) - \frac{\partial v}{\partial \nu}(x - h\nu(x)) \right|^2 ds(x) = 0$$

for the double-layer potential v with density $\varphi \in L^2(\partial D)$. Using Lax's theorem, Hähner [65] has also established that

$$(3.23) \quad \lim_{h \to +0} \int_{\partial D} \left| \operatorname{grad} u(\cdot \pm h\nu) - \int_{\partial D} \operatorname{grad}_x \Phi(\cdot, y)\varphi(y) \, ds(y) \pm \frac{1}{2}\varphi\nu \right|^2 ds = 0$$

for single-layer potentials u with $L^2(\partial D)$ density φ, extending the jump relation (3.7).

3.2 Scattering from a Sound-Soft Obstacle

The scattering of time-harmonic acoustic waves by sound-soft obstacles leads to the following problem.

Direct Acoustic Obstacle Scattering Problem. *Given an entire solution* u^i *to the Helmholtz equation representing an incident field, find a solution*

$$u = u^i + u^s$$

to the Helmholtz equation in $\mathbb{R}^3 \setminus \bar{D}$ *such that the scattered field* u^s *satisfies the Sommerfeld radiation condition and the total field* u *satisfies the boundary condition*

$$u = 0 \quad on \ \partial D.$$

Clearly, after renaming the unknown functions, this direct scattering problem is a special case of the following Dirichlet problem.

Exterior Dirichlet Problem. *Given a continuous function* f *on* ∂D, *find a radiating solution* $u \in C^2(\mathbb{R}^3 \setminus \bar{D}) \cap C(\mathbb{R}^3 \setminus D)$ *to the Helmholtz equation*

$$\triangle u + k^2 u = 0 \quad in \ \mathbb{R}^3 \setminus \bar{D}$$

which satisfies the boundary condition

$$u = f \quad on \ \partial D.$$

We briefly sketch uniqueness, existence and well-posedness for this boundary value problem.

Theorem 3.7 *The exterior Dirichlet problem has at most one solution.*

Proof. We have to show that solutions to the homogeneous boundary value problem $u = 0$ on ∂D vanish identically. If u had a normal derivative in the sense of uniform convergence, we could immediately apply Theorem 2.12 to obtain $u = 0$ in $\mathbb{R}^3 \setminus \bar{D}$. However, in our formulation of the exterior Dirichlet problem we require u only to be continuous up to the boundary which is the natural assumption for posing the Dirichlet boundary condition. There are two possibilities to overcome this difficulty: either we can use the fact that the solution to the Dirichlet problem is in $C^{1,\alpha}(\mathbb{R}^3 \setminus D)$ provided the given boundary data is in $C^{1,\alpha}(\partial D)$ (c.f. [32] or [120]), or we can justify the application of Green's theorem by a more direct argument using convergence theorems for Lebesgue integration. Despite the fact that later we will also need the result on the smoothness of solutions to the exterior Dirichlet problem up to the boundary, we briefly sketch a variant of the second alternative based on an approximation idea due to Heinz (see [55] and also [182] and [76], p. 144). It is more satisfactory since it does not rely on techniques used in the existence results. Thus, we state and prove the following lemma which then justifies the application of Theorem 2.12. Note that this uniqueness result for the Dirichlet problem requires no regularity assumptions on the boundary ∂D.   □

Lemma 3.8 *Let $u \in C^2(\mathbb{R}^3 \setminus \bar{D}) \cap C(\mathbb{R}^3 \setminus D)$ be a solution to the Helmholtz equation in $\mathbb{R}^3 \setminus \bar{D}$ which satisfies the homogeneous boundary condition $u = 0$ on ∂D. Define $D_R := \{y \in \mathbb{R}^3 \setminus \bar{D} : |y| < R\}$ for sufficiently large R. Then $\operatorname{grad} u \in L^2(D_R)$ and*

$$(3.24) \qquad \int_{D_R} |\operatorname{grad} u|^2 \, dx - k^2 \int_{D_R} |u|^2 dx = \int_{|x|=R} u \, \frac{\partial \bar{u}}{\partial \nu} \, ds.$$

Proof. We first assume that u is real valued. We choose an odd function $\psi \in C^1(\mathbb{R})$ such that $\psi(t) = 0$ for $0 \leq t \leq 1$, $\psi(t) = t$ for $t \geq 2$ and $\psi'(t) \geq 0$ for all t, and set $u_n := \psi(nu)/n$. We then have uniform convergence $\|u - u_n\|_\infty \to 0$, $n \to \infty$. Since $u = 0$ on the boundary ∂D, the functions u_n vanish in a neighborhood of ∂D and we can apply Green's theorem (2.2) to obtain

$$\int_{D_R} \operatorname{grad} u_n \cdot \operatorname{grad} u \, dx = k^2 \int_{D_R} u_n u \, dx + \int_{|x|=R} u_n \, \frac{\partial u}{\partial \nu} \, ds.$$

It can be easily seen that

$$0 \leq \operatorname{grad} u_n(x) \cdot \operatorname{grad} u(x) = \psi'(nu(x)) \, |\operatorname{grad} u(x)|^2 \to |\operatorname{grad} u(x)|^2, \quad n \to \infty,$$

for all x not contained in $\{x \in D_R : u(x) = 0, \operatorname{grad} u(x) \neq 0\}$. Since as a consequence of the implicit function theorem the latter set has Lebesgue measure zero, Fatou's lemma tells us that $\operatorname{grad} u \in L^2(D_R)$.

Now assume $u = v + iw$ with real functions v and w. Then, since v and w also satisfy the assumptions of our lemma, we have $\operatorname{grad} v, \operatorname{grad} w \in L^2(D_R)$. From

$$\operatorname{grad} v_n + i \operatorname{grad} w_n = \psi'(nv) \operatorname{grad} v + i \psi'(nw) \operatorname{grad} w$$

we can estimate

$$|(\operatorname{grad} v_n + i \operatorname{grad} w_n) \cdot \operatorname{grad} \bar{u}| \leq 2\|\psi'\|_\infty \left\{ |\operatorname{grad} v|^2 + |\operatorname{grad} w|^2 \right\}.$$

Hence, by the Lebesgue dominated convergence theorem, we can pass to the limit $n \to \infty$ in Green's theorem

$$\int_{D_R} \{(\operatorname{grad} v_n + i \operatorname{grad} w_n) \cdot \operatorname{grad} \bar{u} + (v_n + iw_n) \Delta \bar{u}\} \, dx = \int_{|x|=R} (v_n + i w_n) \frac{\partial \bar{u}}{\partial \nu} \, ds$$

to obtain (3.24).   □

The existence of a solution to the exterior Dirichlet problem can be based on boundary integral equations. In the so-called *layer approach*, we seek the solution in the form of acoustic surface potentials. Here, we choose an approach in the form of a combined acoustic double- and single-layer potential

$$(3.25) \quad u(x) = \int_{\partial D} \left\{ \frac{\partial \Phi(x,y)}{\partial \nu(y)} - i\eta \Phi(x,y) \right\} \varphi(y) \, ds(y), \quad x \in \mathbb{R}^3 \setminus \partial D,$$

with a density $\varphi \in C(\partial D)$ and a real coupling parameter $\eta \neq 0$. Then from the jump relations of Theorem 3.1 we see that the potential u given by (3.25) in $\mathbb{R}^3 \setminus \bar{D}$ solves the exterior Dirichlet problem provided the density is a solution of the integral equation

$$(3.26) \qquad\qquad \varphi + K\varphi - i\eta S\varphi = 2f.$$

Combining Theorems 3.2 and 3.4, the operators $S, K : C(\partial D) \to C(\partial D)$ are seen to be compact. Therefore, the existence of a solution to (3.26) can be established by the Riesz–Fredholm theory for equations of the second kind with a compact operator.

Let φ be a continuous solution to the homogeneous form of (3.26). Then the potential u given by (3.25) satisfies the homogeneous boundary condition $u_+ = 0$ on ∂D whence by the uniqueness for the exterior Dirichlet problem $u = 0$ in $\mathbb{R}^3 \setminus \bar{D}$ follows. The jump relations (3.1)–(3.4) now yield

$$-u_- = \varphi, \qquad -\frac{\partial u_-}{\partial \nu} = i\eta\varphi \quad \text{on } \partial D.$$

Hence, using Green's theorem (2.2), we obtain

$$i\eta \int_{\partial D} |\varphi|^2 ds = \int_{\partial D} \bar{u}_- \frac{\partial u_-}{\partial \nu} \, ds = \int_D \left\{ |\operatorname{grad} u|^2 - k^2 |u|^2 \right\} dx.$$

Taking the imaginary part of the last equation shows that $\varphi = 0$. Thus, we have established uniqueness for the integral equation (3.26), that is, injectivity of the operator $I + K - i\eta S : C(\partial D) \to C(\partial D)$. Then, by the Riesz–Fredholm theory, $I + K - i\eta S$ is bijective and the inverse $(I + K - i\eta S)^{-1} : C(\partial D) \to C(\partial D)$ is bounded. Hence, the inhomogeneous equation (3.26) possesses a solution and this solution depends continuously on f in the maximum norm. From the representation (3.25) of the solution as a combined double- and single-layer potential, with the aid of the regularity estimates in Theorem 3.1, the continuous dependence of the density φ on the boundary data f shows that the exterior Dirichlet problem is well-posed, i.e., small deviations in f in the maximum norm ensure small deviations in u in the maximum norm on $\mathbb{R}^3 \setminus D$ and small deviations of all its derivatives in the maximum norm on closed subsets of $\mathbb{R}^3 \setminus \bar{D}$.

We summarize these results in the following theorem.

Theorem 3.9 *The exterior Dirichlet problem has a unique solution and the solution depends continuously on the boundary data with respect to uniform convergence of the solution on $\mathbb{R}^3 \setminus D$ and all its derivatives on closed subsets of $\mathbb{R}^3 \setminus \bar{D}$.*

Note that for $\eta = 0$ the integral equation (3.26) becomes non-unique if k is a so-called irregular wave number or internal resonance, i.e., if there exist non-trivial solutions u to the Helmholtz equation in the interior domain D satisfying homogeneous Neumann boundary conditions $\partial u / \partial \nu = 0$ on ∂D. The approach (3.25) was introduced independently by Leis [118], Brakhage and Werner [17], and Panich [152] in order to remedy this non-uniqueness deficiency of the classical double-layer approach due to Vekua [179], Weyl [190] and Müller [139]. For an investigation on the proper choice of the coupling parameter η with respect to the condition of the integral equation (3.26), we refer to Kress [103].

In the literature, a variety of other devices have been designed for overcoming the non-uniqueness difficulties of the double-layer integral equation. The combined single- and double-layer approach seems to be the most attractive method from a theoretical point of view since its analysis is straightforward as well as from a numerical point of view since it never fails and can be implemented without additional computational cost as compared with the double-layer approach.

In order to be able to use Green's representation formula for the solution of the exterior Dirichlet problem, we need its normal derivative. However, assuming the given boundary values to be merely continuous means that in general the normal derivative will not exist. Hence, we need to impose some additional smoothness condition on the boundary data.

From Theorems 3.2 and 3.4 we also have compactness of the operators $S, K : C^{1,\alpha}(\partial D) \to C^{1,\alpha}(\partial D)$. Therefore, by the Riesz–Fredholm theory, the injective operator $I + K - i\eta S : C^{1,\alpha}(\partial D) \to C^{1,\alpha}(\partial D)$ again has a bounded inverse $(I+K-i\eta S)^{-1} : C^{1,\alpha}(\partial D) \to C^{1,\alpha}(\partial D)$. Hence, given a right hand side f in $C^{1,\alpha}(\partial D)$, the solution φ of the integral equation (3.26) belongs to $C^{1,\alpha}(\partial D)$ and depends continuously on f in the $\| \cdot \|_{1,\alpha}$ norm. Using the regularity results of Theorem 3.3 for the derivatives of single- and double-layer potentials, from (3.25) we now find that u belongs to $C^{1,\alpha}(\mathbb{R}^3 \setminus D)$ and depends continuously on f. In particular, the normal derivative $\partial u / \partial \nu$ of the solution u exists and belongs to $C^{0,\alpha}(\partial D)$ if $f \in C^{1,\alpha}(\partial D)$ and is given by

$$\frac{\partial u}{\partial \nu} = Af$$

where

$$A := (i\eta I - i\eta K' + T)(I + K - i\eta S)^{-1} : C^{1,\alpha}(\partial D) \to C^{0,\alpha}(\partial D)$$

is bounded. The operator A transfers the boundary values, i.e., the Dirichlet data, into the normal derivative, i.e., the Neumann data, and therefore it is called the *Dirichlet to Neumann map*.

For the sake of completeness, we wish to show that A is bijective and has a bounded inverse. This is equivalent to showing that

$$i\eta I - i\eta K' + T : C^{1,\alpha}(\partial D) \to C^{0,\alpha}(\partial D)$$

is bijective and has a bounded inverse. Since T is not compact, the Riesz–Fredholm theory cannot be employed in a straightforward manner. In order to regularize the operator, we first examine the exterior Neumann problem.

Exterior Neumann Problem. *Given a continuous function g on ∂D, find a radiating solution $u \in C^2(\mathbb{R}^3 \setminus \bar{D}) \cap C(\mathbb{R}^3 \setminus D)$ to the Helmholtz equation*

$$\triangle u + k^2 u = 0 \quad in \ \mathbb{R}^3 \setminus \bar{D}$$

which satisfies the boundary condition

$$\frac{\partial u}{\partial \nu} = g \quad on \ \partial D$$

in the sense of uniform convergence on ∂D.

The exterior Neumann problem describes acoustic scattering from sound-hard obstacles. Uniqueness for the Neumann problem follows from Theorem 2.12. To prove existence we again use a combined single- and double-layer approach. We overcome the problem that the normal derivative of the double-layer potential in general does not exist if the density is merely continuous by incorporating a smoothing operator, that is, we seek the solution in the form

$$(3.27) \quad u(x) = \int_{\partial D} \left\{ \Phi(x,y)\,\varphi(y) + i\eta\, \frac{\partial \Phi(x,y)}{\partial \nu(y)}\,(S_0^2 \varphi)(y) \right\} ds(y), \quad x \notin \partial D,$$

with continuous density φ and a real coupling parameter $\eta \neq 0$. By S_0 we denote the single-layer operator (3.8) in the potential theoretic limit case $k = 0$. Note that by Theorem 3.4 the density $S_0^2 \varphi$ of the double-layer potential belongs to $C^{1,\alpha}(\partial D)$. The idea of using a smoothing operator as in (3.27) was first suggested by Panich [152]. From Theorem 3.1 we see that (3.27) solves the exterior Neumann problem provided the density is a solution of the integral equation

$$(3.28) \qquad\qquad \varphi - K'\varphi - i\eta T S_0^2 \varphi = -2g.$$

By Theorems 3.2 and 3.4 the operators $K' + i\eta T S_0^2 : C(\partial D) \to C(\partial D)$ and $K' + i\eta T S_0^2 : C^{0,\alpha}(\partial D) \to C^{0,\alpha}(\partial D)$ are both compact. Hence, the Riesz–Fredholm theory is available in both spaces.

Let φ be a continuous solution to the homogeneous form of (3.28). Then the potential u given by (3.27) satisfies the homogeneous Neumann boundary condition $\partial u_+/\partial \nu = 0$ on ∂D whence by the uniqueness for the exterior Neumann problem $u = 0$ in $\mathbb{R}^3 \setminus \bar{D}$ follows. The jump relations (3.1)–(3.4) now yield

$$-u_- = i\eta S_0^2 \varphi, \quad -\frac{\partial u_-}{\partial \nu} = -\varphi \quad on \ \partial D$$

and, by interchanging the order of integration and using Green's integral theorem as above in the proof for the Dirichlet problem, we obtain

$$i\eta \int_{\partial D} |S_0 \varphi|^2 ds = i\eta \int_{\partial D} \varphi S_0^2 \bar{\varphi}\, ds = \int_{\partial D} \bar{u}_- \frac{\partial u_-}{\partial \nu}\, ds = \int_D \left\{ |\operatorname{grad} u|^2 - k^2 |u|^2 \right\} dx$$

whence $S_0 \varphi = 0$ on ∂D follows. The single-layer potential w with density φ and wave number $k = 0$ is continuous throughout $\mathbb{R}^3$, harmonic in D and in $\mathbb{R}^3 \setminus \bar{D}$ and vanishes on ∂D and at infinity. Therefore, by the maximum-minimum principle for harmonic functions, we have $w = 0$ in $\mathbb{R}^3$ and the jump relation (3.2) yields $\varphi = 0$. Thus, we have established injectivity of the operator $I - K' - i\eta T S_0^2$ and, by the Riesz–Fredholm theory, $(I - K' - i\eta T S_0^2)^{-1}$ exists and is bounded in both $C(\partial D)$ and $C^{0,\alpha}(\partial D)$. From this we conclude the existence of the solution to the Neumann problem for continuous boundary data g and the continuous dependence of the solution on the boundary data.

Theorem 3.10 *The exterior Neumann problem has a unique solution and the solution depends continuously on the boundary data with respect to uniform convergence of the solution on $\mathbb{R}^3 \setminus D$ and all its derivatives on closed subsets of $\mathbb{R}^3 \setminus \bar{D}$.*

In the case when $g \in C^{0,\alpha}(\partial D)$, the solution φ to the integral equation (3.28) belongs to $C^{0,\alpha}(\partial D)$ and depends continuously on g in the norm of $C^{0,\alpha}(\partial D)$. Using the regularity results of Theorem 3.3 for the single- and double-layer potentials, from (3.27) we now find that u belongs to $C^{1,\alpha}(\mathbb{R}^3 \setminus D)$. In particular, the boundary values u on ∂D are given by

$$u = Bg$$

where

$$B = (i\eta S_0^2 + i\eta K S_0^2 + S)(K' - I + i\eta T S_0^2)^{-1} : C^{0,\alpha}(\partial D) \to C^{1,\alpha}(\partial D)$$

is bounded. Clearly, the operator B is the inverse of A. Thus, we can summarize our regularity analysis as follows.

Theorem 3.11 *The Dirichlet to Neumann map which transfers the boundary values of a radiating solution to the Helmholtz equation into its normal derivative is a bijective bounded operator from $C^{1,\alpha}(\partial D)$ onto $C^{0,\alpha}(\partial D)$ with bounded inverse. The solution to the exterior Dirichlet problem belongs to $C^{1,\alpha}(\mathbb{R}^3 \setminus D)$ if the boundary values are in $C^{1,\alpha}(\partial D)$ and the mapping of the boundary data into the solution is continuous from $C^{1,\alpha}(\partial D)$ into $C^{1,\alpha}(\mathbb{R}^3 \setminus D)$.*

Instead of looking for classical solutions in the spaces of continuous or Hölder continuous functions one can also pose and solve the boundary value problems for the Helmholtz equation in a weak formulation for the boundary condition either in an L^2 sense or in a Sobolev space setting. This then leads to existence results under weaker regularity assumptions on the given boundary data and to continuous dependence in different norms. The latter, in particular, can be useful in the error analysis for approximate solution methods.

A major drawback of the integral equation approach to constructively proving existence of solutions for scattering problems is the relatively strong regularity assumption on the boundary to be of class C^2. It is possible to slightly weaken the regularity and allow *Lyapunov boundaries* instead of C^2 boundaries and still remain within the framework of compact operators. The boundary is said to satisfy a Lyapunov condition if at each point $x \in \partial D$ the normal vector ν to the surface exists and if there are positive constants L and α such that for the angle $\theta(x, y)$ between the normal vectors at x and y there holds $\theta(x, y) \leq L|x - y|^\alpha$ for all $x, y \in \partial D$. For the treatment of the Dirichlet problem for Lyapunov boundaries, which does not differ essentially from that for C^2 boundaries, we refer to Mikhlin [129].

However, the situation changes considerably if the boundary is allowed to have edges and corners since this affects the compactness of the double-layer integral operator. Here, under suitable assumptions on the nature of the edges and corners, the double-layer integral operator can be decomposed into the sum of a compact operator and a bounded operator with norm less than one reflecting the behavior at the edges and corners, and then the Riesz–Fredholm theory still can be employed. For details, we refer to Section 3.5 for the two-dimensional case.

Explicit solutions for the direct scattering problem are only available for special geometries and special incoming fields. In general, to construct a solution one must resort to numerical methods, for example, the numerical solution of the boundary integral equations. An introduction into numerical approximation for integral equations of the second kind by the Nyström method, collocation method and Galerkin method is contained in [106]. We will describe in some detail Nyström methods for the two- and three-dimensional case at the end of this chapter.

For future reference, we present the solution for the scattering of a plane wave

$$u^i(x) = e^{ik\,x\cdot d}$$

by a sound-soft ball of radius R with center at the origin. The unit vector d describes the direction of propagation of the incoming wave. In view of the Jacobi–Anger expansion (2.45) and the boundary condition $u^i + u^s = 0$, we expect the scattered wave to be given by

$$(3.29) \qquad u^s(x) = -\sum_{n=0}^{\infty} i^n (2n+1) \frac{j_n(kR)}{h_n^{(1)}(kR)} \, h_n^{(1)}(k|x|) \, P_n(\cos\theta)$$

where θ denotes the angle between x and d. By the asymptotic behavior (2.37) and (2.38) of the spherical Bessel and Hankel functions for large n, we have

$$\frac{j_n(kR)}{h_n^{(1)}(kR)} \, h_n^{(1)}(k|x|) = O\left(\frac{n!\,(2kR)^n}{(2n+1)!} \, \frac{R^n}{|x|^n} \right), \quad n \to \infty,$$

uniformly on compact subsets of $\mathbb{R}^3 \setminus \{0\}$. Therefore, the series (3.29) is uniformly convergent on compact subsets of $\mathbb{R}^3 \setminus \{0\}$. Hence, by Theorem 2.14 the series represents a radiating field in $\mathbb{R}^3 \setminus \{0\}$, and therefore indeed solves the scattering problem for the sound-soft ball.

For the far field pattern, we see by Theorem 2.15 that

$$(3.30) \qquad u_\infty(\hat{x}) = \frac{i}{k} \sum_{n=0}^{\infty} (2n+1) \frac{j_n(kR)}{h_n^{(1)}(kR)} P_n(\cos\theta).$$

Clearly, as we expect from symmetry reasons, it depends only on the angle θ between the observation direction $\hat{x}$ and the incident direction d.

In general, for the scattering problem the boundary values are as smooth as the boundary since they are given by the restriction of the analytic function u^i to ∂D. In particular, for domains D of class C^2 our regularity analysis shows that the scattered field u^s is in $C^{1,\alpha}(\mathbb{R}^3 \setminus D)$. Therefore, we may apply Green's formula (2.8) with the result

$$u^s(x) = \int_{\partial D} \left\{ u^s(y) \frac{\partial \Phi(x,y)}{\partial \nu(y)} - \frac{\partial u^s}{\partial \nu}(y) \Phi(x,y) \right\} ds(y), \quad x \in \mathbb{R}^3 \setminus \bar{D}.$$

Green's theorem (2.3), applied to the entire solution u^i and $\Phi(x, \cdot)$, gives

$$0 = \int_{\partial D} \left\{ u^i(y) \frac{\partial \Phi(x,y)}{\partial \nu(y)} - \frac{\partial u^i}{\partial \nu}(y) \Phi(x,y) \right\} ds(y), \quad x \in \mathbb{R}^3 \setminus \bar{D}.$$

Adding these two equations and using the boundary condition $u^i + u^s = 0$ on ∂D gives the following theorem. The representation for the far field pattern is obtained with the aid of (2.14).

Theorem 3.12 *For the scattering of an entire field u^i from a sound-soft obstacle D we have*

$$(3.31) \qquad u(x) = u^i(x) - \int_{\partial D} \frac{\partial u}{\partial \nu}(y) \Phi(x,y) \, ds(y), \quad x \in \mathbb{R}^3 \setminus \bar{D},$$

and the far field pattern of the scattered field u^s is given by

$$(3.32) \qquad u_\infty(\hat{x}) = -\frac{1}{4\pi} \int_{\partial D} \frac{\partial u}{\partial \nu}(y) e^{-ik\,\hat{x}\cdot y} \, ds(y), \quad \hat{x} \in \Omega.$$

In physics, the representation (3.31) for the scattered field through the so-called *secondary sources* on the boundary is known as *Huygen's principle*.

We conclude this section by briefly giving the motivation for the *Kirchhoff* or *physical optics approximation* which is frequently used in applications as a physically intuitive procedure to simplify the direct scattering problem. The solution for the scattering of a plane wave with incident direction d at a plane $\Gamma := \{x \in \mathbb{R}^3 : x \cdot \nu = 0\}$ through the origin with normal vector ν is described by

$$u(x) = u^i(x) + u^s(x) = e^{ik\,x\cdot d} - e^{ik\,x\cdot \bar{d}}$$

where $\tilde{d} = d - 2\nu \cdot d\nu$ denotes the reflection of d at the plane Γ. Clearly, $u^i + u^s = 0$ is satisfied on Γ and we evaluate

$$\frac{\partial u}{\partial \nu} = ik\{\nu \cdot d\,u^i + \nu \cdot \tilde{d}\,u^s\} = 2ik\,\nu \cdot d\,u^i = 2\,\frac{\partial u^i}{\partial \nu}.$$

For large wave numbers k, that is for small wavelengths, in a first approximation a convex object D locally may be considered at each point x of ∂D as a plane with normal $\nu(x)$. This leads to setting

$$\frac{\partial u}{\partial \nu} = 2\,\frac{\partial u^i}{\partial \nu}$$

on the region $\partial D_- := \{x \in \partial D : \nu(x) \cdot d < 0\}$ which is illuminated by the plane wave with incident direction d, and

$$\frac{\partial u}{\partial \nu} = 0$$

in the shadow region $\partial D_+ := \{x \in \partial D : \nu(x) \cdot d \geq 0\}$. Thus, the Kirchhoff approximation for the scattering of a plane wave with incident direction d at a convex sound-soft obstacle is given by

$$(3.33) \qquad u(x) = e^{ik\,x\cdot d} - 2\int_{\partial D_-} \frac{\partial e^{ik\,y\cdot d}}{\partial \nu(y)}\, \Phi(x,y)\,ds(y), \quad x \in \mathbb{R}^3 \setminus \bar{D},$$

and the far field pattern of the scattered field is given by

$$(3.34) \qquad u_\infty(\hat{x}) = -\frac{1}{2\pi}\int_{\partial D_-} \frac{\partial e^{ik\,y\cdot d}}{\partial \nu(y)}\, e^{-ik\,\hat{x}\cdot y}\,ds(y), \quad \hat{x} \in \Omega.$$

In this book, the Kirchhoff approximation does not play an important role since we are mainly interested in scattering at low and intermediate values of the wave number.

3.3 The Reciprocity Relation

In the sequel, for an incident plane wave $u^i(x) = u^i(x; d) = e^{ik\,x\cdot d}$ we will indicate the dependence of the scattered field, of the total field and of the far field pattern on the incident direction d by writing, respectively, $u^s(x; d)$, $u(x; d)$ and $u_\infty(\hat{x}; d)$.

Theorem 3.13 *The far field pattern for sound-soft obstacle scattering satisfies the reciprocity relation*

$$(3.35) \qquad u_\infty(\hat{x}; d) = u_\infty(-d; -\hat{x}), \quad \hat{x}, d \in \Omega.$$

Proof. By Green's theorem (2.3), the Helmholtz equation for the incident and the scattered wave and the radiation condition for the scattered wave we find

$$\int_{\partial D} \left\{ u^i(\cdot\,;d) \frac{\partial}{\partial \nu} u^i(\cdot\,;-\hat{x}) - u^i(\cdot\,;-\hat{x}) \frac{\partial}{\partial \nu} u^i(\cdot\,;d) \right\} ds = 0$$

and

$$\int_{\partial D} \left\{ u^s(\cdot\,;d) \frac{\partial}{\partial \nu} u^s(\cdot\,;-\hat{x}) - u^s(\cdot\,;-\hat{x}) \frac{\partial}{\partial \nu} u^s(\cdot\,;d) \right\} ds = 0.$$

From (2.13) we deduce that

$$4\pi u_\infty(\hat{x};d) = \int_{\partial D} \left\{ u^s(\cdot\,;d) \frac{\partial}{\partial \nu} u^i(\cdot\,;-\hat{x}) - u^i(\cdot\,;-\hat{x}) \frac{\partial}{\partial \nu} u^s(\cdot\,;d) \right\} ds$$

and, interchanging the roles of $\hat{x}$ and d,

$$4\pi u_\infty(-d;-\hat{x}) = \int_{\partial D} \left\{ u^s(\cdot\,;-\hat{x}) \frac{\partial}{\partial \nu} u^i(\cdot\,;d) - u^i(\cdot\,;d) \frac{\partial}{\partial \nu} u^s(\cdot\,;-\hat{x}) \right\} ds.$$

We now subtract the last equation from the sum of the three preceding equations to obtain

$$4\pi \{ u_\infty(\hat{x};d) - u_\infty(-d;-\hat{x}) \}$$

(3.36)

$$= \int_{\partial D} \left\{ u(\cdot\,;d) \frac{\partial}{\partial \nu} u(\cdot\,;-\hat{x}) - u(\cdot\,;-\hat{x}) \frac{\partial}{\partial \nu} u(\cdot\,;d) \right\} ds$$

whence (3.35) follows by using the boundary condition $u(\cdot\,;d) = u(\cdot\,;-\hat{x}) = 0$ on ∂D.   □

In the derivation of (3.36), we only used the Helmholtz equation for the incident field in $\mathbb{R}^3$ and for the scattered field in $\mathbb{R}^3 \setminus \bar{D}$ and the radiation condition. Therefore, we can conclude that the reciprocity relation (3.35) is also valid for the sound-hard, impedance and transmission boundary conditions.

We now ask the question if the far field patterns for a fixed sound-soft obstacle D and all incident plane waves are complete in $L^2(\Omega)$. We call a subset U of a Hilbert space X *complete* if the linear combinations of elements from U are dense in X, that is, if $X = \overline{\text{span } U}$. Recall that U is complete in the Hilbert space X if and only if $(u, \varphi) = 0$ for all $u \in U$ implies that $\varphi = 0$ (see [49]).

Definition 3.14 *A* Herglotz wave function *is a function of the form*

(3.37)
$$v(x) = \int_\Omega e^{ik\,x\cdot d} g(d)\, ds(d), \quad x \in \mathbb{R}^3,$$

where $g \in L^2(\Omega)$. *The function* g *is called the* Herglotz kernel *of* v.

Herglotz wave functions are clearly entire solutions to the Helmholtz equation. We note that for a given $g \in L^2(\Omega)$ the function

$$v(x) = \int_\Omega e^{-ik\,x\cdot d} g(d)\, ds(d), \quad x \in \mathbb{R}^3,$$

also defines a Herglotz wave function. The following theorem establishes a one-to-one correspondence between Herglotz wave functions and their kernels.

Theorem 3.15 *Assume that the Herglotz wave function v with kernel g vanishes in all of $\mathbb{R}^3$. Then $g = 0$.*

Proof. From $v(x) = 0$ for all $x \in \mathbb{R}^3$ and the Funk–Hecke formula (2.44), we see that

$$\int_\Omega g\, Y_n\, ds = 0$$

for all spherical harmonics Y_n of order $n = 0, 1, \dots$. Now $g = 0$ follows from the completeness of the spherical harmonics (Theorem 2.7).   □

Lemma 3.16 *For a given function $g \in L^2(\Omega)$ the solution to the scattering problem for the incident wave*

$$v^i(x) = \int_\Omega e^{ik\,x\cdot d} g(d)\, ds(d), \quad x \in \mathbb{R}^3,$$

is given by

$$v^s(x) = \int_\Omega u^s(x; d) g(d)\, ds(d), \quad x \in \mathbb{R}^3 \setminus \bar{D},$$

and has the far field pattern

$$v_\infty(\hat{x}) = \int_\Omega u_\infty(\hat{x}; d) g(d)\, ds(d), \quad \hat{x} \in \Omega.$$

Proof. Multiply (3.25) and (3.26) by g, integrate with respect to d over Ω and interchange orders of integration.   □

Now, the rather surprising answer to our completeness question, due to Colton and Kirsch [26], will be that the far field patterns are complete in $L^2(\Omega)$ if and only if there does not exist a nontrivial Herglotz wave function v that vanishes on ∂D. A nontrivial Herglotz wave function that vanishes on ∂D, of course, is a Dirichlet eigenfunction, i.e., a solution to the Dirichlet problem in D with zero boundary condition, and this is peculiar since from physical considerations the eigenfunctions corresponding to the *Dirichlet eigenvalues* of the negative Laplacian in D should have nothing to do with the exterior scattering problem at all.

Theorem 3.17 *Let (d_n) be a sequence of unit vectors that is dense on Ω and define the set $\mathcal{F}$ of far field patterns by*

$$\mathcal{F} := \{u_\infty(\cdot\,; d_n) : n = 1, 2, \dots\}.$$

Then $\mathcal{F}$ is complete in $L^2(\Omega)$ if and only if there does not exist a Dirichlet eigenfunction for D which is a Herglotz wave function.

Proof. Deviating from the original proof by Colton and Kirsch [26], we make use of the reciprocity relation. By the continuity of u_∞ as a function of d and Theorem 3.13, the completeness condition

$$\int_\Omega u_\infty(\hat{x}; d_n) h(\hat{x})\, ds(\hat{x}) = 0, \quad n = 1, 2, \ldots,$$

for a function $h \in L^2(\Omega)$ is equivalent to the condition

(3.38) $$\int_\Omega u_\infty(\hat{x}; d) g(d)\, ds(d) = 0, \quad \hat{x} \in \Omega,$$

for $g \in L^2(\Omega)$ with $g(d) = h(-d)$.

By Theorem 3.15 and Lemma 3.16, the existence of a nontrivial function g satisfying (3.38) is equivalent to the existence of a nontrivial Herglotz wave function v^i (with kernel g) for which the far field pattern of the corresponding scattered wave v^s is $v_\infty = 0$. By Theorem 2.13, the vanishing far field $v_\infty = 0$ on Ω is equivalent to $v^s = 0$ in $\mathbb{R}^3 \setminus \bar{D}$. By the boundary condition $v^i + v^s = 0$ on ∂D and the uniqueness for the exterior Dirichlet problem, this is equivalent to $v^i = 0$ on ∂D and the proof is finished. $\qquad\square$

Clearly, by the Funk–Hecke formula (2.44), the spherical wave functions

$$u_n(x) = j_n(k|x|)\, Y_n(\hat{x})$$

provide examples of Herglotz wave functions. The spherical wave functions also describe Dirichlet eigenfunctions for a ball of radius R centered at the origin with the eigenvalues k^2 given in terms of the zeros $j_n(kR) = 0$ of the spherical Bessel functions. By arguments similar to those used in the proof of Rellich's Lemma 2.11, an expansion with respect to spherical harmonics shows that all the eigenfunctions for a ball are indeed spherical wave functions. Therefore, the eigenfunctions for balls are always Herglotz wave functions and by Theorem 3.17 the far field patterns for plane waves are not complete for a ball D when k^2 is a Dirichlet eigenvalue.

We can also express the result of Theorem 3.17 in terms of a far field operator.

Corollary 3.18 *The operator $F : L^2(\Omega) \to L^2(\Omega)$ defined by*

$$(Fg)(\hat{x}) := \int_\Omega u_\infty(\hat{x}; d) g(d)\, ds(d), \quad \hat{x} \in \Omega,$$

is injective and has dense range if and only if there does not exist a Dirichlet eigenfunction for D which is a Herglotz wave function.

Proof. For the L^2 adjoint $F^* : L^2(\Omega) \to L^2(\Omega)$, given by

$$(F^*h)(d) := \int_\Omega \overline{u_\infty(\hat{x}; d)} h(\hat{x})\, ds(\hat{x}), \quad d \in \Omega,$$

the reciprocity relation implies that

$$(F^*h)(d) = \overline{(Fg)(-d)}, \quad d \in \Omega,$$

where $g(\hat{x}) = \overline{h(-\hat{x})}$. Hence, the operator F is injective if and only if its adjoint F^* is injective. Observing that in a Hilbert space we have $N(F^*)^\perp = \overline{F(L^2(\Omega))}$ for bounded operators F (see Theorem 4.6), the statement of the corollary is indeed seen to be a reformulation of the preceding theorem. □

The corresponding completeness results for the transmission problem were given by Kirsch [87] and for the resistive boundary condition by Hettlich [69]. For extensions to Sobolev and Hölder norms we refer to Kirsch [88].

The question of when we can find a superposition of incident plane waves such that the resulting far field pattern coincides with a prescribed far field is answered in terms of a solvability condition for an integral equation of the first kind in the following theorem.

Theorem 3.19 *Let v^s be a radiating solution to the Helmholtz equation with far field pattern v_∞. Then the integral equation of the first kind*

$$(3.39) \qquad \int_\Omega u_\infty(\hat{x}; d) g(d) \, ds(d) = v_\infty(\hat{x}), \quad \hat{x} \in \Omega,$$

possesses a solution $g \in L^2(\Omega)$ if and only if v^s is defined in $\mathbb{R}^3 \setminus \bar{D}$, is continuous in $\mathbb{R}^3 \setminus D$ and the interior Dirichlet problem for the Helmholtz equation

$$(3.40) \qquad \Delta v^i + k^2 v^i = 0 \quad in \ D,$$

$$(3.41) \qquad v^i + v^s = 0 \quad on \ \partial D$$

is solvable with a solution v^i being a Herglotz wave function.

Proof. By Theorem 3.15 and Lemma 3.16, the solvability of the integral equation (3.39) for g is equivalent to the existence of a Herglotz wave function v^i (with kernel g) for which the far field pattern for the scattering by the obstacle D coincides with the given v_∞, that is, the scattered wave coincides with the given v^s. This completes the proof. □

Special cases of Theorem 3.19 include the radiating spherical wave function

$$v^s(x) = h_n^{(1)}(k|x|) Y_n(\hat{x})$$

of order n with far field pattern

$$v_\infty = \frac{1}{ki^{n+1}} Y_n.$$

Here, for solvability of (3.39) it is necessary that the origin is contained in D.

The integral equation (3.39) will play a role in our investigation of the inverse scattering problem in Section 5.5. By reciprocity, the solvability of (3.39) is equivalent to the solvability of

$$(3.42) \qquad \int_\Omega u_\infty(\hat{x}; d) h(\hat{x})\, ds(\hat{x}) = v_\infty(-d), \quad d \in \Omega,$$

where $h(\hat{x}) = g(-\hat{x})$. Since the Dirichlet problem (3.40), (3.41) is solvable provided k^2 is not a Dirichlet eigenvalue, the crucial condition in Theorem 3.19 is the property of the solution to be a Herglotz wave function, that is, a strong regularity condition. In the special case $v_\infty = 1$, the connection between the solution to the integral equation (3.42) and the interior Dirichlet problem (3.40), (3.41) as described in Theorem 3.19 was first established by Colton and Monk [37] without, however, making use of the reciprocity Theorem 3.13.

The original proof for Theorem 3.17 by Colton and Kirsch [26] is based on the following completeness result which we include for its own interest.

Theorem 3.20 *Let (d_n) be a sequence of unit vectors that is dense on Ω. Then the normal derivatives of the total fields*

$$\left\{ \frac{\partial}{\partial \nu} u(\cdot\,; d_n) : n = 1, 2, \ldots \right\}$$

corresponding to incident plane waves with directions (d_n) are complete in $L^2(\partial D)$.

Proof. The weakly singular operators $K - iS$ and $K' - iS$ are both compact from $C(\partial D)$ into $C(\partial D)$ and from $L^2(\partial D)$ into $L^2(\partial D)$ and they are adjoint with respect to the L^2 bilinear form, that is,

$$\int_{\partial D} (K - iS)\varphi\,\psi\, ds = \int_{\partial D} \varphi\,(K' - iS)\psi\, ds$$

for all $\varphi, \psi \in L^2(\partial D)$. From the proof of Theorem 3.9, we know that the operator $I + K - iS$ has a trivial nullspace in $C(\partial D)$. Therefore, by the Fredholm alternative applied in the dual system $\langle C(\partial D), L^2(\partial D) \rangle$ with the L^2 bilinear form, the adjoint operator $I + K' - iS$ has a trivial nullspace in $L^2(\partial D)$. Again by the Fredholm alternative, but now applied in the dual system $\langle L^2(\partial D), L^2(\partial D) \rangle$ with the L^2 bilinear form, the operator $I + K - iS$ also has a trivial nullspace in $L^2(\partial D)$. Hence, by the Riesz–Fredholm theory for compact operators, both the operators $I + K - iS : L^2(\partial D) \to L^2(\partial D)$ and $I + K' - iS : L^2(\partial D) \to L^2(\partial D)$ are bijective and have a bounded inverse. This idea to employ the Fredholm alternative in two different dual systems for showing that the dimensions of the nullspaces for weakly singular integral operators of the second kind in the space of continuous functions and in the L^2 space coincide is due to Hähner [66].

From the representation (3.31), the boundary condition $u = 0$ on ∂D and the jump relations of Theorem 3.1 we deduce that

$$\frac{\partial u}{\partial \nu} + K' \frac{\partial u}{\partial \nu} - iS \frac{\partial u}{\partial \nu} = 2\frac{\partial u^i}{\partial \nu} - 2iu^i.$$

Now let $g \in L^2(\partial D)$ satisfy

$$\int_{\partial D} g \, \frac{\partial u(\cdot\,; d_n)}{\partial \nu} \, ds = 0, \quad n = 1, 2, \dots.$$

This, by the continuity of the Dirichlet to Neumann map (Theorem 3.11), implies

$$\int_{\partial D} g \, \frac{\partial u(\cdot\,; d)}{\partial \nu} \, ds = 0$$

for all $d \in \Omega$. Then from

$$\frac{\partial u}{\partial \nu} = 2(I + K' - iS)^{-1} \left\{ \frac{\partial u^i}{\partial \nu} - iu^i \right\}$$

we obtain

$$\int_{\partial D} g \, (I + K' - iS)^{-1} \left\{ \frac{\partial}{\partial \nu} \, u^i(\cdot\,; d) - iu^i(\cdot\,; d) \right\} ds = 0$$

for all $d \in \Omega$, and consequently

$$\int_{\partial D} \varphi(y) \left\{ \frac{\partial}{\partial \nu(y)} \, e^{ik\,y\cdot d} - ie^{ik\,y\cdot d} \right\} ds(y) = 0$$

for all $d \in \Omega$ where we have set

$$\varphi := (I + K - iS)^{-1} g.$$

Therefore, since $I + K - iS$ is bijective, our proof will be finished by showing that $\varphi = 0$. To this end, by (2.14) and (2.15), we deduce from the last equation that the combined single- and double-layer potential

$$v(x) := \int_{\partial D} \varphi(y) \left\{ \frac{\partial \Phi(x, y)}{\partial \nu(y)} - i\Phi(x, y) \right\} ds(y), \quad x \in \mathbb{R}^3 \setminus \bar{D},$$

has far field pattern

$$v_\infty(\hat{x}) = \frac{1}{4\pi} \int_{\partial D} \varphi(y) \left\{ \frac{\partial}{\partial \nu(y)} \, e^{-ik\,y\cdot \hat{x}} - ie^{-ik\,y\cdot \hat{x}} \right\} ds(y) = 0, \quad \hat{x} \in \Omega.$$

By Theorem 2.13, this implies $v = 0$ in $\mathbb{R}^3 \setminus \bar{D}$, and letting x tend to the boundary ∂D with the help of the L^2 jump relations (3.19) and (3.21) yields

$$\varphi + K\varphi - iS\varphi = 0,$$

whence $\varphi = 0$ follows. □

With the tools involved in the proof of Theorem 3.20, we can establish the following result which we shall also need in our analysis of the inverse problem in Section 5.5.

Theorem 3.21 *The operator* $A : C(\partial D) \to L^2(\Omega)$ *which maps the boundary values of radiating solutions* $u \in C^2(\mathbb{R}^3 \setminus \bar{D}) \cap C(\mathbb{R}^3 \setminus D)$ *to the Helmholtz equation onto the far field pattern* u_∞ *can be extended to an injective bounded linear operator* $A : L^2(\partial D) \to L^2(\Omega)$.

Proof. From the solution (3.25) to the exterior Dirichlet problem, for $\hat{x} \in \Omega$ we derive

$$u_\infty(\hat{x}) = \frac{1}{2\pi} \int_{\partial D} \left\{ \frac{\partial}{\partial \nu(y)} e^{-iky \cdot \hat{x}} - i e^{-iky \cdot \hat{x}} \right\} \left((I + K - iS)^{-1} f \right)(y) \, ds(y)$$

with the boundary values $u = f$ on ∂D. From this, given the boundedness of $(I + K - iS)^{-1} : L^2(\partial D) \to L^2(\partial D)$ from the proof of Theorem 3.20, it is obvious that A is bounded from $L^2(\partial D) \to L^2(\Omega)$. The injectivity of A is also immediate from the proof of Theorem 3.20.   □

We now wish to study Herglotz wave functions more closely. The concept of the growth condition in the following theorem for solutions to the Helmholtz equation was introduced by Herglotz in a lecture in 1945 in Göttingen and was studied further by Magnus [123] and Müller [138]. The equivalence stated in the theorem was shown by Hartman and Wilcox [68].

Theorem 3.22 *An entire solution* v *to the Helmholtz equation possesses the growth property*

(3.43)
$$\sup_{R>0} \frac{1}{R} \int_{|x| \leq R} |v(x)|^2 dx < \infty$$

if and only if it is a Herglotz wave function, that is, if and only if there exists a function $g \in L^2(\Omega)$ *such that* v *can be represented in the form (3.37).*

Proof. Before we can prove this result, we need to note two properties for integrals containing spherical Bessel functions. From the asymptotic behavior (2.41), that is, from

$$j_n(t) = \frac{1}{t} \cos\left(t - \frac{n\pi}{2} - \frac{\pi}{2} \right) \left\{ 1 + O\left(\frac{1}{t} \right) \right\}, \quad t \to \infty,$$

we readily find that

(3.44)
$$\lim_{T \to \infty} \frac{1}{T} \int_0^T t^2 [j_n(t)]^2 dt = \frac{1}{2}, \quad n = 0, 1, 2, \ldots.$$

We now want to establish that the integrals in (3.44) are uniformly bounded with respect to T and n. This does not follow immediately since the asymptotic behavior for the spherical Bessel functions is not uniformly valid with respect to the order n. If we multiply the differential formula (2.34) rewritten in the form

$$j_{n+1}(t) = -\frac{1}{\sqrt{t}} \frac{d}{dt} \sqrt{t} \, j_n(t) + \left(n + \frac{1}{2} \right) \frac{1}{t} j_n(t)$$

by two and subtract it from the recurrence relation (2.33), that is, from

$$j_{n-1}(t) + j_{n+1}(t) = \frac{2n+1}{t}\, j_n(t),$$

we obtain

$$j_{n-1}(t) - j_{n+1}(t) = \frac{2}{\sqrt{t}}\frac{d}{dt}\sqrt{t}\, j_n(t).$$

Hence, from the last two equations we get

$$\int_0^T t^2 \left\{ [j_{n-1}(t)]^2 - [j_{n+1}(t)]^2 \right\} dt = (2n+1)T\, [j_n(T)]^2$$

for $n = 1, 2, \ldots$ and all $T > 0$. From this monotonicity, together with (3.44) for $n = 0$ and $n = 1$, it is now obvious that

$$(3.45) \qquad \sup_{\substack{T>0 \\ n=0,1,2,\ldots}} \frac{1}{T}\int_0^T t^2[j_n(t)]^2 dt < \infty.$$

For the proof of the theorem, we first observe that any entire solution v of the Helmholtz equation can be expanded in a series

$$(3.46) \qquad v(x) = 4\pi \sum_{n=0}^{\infty} \sum_{m=-n}^{n} i^n\, a_n^m\, j_n(k|x|)\, Y_n^m(\hat{x})$$

where $\hat{x} = x/|x|$ and the series converges uniformly on compact subsets of $\mathbb{R}^3$. This follows from Green's representation formula (2.5) for v in a ball with radius R and center at the origin and inserting the addition theorem (2.42) with the roles of x and y interchanged, that is,

$$\Phi(x,y) = ik \sum_{n=0}^{\infty} \sum_{m=-n}^{n} j_n(k|x|)\, Y_n^m(\hat{x})\, h_n^{(1)}(k|y|)\, \overline{Y_n^m(\hat{y})}, \quad |x| < |y|.$$

Since the expansion derived for two different radii represent the same function in the ball with the smaller radius, the coefficients a_n^m do not depend on the radius R. Because of the uniform convergence, we can integrate term by term and use the orthonormality of the Y_n^m to find that

$$(3.47) \qquad \frac{1}{R}\int_{|x|\leq R} |v(x)|^2 dx = \frac{16\pi^2}{R} \sum_{n=0}^{\infty} \int_0^R r^2[j_n(kr)]^2 dr \sum_{m=-n}^{n} |a_n^m|^2.$$

Now assume that v satisfies

$$\frac{1}{R}\int_{|x|\leq R} |v(x)|^2 dx \leq C$$

for all $R > 0$ and some constant $C > 0$. This, by (3.47), implies that

$$\frac{16\pi^2}{R} \sum_{n=0}^{N} \int_0^R r^2[j_n(kr)]^2 dr \sum_{m=-n}^{n} |a_n^m|^2 \leq C$$

for all $R > 0$ and all $N \in \mathbb{N}$. Hence, by first passing to the limit $R \to \infty$ with the aid of (3.44) and then letting $N \to \infty$ we obtain

$$\sum_{n=0}^{\infty} \sum_{m=-n}^{n} |a_n^m|^2 \leq \frac{k^2 C}{8\pi^2}.$$

Therefore,

$$g := \sum_{n=0}^{\infty} \sum_{m=-n}^{n} a_n^m Y_n^m$$

defines a function $g \in L^2(\Omega)$. From the Jacobi–Anger expansion (2.45) and the addition theorem (2.29), that is, from

$$e^{ik\,x\cdot d} = 4\pi \sum_{n=0}^{\infty} \sum_{m=-n}^{n} i^n j_n(k|x|) Y_n^m(\hat{x}) \overline{Y_n^m(d)}$$

we now derive

$$\int_{\Omega} g(d) e^{ik\,x\cdot d} ds(d) = 4\pi \sum_{n=0}^{\infty} \sum_{m=-n}^{n} i^n a_n^m j_n(k|x|) Y_n^m(\hat{x}) = v(x)$$

for all $x \in \mathbb{R}^3$, that is, we have shown that v can be represented in the form (3.37).

Conversely, for a given $g \in L^2(\Omega)$ we have an expansion

$$g = \sum_{n=0}^{\infty} \sum_{m=-n}^{n} a_n^m Y_n^m,$$

where, by Parseval's equality, the coefficients satisfy

(3.48) $$\|g\|_{L^2(\Omega)}^2 = \sum_{n=0}^{\infty} \sum_{m=-n}^{n} |a_n^m|^2 < \infty.$$

Then for the entire solution v to the Helmholtz equation defined by

$$v(x) := \int_{\Omega} e^{ik\,x\cdot d} g(d)\, ds(d), \quad x \in \mathbb{R}^3,$$

we again see by the Jacobi–Anger expansion that

(3.49) $$v(x) = 4\pi \sum_{n=0}^{\infty} \sum_{m=-n}^{n} i^n a_n^m j_n(k|x|) Y_n^m(\hat{x})$$

and from (3.45), (3.47) and (3.49) we conclude that the growth condition (3.43) is fulfilled for v. The proof is now complete.   □

With the help of (3.45), we observe that the series (3.47) has a convergent majorant independent of R. Hence, it is uniformly convergent for all $R > 0$ and we may interchange the limit $R \to \infty$ with the series and use (3.44) and (3.48) to obtain that for the Herglotz wave function v with kernel g we have

$$\lim_{R \to \infty} \frac{1}{R} \int_{|x| \leq R} |v(x)|^2 dx = \frac{8\pi^2}{k^2} \|g\|_{L^2(\Omega)}^2.$$

3.4 The Two-Dimensional Case

The scattering from infinitely long cylindrical obstacles leads to exterior bound-ary value problems for the Helmholtz equation in $\mathbb{R}^2$. The two-dimensional case can be used as an approximation for the scattering from finitely long cylinders, and more important, it can serve as a model case for testing numerical approx-imation schemes in direct and inverse scattering. Without giving much of the details, we would like to show how all the results of this chapter remain valid in two dimensions after appropriate modifications of the fundamental solution, the radiation condition and the spherical wave functions.

We note that in two dimensions there exist two linearly independent spher-ical harmonics of order n which can be represented by $e^{\pm in\varphi}$. Correspondingly, looking for solutions to the Helmholtz equation of the form

$$u(x) = f(kr)\, e^{\pm in\varphi}$$

in polar coordinates (r, φ) leads to the *Bessel differential equation*

$$(3.50) \qquad t^2 f''(t) + t f'(t) + [t^2 - n^2] f(t) = 0$$

with integer order $n = 0, 1, \ldots$. The analysis of the Bessel equation which is required for the study of the two-dimensional Helmholtz equation, in particular the asymptotics of the solutions for large argument, is more involved than the corresponding analysis for the spherical Bessel equation (2.30). Therefore, here we will list only the relevant results without proofs. For a concise treatment of the Bessel equation for the purpose of scattering theory, we refer to Colton [25] or Lebedev [117].

By direct calculations and the ratio test, we can easily verify that for $n = 0, 1, 2, \ldots$ the functions

$$(3.51) \qquad J_n(t) := \sum_{p=0}^{\infty} \frac{(-1)^p}{p!\,(n+p)!} \left(\frac{t}{2}\right)^{n+2p}$$

represent solutions to Bessel's equation which are analytic for all $t \in \mathbb{R}$ and these are known as *Bessel functions* of order n. As opposed to the spherical Bessel equation, here it is more complicated to construct a second linearly inde-pendent solution. Patient, but still straightforward, calculations together with the ratio test show that

$$
\begin{aligned}
Y_n(t) \; := \; & \frac{2}{\pi}\left\{\ln\frac{t}{2} + C\right\} J_n(t) - \frac{1}{\pi}\sum_{p=0}^{n-1}\frac{(n-1-p)!}{p!}\left(\frac{2}{t}\right)^{n-2p} \\[2mm]
& - \frac{1}{\pi}\sum_{p=0}^{\infty}\frac{(-1)^p}{p!\,(n+p)!}\left(\frac{t}{2}\right)^{n+2p}\{\psi(p+n)+\psi(p)\}
\end{aligned}
$$

(3.52)

for $n = 0, 1, 2, \ldots$ provide solutions to Bessel's equation which are analytic for all $t \in (0, \infty)$. Here, we define $\psi(0) := 0$,

$$\psi(p) := \sum_{m=1}^{p} \frac{1}{m}, \quad p = 1, 2, \ldots,$$

let

$$C := \lim_{p \to \infty} \left\{ \sum_{m=1}^{p} \frac{1}{m} - \ln p \right\}$$

denote Euler's constant, and if $n = 0$ the finite sum in (3.52) is set equal to zero. The functions Y_n are called *Neumann functions* of order n and the linear combinations

$$H_n^{(1,2)} := J_n \pm iY_n$$

are called *Hankel functions* of the first and second kind of order n respectively.

From the series representation (3.51) and (3.52), by equating powers of t, it is readily verified that both $f_n = J_n$ and $f_n = Y_n$ satisfy the recurrence relation

$$(3.53) \qquad f_{n+1}(t) + f_{n-1}(t) = \frac{2n}{t} f_n(t), \quad n = 1, 2, \ldots.$$

Straightforward differentiation of the series (3.51) and (3.52) shows that both $f_n = J_n$ and $f_n = Y_n$ satisfy the differentiation formulas

$$(3.54) \qquad f_{n+1}(t) = -t^n \frac{d}{dt} \left\{ t^{-n} f_n(t) \right\}, \quad n = 0, 1, 2, \ldots,$$

and

$$(3.55) \qquad t^{n+1} f_{n-1}(t) = \frac{d}{dt} \left\{ t^n f_n(t) \right\}, \quad n = 1, 2, \ldots.$$

The Wronskian

$$W(J_n(t), Y_n(t)) := J_n(t)Y_n'(t) - Y_n(t)J_n'(t)$$

satisfies

$$W' + \frac{1}{t} W = 0.$$

Therefore, $W(J_n(t), Y_n(t)) = C/t$ for some constant C and by passing to the limit $t \to 0$ it follows that

$$(3.56) \qquad J_n(t)Y_n'(t) - J_n'(t)Y_n(t) = \frac{2}{\pi t}.$$

From the series representation of the Bessel and Neumann functions, it is obvious that

$$(3.57) \qquad J_n(t) = \frac{t^n}{2^n n!} \left(1 + O\left(\frac{1}{n}\right) \right), \quad n \to \infty,$$

uniformly on compact subsets of $\mathbb{R}$ and

$$(3.58) \qquad H_n^{(1)}(t) = \frac{2^n (n-1)!}{\pi i t^n} \left(1 + O\left(\frac{1}{n}\right)\right), \qquad n \to \infty,$$

uniformly on compact subsets of $(0, \infty)$.

For large arguments, we have the following asymptotic behavior of the Hankel functions

$$H_n^{(1,2)}(t) = \sqrt{\frac{2}{\pi t}}\, e^{\pm i\left(t - \frac{n\pi}{2} - \frac{\pi}{4}\right)} \left\{1 + O\left(\frac{1}{t}\right)\right\}, \qquad t \to \infty,$$

$$(3.59)$$

$$H_n^{(1,2)\prime}(t) = \sqrt{\frac{2}{\pi t}}\, e^{\pm i\left(t - \frac{n\pi}{2} + \frac{\pi}{4}\right)} \left\{1 + O\left(\frac{1}{t}\right)\right\}, \qquad t \to \infty.$$

For a proof, we refer to Lebedev [117]. Taking the real and the imaginary part of (3.59) we also have asymptotic formulas for the Bessel and Neumann functions.

Now we have listed all the necessary tools for carrying over the analysis of Chapters 2 and 3 for the Helmholtz equation from three to two dimensions. The fundamental solution to the Helmholtz equation in two dimensions is given by

$$(3.60) \qquad \Phi(x, y) := \frac{i}{4} H_0^{(1)}(k|x - y|), \qquad x \neq y.$$

For fixed $y \in \mathbb{R}^2$, it satisfies the Helmholtz equation in $\mathbb{R}^2 \setminus \{y\}$. From the expansions (3.51) and (3.52), we deduce that

$$(3.61) \quad \Phi(x, y) = \frac{1}{2\pi} \ln \frac{1}{|x - y|} + \frac{i}{4} - \frac{1}{2\pi} \ln \frac{k}{2} - \frac{C}{2\pi} + O\left(|x - y|^2 \ln \frac{1}{|x - y|}\right)$$

for $|x - y| \to 0$. Therefore, the fundamental solution to the Helmholtz equation in two dimensions has the same singular behavior as the fundamental solution of Laplace's equation. As a consequence, Green's formula (2.5) and the jump relations and regularity results on single- and double-layer potentials of Theorems 3.1 and 3.3 can be carried over to two dimensions. From (3.61) we note that, in contrast to three dimensions, the fundamental solution does not converge for $k \to 0$ to the fundamental solution for the Laplace equation. This leads to some difficulties in the investigation of the convergence of the solution to the exterior Dirichlet problem as $k \to 0$ (see Werner [186] and Kress [105]).

In $\mathbb{R}^2$ the Sommerfeld radiation condition has to be replaced by

$$(3.62) \qquad \lim_{r \to \infty} \sqrt{r} \left(\frac{\partial u}{\partial r} - iku\right) = 0, \qquad r = |x|,$$

uniformly for all directions $x/|x|$. From (3.59) it is obvious that the fundamental solution satisfies the radiation condition uniformly with respect to y on compact sets. Therefore, Green's representation formula (2.8) can be shown to be valid for two-dimensional radiating solutions. According to the form (3.62) of the radiation condition, the definition of the far field pattern (2.12) has to be replaced by

$$(3.63) \qquad u(x) = \frac{e^{ik|x|}}{\sqrt{|x|}} \left\{ u_\infty(\hat{x}) + O\left(\frac{1}{|x|}\right) \right\}, \qquad |x| \to \infty,$$

and, due to (3.59), the representation (2.13) has to be replaced by

$$(3.64) \qquad u_\infty(\hat{x}) = \frac{e^{i\frac{\pi}{4}}}{\sqrt{8\pi k}} \int_{\partial D} \left\{ u(y) \frac{\partial e^{-ik\hat{x}\cdot y}}{\partial \nu(y)} - \frac{\partial u}{\partial \nu}(y)\, e^{-ik\hat{x}\cdot y} \right\} ds(y)$$

for $|\hat{x}| = x/|x|$. We explicitly write out the addition theorem

$$(3.65) \quad H_0^{(1)}(k|x-y|) = H_0^{(1)}(k|x|)\, J_0(k|y|) + 2 \sum_{n=1}^{\infty} H_n^{(1)}(k|x|)\, J_n(k|y|) \cos n\theta$$

which is valid for $|x| > |y|$ in the sense of Theorem 2.10 and where θ denotes the angle between x and y. The proof is analogous to that of Theorem 2.10. We note that the entire spherical wave functions in $\mathbb{R}^2$ are given by $J_n(kr)e^{\pm in\varphi}$ and the radiating spherical wave functions by $H_n^{(1)}(kr)e^{\pm in\varphi}$. Similarly, the Jacobi–Anger expansion (2.45) assumes the form

$$(3.66) \qquad e^{ik\,x\cdot d} = J_0(k|x|) + 2 \sum_{n=1}^{\infty} i^n J_n(k|x|) \cos n\theta, \qquad x \in \mathbb{R}^2.$$

With all these prerequisites, it is left as an excercise to establish that, with minor adjustments in the proofs, all the results of Sections 2.5, 3.2 and 3.3 remain valid in two dimensions.

3.5 On the Numerical Solution in $\mathbb{R}^2$

We would like to include in our presentation an advertisement for what we think is the most efficient method for the numerical solution of the boundary integral equations for two-dimensional problems. Since it seems to be safe to state that the boundary curves in most practical applications are either analytic or piecewise analytic with corners, we restrict our attention to approximation schemes which are the most appropriate under these regularity assumptions.

We begin with the analytic case where we recommend the Nyström method based on appropriately weighted numerical quadratures on an equidistant mesh. We first describe the necessary parametrization of the integral equation (3.26) in the two-dimensional case. We assume that the boundary curve ∂D possesses a regular analytic and 2π–periodic parametric representation of the form

$$(3.67) \qquad x(t) = (x_1(t), x_2(t)), \qquad 0 \le t \le 2\pi,$$

in counterclockwise orientation satisfying $[x_1'(t)]^2 + [x_2'(t)]^2 > 0$ for all t. Then, by straightforward calculations using $H_1^{(1)} = -H_0^{(1)\prime}$, we transform (3.26) into the parametric form

$$\psi(t) - \int_0^{2\pi} \{L(t,\tau) + i\eta M(t,\tau)\}\,\psi(\tau)\,d\tau = g(t), \quad 0 \le t \le 2\pi,$$

where we have set $\psi(t) := \varphi(x(t))$, $g(t) := 2f(x(t))$ and the kernels are given by

$$L(t,\tau) := \frac{ik}{2}\,\{x_2'(\tau)[x_1(\tau) - x_1(t)] - x_1'(\tau)[x_2(\tau) - x_2(t)]\}\,\frac{H_1^{(1)}(kr(t,\tau))}{r(t,\tau)},$$

$$M(t,\tau) := \frac{i}{2}\,H_0^{(1)}(kr(t,\tau))\{[x_1'(\tau)]^2 + [x_2'(\tau)]^2\}^{1/2}$$

for $t \ne \tau$. Here, we have set

$$r(t,\tau) := \{[x_1(t) - x_1(\tau)]^2 + [x_2(t) - x_2(\tau)]^2\}^{1/2}.$$

From the expansion (3.52) for the Neumann functions, we see that the kernels L and M have logarithmic singularities at $t = \tau$. Hence, for their proper numerical treatment, following Martensen [125] and Kussmaul [113], we split the kernels into

$$L(t,\tau) = L_1(t,\tau)\ln\left(4\sin^2\frac{t-\tau}{2}\right) + L_2(t,\tau),$$

$$M(t,\tau) = M_1(t,\tau)\ln\left(4\sin^2\frac{t-\tau}{2}\right) + M_2(t,\tau),$$

where

$$L_1(t,\tau) := \frac{k}{2\pi}\,\{x_2'(\tau)[x_1(t) - x_1(\tau)] - x_1'(\tau)[x_2(t) - x_2(\tau)]\}\,\frac{J_1(kr(t,\tau))}{r(t,\tau)},$$

$$L_2(t,\tau) := L(t,\tau) - L_1(t,\tau)\ln\left(4\sin^2\frac{t-\tau}{2}\right),$$

$$M_1(t,\tau) := -\frac{1}{2\pi}\,J_0(kr(t,\tau))\,\{[x_1'(\tau)]^2 + [x_2'(\tau)]^2\}^{1/2},$$

$$M_2(t,\tau) := M(t,\tau) - M_1(t,\tau)\ln\left(4\sin^2\frac{t-\tau}{2}\right).$$

The kernels L_1, L_2, M_1, and M_2 turn out to be analytic. In particular, using the expansions (3.51) and (3.52) we can deduce the diagonal terms

$$L_2(t,t) = L(t,t) = \frac{1}{2\pi}\,\frac{x_1'(t)x_2''(t) - x_2'(t)x_1''(t)}{[x_1'(t)]^2 + [x_2'(t)]^2}$$

and

$$M_2(t,t) = \left\{ \frac{i}{2} - \frac{C}{\pi} - \frac{1}{2\pi} \ln \left(\frac{k^2}{4} \{[x_1'(t)]^2 + [x_2'(t)]^2\} \right) \right\} \{[x_1'(t)]^2 + [x_2'(t)]^2\}^{1/2}$$

for $0 \leq t \leq 2\pi$. We note that despite the continuity of the kernel L, for numerical accuracy it is advantageous to separate the logarithmic part of L since the derivatives of L fail to be continuous at $t = \tau$.

Hence, we have to numerically solve an integral equation of the form

$$(3.68) \qquad \psi(t) - \int_0^{2\pi} K(t,\tau)\psi(\tau)\, d\tau = g(t), \quad 0 \leq t \leq 2\pi,$$

where the kernel can be written in the form

$$(3.69) \qquad K(t,\tau) = K_1(t,\tau) \ln \left(4 \sin^2 \frac{t-\tau}{2} \right) + K_2(t,\tau)$$

with analytic functions K_1 and K_2 and with an analytic right hand side g. Here we wish to point out that it is essential to split off the logarithmic singularity in a fashion which preserves the 2π–periodicity for the kernels K_1 and K_2.

For the numerical solution of integral equations of the second kind, in principle, there are three basic methods available, the *Nyström method*, the *collocation method* and the *Galerkin method*. In the case of one-dimensional integral equations, the Nyström method is more practical than the collocation and Galerkin method since it requires the least computational effort. In each of the three methods, the approximation requires the solution of a finite dimensional linear system. In the Nyström method, for the evaluation of each of the matrix elements of this linear system only an evaluation of the kernel function is needed, whereas in the collocation and Galerkin method the matrix elements are simple or double integrals demanding numerical quadratures. In addition, the Nyström method is generically stable in the sense that it preserves the condition of the integral equation whereas in the collocation and Galerkin method the condition can be disturbed by a poor choice of the basis (see [106]).

In the case of integral equations for periodic analytic functions, using global approximations via trigonometric polynomials is superior to using local approximations via low order polynomial splines since the trigonometric approximations yield much better convergence. By choosing the appropriate basis, the computational effort for the global approximation is comparable to that for local approximations.

The Nyström method consists in the straightforward approximation of the integrals by quadrature formulas. In our case, for the 2π–periodic integrands, we choose an equidistant set of knots $t_j := \pi j / n$, $j = 0, \ldots, 2n - 1$, and use the quadrature rule

$$(3.70) \qquad \int_0^{2\pi} \ln \left(4 \sin^2 \frac{t-\tau}{2} \right) f(\tau) d\tau \approx \sum_{j=0}^{2n-1} R_j^{(n)}(t) f(t_j), \quad 0 \leq t \leq 2\pi,$$

with the quadrature weights given by

$$R_j^{(n)}(t) := -\frac{2\pi}{n} \sum_{m=1}^{n-1} \frac{1}{m} \cos m(t - t_j) - \frac{\pi}{n^2} \cos n(t - t_j), \quad j = 0, \ldots, 2n - 1,$$

and the trapezoidal rule

(3.71) $$\int_0^{2\pi} f(\tau)d\tau \approx \frac{\pi}{n} \sum_{j=0}^{2n-1} f(t_j).$$

Both these numerical integration formulas are obtained by replacing the integrand f by its trigonometric interpolation polynomial and then integrating exactly. The quadrature formula (3.70) was first used by Martensen [125] and Kussmaul [113]. Provided f is analytic, according to derivative-free error estimates for the remainder term in trigonometric interpolation for periodic analytic functions (see [101, 106]), the errors for the quadrature rules (3.70) and (3.71) decrease at least exponentially when the number $2n$ of knots is increased. More precisely, the error is of order $O(\exp(-n\sigma))$ where σ denotes half of the width of a parallel strip in the complex plane into which the real analytic function f can be holomorphically extended.

Of course, it is also possible to use quadrature rules different from (3.70) and (3.71) obtained from other approximations for the integrand f. However, due to their simplicity and high approximation order we strongly recommend the application of (3.70) and (3.71).

In the Nyström method, the integral equation (3.68) is replaced by the approximating equation

(3.72) $$\psi^{(n)}(t) - \sum_{j=0}^{2n-1} \left\{ R_j^{(n)}(t)K_1(t, t_j) + \frac{\pi}{n} K_2(t, t_j) \right\} \psi^{(n)}(t_j) = g(t)$$

for $0 \leq t \leq 2\pi$. Equation (3.72) is obtained from (3.68) by applying the quadrature rule (3.70) to $f = K_1(t, .)\psi$ and (3.71) to $f = K_2(t, .)\psi$. The solution of (3.72) reduces to solving a finite dimensional linear system. In particular, for any solution of (3.72) the values $\psi_i^{(n)} = \psi^{(n)}(t_i)$, $i = 0, \ldots, 2n - 1$, at the quadrature points trivially satisfy the linear system

(3.73) $$\psi_i^{(n)} - \sum_{j=0}^{2n-1} \left\{ R_{|i-j|}^{(n)} K_1(t_i, t_j) + \frac{\pi}{n} K_2(t_i, t_j) \right\} \psi_j^{(n)} = g(t_i)$$

for $i = 0, \ldots, 2n - 1$, where

$$R_j^{(n)} := R_j^{(n)}(0) = -\frac{2\pi}{n} \sum_{m=1}^{n-1} \frac{1}{m} \cos \frac{mj\pi}{n} - \frac{(-1)^j \pi}{n^2}, \quad j = 0, \ldots, 2n - 1.$$

Conversely, given a solution $\psi_i^{(n)}$, $i = 0, \ldots, 2n - 1$, of the system (3.73), the function $\psi^{(n)}$ defined by

(3.74) $$\psi^{(n)}(t) := \sum_{j=0}^{2n-1} \left\{ R_j^{(n)}(t)K_1(t, t_j) + \frac{\pi}{n} K_2(t, t_j) \right\} \psi_j^{(n)} + g(t)$$

for $0 \le t \le 2\pi$ is readily seen to satisfy the approximating equation (3.72). The formula (3.74) may be viewed as a natural interpolation of the values $\psi_i^{(n)}$, $i = 0, \ldots, 2n-1$, at the quadrature points to obtain the approximating function $\psi^{(n)}$ and goes back to Nyström.

For the solution of the large linear system (3.73), we recommend the use of the fast iterative two-grid or multi-grid methods as described in [106] or, in more detail, in [62].

Provided the integral equation (3.68) itself is uniquely solvable and the kernels K_1 and K_2 and the right hand side g are continuous, a rather involved error analysis (for the details we refer to [106]) shows that

1. the approximating linear system (3.73), i.e., the approximating equation (3.72), is uniquely solvable for all sufficiently large n;
2. as $n \to \infty$ the approximate solutions $\psi^{(n)}$ converge uniformly to the solution ψ of the integral equation;
3. the convergence order of the quadrature errors for (3.70) and (3.71) carries over to the error $\psi^{(n)} - \psi$.

The latter, in particular, means that in the case of analytic kernels K_1 and K_2 and analytic right hand sides g the approximation error decreases exponentially, i.e., there exist positive constants C and σ such that

$$(3.75) \qquad |\psi^{(n)}(t) - \psi(t)| \le C\,e^{-n\sigma}, \quad 0 \le t \le 2\pi,$$

for all n. In principle, the constants in (3.75) are computable but usually they are difficult to evaluate. In most practical cases, it is sufficient to judge the accuracy of the computed solution by doubling the number $2n$ of knots and then comparing the results for the coarse and the fine grid with the aid of the exponential convergence order, i.e., by the fact that doubling the number $2n$ of knots will double the number of correct digits in the approximate solution.

Fig. 3.1. Kite-shaped domain for numerical example

For a numerical example, we consider the scattering of a plane wave by a cylinder with a non-convex kite-shaped cross section with boundary ∂D illustrated in Fig. 3.1 and described by the parametric representation

$$x(t) = (\cos t + 0.65 \cos 2t - 0.65 \,,\; 1.5 \sin t), \quad 0 \le t \le 2\pi.$$

From the asymptotics (3.59) for the Hankel functions, analogous to (3.64) it can be deduced that the far field pattern of the combined potential (3.25) in two dimensions is given by

$$u_\infty(\hat{x}) = \frac{e^{-i\frac{\pi}{4}}}{\sqrt{8\pi k}} \int_{\partial D} \{k\,\nu(y) \cdot \hat{x} + \eta\} e^{-ik\hat{x}\cdot y} \varphi(y)\, ds(y), \quad |\hat{x}| = 1,$$

which can be evaluated again by the trapezoidal rule after solving the integral equation for φ. Table 3.1 gives some approximate values for the far field pattern $u_\infty(d)$ and $u_\infty(-d)$ in the forward direction d and the backward direction $-d$. The direction d of the incident wave is $d = (1,0)$ and, as recommended in [103], the coupling parameter is $\eta = k$. Note that the exponential convergence is clearly exhibited.

Table 3.1. Numerical results for Nyström's method

	n	$\mathrm{Re}\,u_\infty(d)$	$\mathrm{Im}\,u_\infty(d)$	$\mathrm{Re}\,u_\infty(-d)$	$\mathrm{Im}\,u_\infty(-d)$
$k = 1$	8	-1.62642413	0.60292714	1.39015283	0.09425130
	16	-1.62745909	0.60222343	1.39696610	0.09499454
	32	-1.62745750	0.60222591	1.39694488	0.09499635
	64	-1.62745750	0.60222591	1.39694488	0.09499635
$k = 5$	8	-2.30969119	1.52696566	-0.30941096	0.11503232
	16	-2.46524869	1.67777368	-0.19932343	0.06213859
	32	-2.47554379	1.68747937	-0.19945788	0.06015893
	64	-2.47554380	1.68747937	-0.19945787	0.06015893

For domains D with corners, a uniform mesh yields only poor convergence and therefore has to be replaced by a graded mesh. We suggest to base this grading upon the idea of substituting an appropriate new variable and then using the Nyström method as described above for the transformed integral equation. With a suitable choice for the substitution, this will lead to high order convergence.

Without loss of generality, we confine our presentation to a boundary curve ∂D with one corner at the point x_0 and assume $\partial D \setminus \{x_0\}$ to be C^2 and piecewise analytic. We do not allow cusps in our analysis, i.e., the angle γ at the corner is assumed to satisfy $0 < \gamma < 2\pi$.

Using the fundamental solution

$$\Phi_0(x,y) := \frac{1}{2\pi} \ln \frac{1}{|x-y|}, \quad x \neq y,$$

to the Laplace equation in $\mathbb{R}^2$ to subtract a vanishing term, we rewrite the combined double- and single-layer potential (3.25) in the form

$$u(x) = \int_{\partial D} \left[\left\{ \frac{\partial \Phi(x,y)}{\partial \nu(y)} - i\eta \Phi(x,y) \right\} \varphi(y) - \frac{\partial \Phi_0(x,y)}{\partial \nu(y)} \varphi(x_0) \right] ds(y).$$

for $x \in \mathbb{R}^2 \setminus \bar{D}$. This modification is notationally advantageous for the corner case and it makes the error analysis for the Nyström method work. The integral equation (3.26) now becomes

(3.76)
$$\varphi(x) - \varphi(x_0) + 2 \int_{\partial D} \left\{ \frac{\partial \Phi(x,y)}{\partial \nu(y)} - i\eta \Phi(x,y) \right\} \varphi(y) \, ds(y)$$
$$- 2 \int_{\partial D} \frac{\partial \Phi_0(x,y)}{\partial \nu(y)} \varphi(x_0) \, ds(y) = 2f(x), \quad x \in \partial D.$$

Despite the corner at x_0, there is no change in the residual term in the jump relations since the density $\varphi - \varphi(x_0)$ of the leading term in the singularity vanishes at the corner. However, the kernel of the integral equation (3.76) at the corner no longer remains weakly singular. For a C^2 boundary, the weak singularity of the kernel of the double-layer operator rests on the inequality

(3.77)
$$|\nu(y) \cdot (x - y)| \leq L|x - y|^2, \quad x, y \in \partial D,$$

for some positive constant L. This inequality expresses the fact that the vector $x - y$ for x close to y is almost orthogonal to the normal vector $\nu(y)$. For a proof, we refer to [32]. However, in the vicinity of a corner (3.77) does not remain valid.

After splitting off the operator $K_0 : C(\partial D) \to C(\partial D)$ defined by

$$(K_0 \varphi)(x) := 2 \int_{\partial D} \frac{\partial \Phi_0(x,y)}{\partial \nu(y)} [\varphi(y) - \varphi(x_0)] \, ds(y), \quad x \in \partial D,$$

from (3.61) we see that the remaining integral operator in (3.76) has a weakly singular kernel and therefore is compact. For the further investigation of the non-compact part K_0, we choose a sufficiently small positive number r and denote the two arcs of the boundary ∂D contained in the disk of radius r and center at the corner x_0 by A and B. These arcs intersect at x_0 with an angle γ and without loss of generality we restrict our presentation to the case where $\gamma < \pi$. By elementary geometry and continuity, we can assume that r is chosen such that both A and B have length less than $2r$ and for the angle $\alpha(x, B)$ between the two straight lines connecting the points $x \in A \setminus \{x_0\}$ with the two endpoints of the arc B we have

$$0 < \alpha(x, B) \leq \pi - \frac{1}{2}\gamma, \quad x \in A \setminus x_0,$$

and analogously with the roles of A and B interchanged. For the sake of brevity, we confine ourselves to the case where the boundary ∂D in a neighborhoud of the corner x_0 consists of two straight lines intersecting at x_0. Then we can assume that r is chosen such that the function $(x, y) \mapsto \nu(y) \cdot (y - x)$ does not change its sign for all $(x, y) \in A \times B$ and all $(x, y) \in B \times A$. Finally, for the two C^2 arcs A and B, there exists a constant L independent of r such that the estimate (3.77) holds for all $(x, y) \in A \times A$ and all $(x, y) \in B \times B$.

We now choose a continuous cut-off function $\psi : \mathbb{R}^2 \to [0,1]$ such that $\psi(x) = 1$ for $0 \le |x - x_0| \le r/2$, $\psi(x) = 0$ for $r \le |x - x_0| < \infty$ and define $K_{0,r} : C(\partial D) \to C(\partial D)$ by

$$K_{0,r}\varphi := \psi K_0(\psi \varphi).$$

Then, the kernel of $K_0 - K_{0,r}$ vanishes in a neighborhood of (x_0, x_0) and therefore is weakly singular.

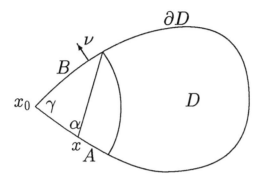

Fig. 3.2. Domain with a corner

We introduce the norm

$$\|\varphi\|_{\infty,0} := \max_{x \in \partial D} |\varphi(x) - \varphi(x_0)| + |\varphi(x_0)|$$

which obviously is equivalent to the maximum norm. We now show that r can be chosen such that $\|K_{0,r}\|_{\infty,0} < 1$. Then, by the Neumann series, the operator $I + K_{0,r}$ has a bounded inverse and the results of the Riesz–Fredholm theory are available for the corner integral equation (3.76).

By our assumptions on the choice of r, we can estimate

$$(3.78) \qquad |(K_{0,r}\varphi)(x_0)| \le \frac{4Lr}{\pi} \|\varphi\|_{\infty,0}$$

since (3.77) holds for $x = x_0$ and all $y \in A \cup B$. For $x \in A \setminus \{x_0\}$ we split the integral into the parts over A and over B and evaluate the second one by using Green's integral theorem and our assumptions on the geometry to obtain

$$2 \int_B \left| \frac{\partial \Phi_0(x,y)}{\partial \nu(y)} \right| ds(y) = 2 \left| \int_B \frac{\partial \Phi_0(x,y)}{\partial \nu(y)} ds(y) \right| = \frac{\alpha(x,B)}{\pi}, \qquad x \in A \setminus \{x_0\},$$

and consequently

$$(3.79) \qquad |(K_{0,r}\varphi)(x)| \le \left\{ \frac{2Lr}{\pi} + 1 - \frac{\gamma}{2\pi} \right\} \|\varphi\|_{\infty,0},$$

which by symmetry is valid for all $x \in A \cup B \setminus \{x_0\}$. Summarizing, from the inequalities (3.78) and (3.79) we deduce that we can choose r small enough such that $\|K_{0,r}\|_{\infty,0} < 1$. For an analysis for more general domains with corners we refer to Ruland [163] and the literature therein.

The above analysis establishes the existence of a continuous solution to the integral equation (3.76). However, due to the singularities of elliptic boundary value problems in domains with corners (see [60]), this solution will have singularities in the derivatives at the corner. To take proper care of this corner singularity, we replace our equidistant mesh by a graded mesh through substituting a new variable in such a way that the derivatives of the new integrand vanish up to a certain order at the endpoints and then use the quadrature rules (3.70) and (3.71) for the transformed integrals.

We describe this numerical quadrature rule for the integral $\int_0^{2\pi} f(t)\, dt$ where the integrand f is analytic in $(0, 2\pi)$ but has singularities at the endpoints $t = 0$ and $t = 2\pi$. Let the function $w : [0, 2\pi] \to [0, 2\pi]$ be one-to-one, strictly monotonically increasing and infinitely differentiable. We assume that the derivatives of w at the endpoints $t = 0$ and $t = 2\pi$ vanish up to an order $p \in \mathbb{N}$. We then substitute $t = w(s)$ to obtain

$$\int_0^{2\pi} f(t)\, dt = \int_0^{2\pi} w'(s)\, f(w(s))\, ds.$$

Applying the trapezoidal rule to the transformed integral now yields the quadrature formula

$$(3.80) \qquad \int_0^{2\pi} f(t)\, dt \approx \frac{\pi}{n} \sum_{j=1}^{2n-1} a_j\, f(s_j)$$

with the weights and mesh points given by

$$a_j = w'\left(\frac{j\pi}{n}\right), \quad s_j = w\left(\frac{j\pi}{n}\right), \quad j = 1, \ldots, 2n - 1.$$

A typical example for such a substitution is given by

$$(3.81) \qquad w(s) = 2\pi \frac{[v(s)]^p}{[v(s)]^p + [v(2\pi - s)]^p}, \quad 0 \le s \le 2\pi,$$

where

$$v(s) = \left(\frac{1}{p} - \frac{1}{2}\right)\left(\frac{\pi - s}{\pi}\right)^3 + \frac{1}{p}\frac{s - \pi}{\pi} + \frac{1}{2}$$

and $p \ge 2$. Note that the cubic polynomial v is chosen such that $v(0) = 0$, $v(2\pi) = 1$ and $w'(\pi) = 2$. The latter property ensures, roughly speaking, that one half of the grid points is equally distributed over the total interval, whereas the other half is accumulated towards the two end points.

For an error analysis for the quadrature rule (3.80) with substitutions of the form described above and using the Euler–MacLaurin expansion, we refer to Kress [107]. Assume f is $2q + 1$–times continuously differentiable on $(0, 2\pi)$ such that for some $0 < \alpha < 1$ with $\alpha p \geq 2q + 1$ the integrals

$$\int_0^{2\pi} \left[\sin \frac{t}{2} \right]^{m-\alpha} |f^{(m)}(t)| \, dt$$

exist for $m = 0, 1, \ldots, 2q + 1$. The error $E^{(n)}(f)$ in the quadrature (3.80) can then be estimated by

(3.82)
$$|E^{(n)}(f)| \leq \frac{C}{n^{2q+1}}$$

with some constant C. Thus, by choosing p large enough, we can obtain almost exponential convergence behavior.

For the numerical solution of the corner integral equation (3.76), we choose a parametric representation of the form (3.67) such that the corner x_0 corresponds to the parameter $t = 0$ and rewrite (3.76) in the parametrized form

(3.83)
$$\psi(t) - \psi(0) - \int_0^{2\pi} K(t, \tau) \psi(\tau) \, d\tau$$
$$- \int_0^{2\pi} H(t, \tau) \psi(0) \, d\tau = g(t), \quad 0 \leq t \leq 2\pi,$$

where K is given as above in the analytic case and where

$$H(t, \tau) = \begin{cases} \dfrac{1}{\pi} \dfrac{x_2'(\tau)[x_1(t) - x_1(\tau)] - x_1'(\tau)[x_2(t) - x_2(\tau)]}{[x_1(t) - x_1(\tau)]^2 + [x_2(t) - x_2(\tau)]^2}, & t \neq \tau, \\[4mm] \dfrac{1}{2\pi} \dfrac{x_2'(t)x_1''(t) - x_1'(t)x_2''(t)}{[x_1'(t)]^2 + [x_2'(t)]^2}, & t = \tau, \ t \neq 0, 2\pi, \end{cases}$$

corresponds to the additional term in (3.76). For the numerical solution of the integral equation (3.83) by Nyström's method on the graded mesh, we also have to take into account the logarithmic singularity. We set $t = w(s)$ and $\tau = w(\sigma)$ to obtain

$$\int_0^{2\pi} K(t, \tau) \, \psi(\tau) \, d\tau = \int_0^{2\pi} K(w(s), w(\sigma)) \, w'(\sigma) \psi(w(\sigma)) \, d\sigma$$

and then write

$$K(w(s), w(\sigma)) = \tilde{K}_1(s, \sigma) \ln \left(4 \sin^2 \frac{s - \sigma}{2} \right) + \tilde{K}_2(s, \sigma).$$

This decomposition is related to (3.69) by

$$\tilde{K}_1(s, \sigma) = K_1(w(s), w(\sigma))$$

and

$$\tilde{K}_2(s, \sigma) = K(w(s), w(\sigma)) - \tilde{K}_1(s, \sigma) \ln \left(4 \sin^2 \frac{s - \sigma}{2} \right), \quad s \neq \sigma.$$

From

$$K_2(s,s) = \lim_{\sigma \to s} \left[K(s,\sigma) - K_1(s,\sigma) \ln \left(4 \sin^2 \frac{s-\sigma}{2} \right) \right]$$

we deduce the diagonal term

$$\tilde{K}_2(s,s) = K_2(w(s), w(s)) + 2 \ln w'(s) \, K_1(w(s), w(s)).$$

Now, proceeding as in the derivation of (3.73), for the approximate values $\psi_i^{(n)} = \psi^{(n)}(s_i)$ at the quadrature points s_i for $i = 1, \ldots, 2n - 1$ and $\psi_0^{(n)} = \psi^{(n)}(0)$ at the corner $s_0 = 0$ we arrive at the linear system

(3.84)

$$\psi_i^{(n)} - \psi_0^{(n)} - \sum_{j=1}^{2n-1} \left\{ R_{|i-j|}^{(n)} \tilde{K}_1(s_i, s_j) + \frac{\pi}{n} \tilde{K}_2(s_i, s_j) \right\} a_j \psi_j^{(n)}$$

$$- \sum_{j=1}^{2n-1} \frac{\pi}{n} H(s_i, s_j) \, a_j \, \psi_0^{(n)} = g(s_i), \quad i = 0, \ldots, 2n - 1.$$

A rigorous error analysis carrying over the error behavior (3.82) to the approximate solution of the integral equation obtained from (3.84) for the potential theoretic case $k = 0$ has been worked out by Kress [107].

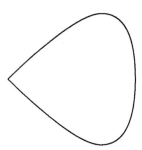

Fig. 3.3. Drop-shaped domain for numerical example

For a numerical example, we used the substitution (3.81) with order $p = 8$. We consider a drop-shaped domain with the boundary curve ∂D illustrated by Fig. 3.3 and given by the parametric representation

$$x(t) = \left(2 \sin \frac{t}{2}, \ - \sin t \right), \quad 0 \le t \le 2\pi.$$

It has a corner at $t = 0$ with interior angle $\gamma = \pi/2$. The direction d of the incoming plane wave and the coupling parameter η are chosen as in our previous example. Table 3.2 clearly exhibits the fast convergence of the method.

Table 3.2. Nyström's method for a domain with corner

	n	$\operatorname{Re} u_\infty(d)$	$\operatorname{Im} u_\infty(d)$	$\operatorname{Re} u_\infty(-d)$	$\operatorname{Im} u_\infty(-d)$
$k = 1$	16	-1.28558226	0.30687170	-0.53002440	-0.41033666
	32	-1.28549613	0.30686638	0.53020518	-0.41094518
	64	-1.28549358	0.30686628	-0.53021014	-0.41096324
	128	-1.28549353	0.30686627	-0.53021025	-0.41096364
$k = 5$	16	-1.73779647	1.07776749	-0.18112826	-0.20507986
	32	-1.74656264	1.07565703	-0.19429063	-0.19451172
	64	-1.74656303	1.07565736	-0.19429654	-0.19453324
	128	-1.74656304	1.07565737	-0.19429667	-0.19453372

3.6 On the Numerical Solution in $\mathbb{R}^3$

In three dimensions, for the numerical solution of the boundary integral equation (3.26) the Nyström, collocation and Galerkin methods are still available. However, for surface integral equations we have to modify our statements on comparing the efficiency of the three methods. Firstly, there is no straightforward simple quadrature rule analogous to (3.70) available that deals appropriately with the singularity of the three-dimensional fundamental solution. Hence, the Nyström method looses some of its attraction. Secondly, for the surface integral equations there is no immediate choice for global approximations like the trigonometric polynomials in the one-dimensional periodic case. Therefore, local approximations by low order polynomial splines have been more widely used and the collocation method is the most important numerical approximation method. To implement the collocation method, the boundary surface is first subdivided into a finite number of segments, like curved triangles and squares. The approximation space is then chosen to consist of low order polynomial splines with respect to these surface elements. The simplest choices are piecewise constants or piecewise linear functions. Within each segment, depending on the degree of freedom in the chosen splines, a number of collocation points is selected. Then, the integrals for the matrix elements in the collocation system are evaluated using numerical integration. Due to the weak singularity of the kernels, the calculation of the improper integrals for the diagonal elements of the matrix, where the collocation points and the surface elements coincide, needs special attention. For a detailed description of this so-called *boundary element method* we refer to Brebbia, Telles and Wrobel [18].

Besides such local approximations there are also a few global approaches available. For surfaces which can be mapped onto spheres, Atkinson [10] has developed a Galerkin method for the Laplace equation using spherical harmonics as a possible counterpart of the trigonometric polynomials. This method has been extended to the Helmholtz equation by Lin [121]. Based on spherical harmonics and transforming the boundary surface to a sphere as in Atkinson's method, Wienert [192] has developed a Nyström type method for the boundary integral equations for three-dimensional Helmholtz problems which is exponentially convergent for analytic boundary surfaces. We conclude this chapter by introducing the main ideas of Wienert's method.

We begin by describing a numerical quadrature scheme for the integration of analytic functions over closed analytic surfaces Γ in $\mathbb{R}^3$ which are homeomorphic to the unit sphere Ω and then we proceed to corresponding quadratures for acoustic single- and double-layer potentials. To this end, we first introduce a suitable projection operator Q_N onto the linear space H_{N-1} of all spherical harmonics of order less than N. We denote by $-1 < t_1 < t_2 < \cdots < t_N < 1$ the zeros of the Legendre polynomial P_N (the existence of N distinct zeros of P_N in the interval $(-1,1)$ is a consequence of the orthogonality relation (2.24) – see [49], p. 236) and by

$$\alpha_j := \frac{2(1 - t_j^2)}{[N P_{N-1}(t_j)]^2} , \quad j = 1, \ldots, N,$$

the weights of the Gauss–Legendre quadrature rule which are uniquely determined by the property

$$(3.85) \qquad \int_{-1}^{1} p(t)\, dt = \sum_{j=1}^{N} \alpha_j p(t_j)$$

for all polynomials p of degree less than or equal to $2N - 1$ (see [50], p. 89). We then choose a set of points x_{jk} on the unit sphere Ω given in polar coordinates by

$$x_{jk} := (\sin \theta_j \cos \varphi_k, \sin \theta_j \sin \varphi_k, \cos \theta_j)$$

for $j = 1, \ldots, N$ and $k = 0, \ldots, 2N - 1$ where $\theta_j := \arccos t_j$ and $\varphi_k = \pi k / N$ and define $Q_N : C(\Omega) \to H_{N-1}$ by

$$(3.86) \qquad Q_N f := \frac{\pi}{N} \sum_{j=1}^{N} \sum_{k=0}^{2N-1} \alpha_j f(x_{jk}) \sum_{n=0}^{N-1} \sum_{m=-n}^{n} Y_n^{-m}(x_{jk})\, Y_n^m$$

where the spherical harmonics Y_n^m are given by (2.27). By orthogonality we clearly have

$$(3.87) \qquad \int_{\Omega} Q_N f\, Y_n^{-m}\, ds = \frac{\pi}{N} \sum_{j=1}^{N} \sum_{k=0}^{2N-1} \alpha_j f(x_{jk}) Y_n^{-m}(x_{jk})$$

for $|m| \leq n < N$. Since the trapezoidal rule with $2N$ knots integrates trigonometric polynomials of degree less than N exactly, we have

$$\frac{\pi}{N} \sum_{k=0}^{2N-1} Y_n^m(x_{jk}) Y_{n'}^{-m'}(x_{jk}) = \int_{0}^{2\pi} Y_n^m(\theta_j, \varphi)\, Y_{n'}^{-m'}(\theta_j, \varphi)\, d\varphi$$

for $|m|, |m'| \leq n < N$ and these integrals, in view of (2.27), vanish if $m \neq m'$. For $m = m'$, by (2.26) and (2.27), $Y_n^m Y_{n'}^{-m}$ is a polynomial of degree less than

$2N$ in $\cos\theta$. Hence, by the property (3.85) of the Gauss–Legendre quadrature rule, summing the previous equation we find

$$\frac{\pi}{N}\sum_{j=1}^{N}\sum_{k=0}^{2N-1}\alpha_j Y_n^m(x_{jk})\,Y_{n'}^{-m'}(x_{jk}) = \int_\Omega Y_n^m\,Y_{n'}^{-m'}\,ds,$$

that is, $Q_N Y_n^m = Y_n^m$ for $|m| \le n < N$ and therefore Q_N is indeed a projection operator onto H_{N-1}. We note that Q_N is not an interpolation operator since by Theorem 2.6 we have $\dim H_{N-1} = N^2$ whereas we have $2N^2$ points x_{jk}. With the aid of (2.21), the addition theorem (2.29) and (3.85) we can estimate

$$\|Q_N f\|_\infty \le \frac{1}{4N}\sum_{j=1}^{N}\sum_{k=0}^{2N-1}\alpha_j\sum_{n=0}^{N-1}(2n+1)\|f\|_\infty = N^2\|f\|_\infty$$

whence

(3.88)
$$\|Q_N\|_\infty \le N^2$$

follows.

For analytic functions $f : \Omega \to \mathbb{C}$, Wienert [192] has shown that the approximation error $f - Q_N f$ decreases exponentially, that is, there exist positive constants C and σ depending on f such that

(3.89)
$$\|f - Q_N f\|_{\infty,\Omega} \le C e^{-N\sigma}$$

for all $N \in \mathbb{N}$.

Integrating the approximation $Q_N f$ instead of f we obtain the so-called Gauss trapezoidal product rule

(3.90)
$$\int_\Omega f\,ds \approx \frac{\pi}{N}\sum_{j=1}^{N}\sum_{k=0}^{2N-1}\alpha_j f(x_{jk})$$

for the numerical integration over the unit sphere. For analytic surfaces Γ which can be mapped bijectively through an analytic function $q : \Omega \to \Gamma$ onto the unit sphere, (3.90) can also be used after the substitution

$$\int_\Gamma g(\xi)\,ds(\xi) = \int_\Omega g(q(x))J_q(x)\,ds(x)$$

where J_q stands for the Jacobian of the mapping q. For analytic functions, the exponential convergence (3.89) carries over to the quadrature (3.90).

By passing to the limit $k \to 0$ in (2.43), with the help of (2.31) and (2.32), we find

$$\int_\Omega \frac{Y_n(y)}{|x-y|}\,ds(y) = \frac{4\pi}{2n+1}\,Y_n(x), \quad x \in \Omega,$$

for spherical harmonics Y_n of order n. This can be used together with the addition formula (2.29) to obtain the approximation

$$\int_\Omega \frac{f(y)}{|x-y|}\,ds(y) \approx \frac{\pi}{N}\sum_{j=1}^{N}\sum_{k=0}^{2N-1}\alpha_j f(x_{jk})\sum_{n=0}^{N-1}P_n(x_{jk}\cdot x), \quad x \in \Omega,$$

which again is based on replacing f by $Q_N f$. In particular, for the north pole $x_0 = (0, 0, 1)$ this reads

$$(3.91) \qquad \int_\Omega \frac{f(y)}{|x_0 - y|} \, ds(y) \approx \sum_{j=1}^{N} \sum_{k=0}^{2N-1} \beta_j f(x_{jk})$$

where

$$\beta_j := \frac{\pi \alpha_j}{N} \sum_{n=0}^{N-1} P_n(t_j), \quad j = 1, \ldots, N.$$

The exponential convergence for analytic densities $f : \Omega \to \mathbb{C}$ again carries over from (3.89) to the numerical quadrature (3.91) of the harmonic single-layer potential.

For the extension of this quadrature scheme to more general surfaces Γ, we need to allow more general densities and we can do this without losing the rapid convergence order. Denote by $\tilde{\Omega}$ the cylinder

$$\tilde{\Omega} := \{(\cos \varphi, \sin \varphi, \theta) : 0 \leq \theta \leq \pi, \, 0 \leq \varphi \leq 2\pi\}.$$

Then we can identify functions defined on $\tilde{\Omega}$ with functions on Ω through the mapping

$$(\cos \varphi, \sin \varphi, \theta) \mapsto (\sin \theta \cos \varphi, \sin \theta \sin \varphi, \cos \theta)$$

and, loosely speaking, in the sequel we refer to functions on $\tilde{\Omega}$ as functions on Ω depending on the azimuth φ at the poles. As Wienert [192] has shown, the exponential convergence is still true for the application of (3.91) to analytic functions $f : \tilde{\Omega} \to \mathbb{C}$.

For the general surface Γ as above, we write

$$\int_\Gamma \frac{g(\eta)}{|q(x) - \eta|} \, ds(\eta) = \int_\Omega \frac{F(x, y) f(y)}{|x - y|} \, ds(y)$$

where we have set $f(y) := g(q(y)) J_q(y)$ and

$$(3.92) \qquad F(x, y) := \frac{|x - y|}{|q(x) - q(y)|}, \quad x \neq y.$$

Unfortunately, as can be seen from simple examples, the function F in general cannot be extended as a continuous function on $\Omega \times \Omega$. However, since on the unit sphere we have

$$|x - y|^2 = 2(1 - x \cdot y)$$

from the estimate (see the proof of Theorem 2.2 in [32])

$$c_1 |x - y|^2 \leq |q(x) - q(y)|^2 \leq c_2 |x - y|^2$$

which is valid for all $x, y \in \Omega$ and some constants $0 < c_1 < c_2$ it can be seen that $F(x_0, \cdot)$ is analytic on $\tilde{\Omega}$.

For $\psi \in \mathbb{R}$, we define the orthogonal transformations

$$D_P(\psi) := \begin{pmatrix} \cos \psi & -\sin \psi & 0 \\ \sin \psi & \cos \psi & 0 \\ 0 & 0 & 1 \end{pmatrix}$$

and

$$D_T(\psi) := \begin{pmatrix} \cos \psi & 0 & -\sin \psi \\ 0 & 1 & 0 \\ \sin \psi & 0 & \cos \psi \end{pmatrix}.$$

Then for $x \in \Omega$ with polar coordinates θ and φ the orthogonal transformation

$$T_x := D_P(\varphi)D_T(\theta)D_P(-\varphi)$$

has the property

$$T_x x = (0, 0, 1), \quad x \in \Omega.$$

Therefore

$$(3.93) \qquad \int_\Omega \frac{F(x,y)f(y)}{|x-y|} \, ds(y) \approx \sum_{j=1}^{N} \sum_{k=0}^{2N-1} \beta_j F(x, T_x^{-1} x_{jk}) f(T_x^{-1} x_{jk})$$

is exponentially convergent for analytic densities f in the sense of (3.89) since x is the north pole for the set of quadrature points $T_x^{-1} x_{jk}$. It can be shown that the exponential convergence is uniform with respect to $x \in \Omega$.

By decomposing

$$\frac{e^{ik|x-y|}}{|x-y|} = \frac{\cos k|x-y|}{|x-y|} + i \, \frac{\sin k|x-y|}{|x-y|},$$

we see that the integral equation (3.26) for the exterior Dirichlet problem is of the form

$$g(\xi) - \int_\Gamma \left\{ \frac{h_1(\xi, \eta)}{|\xi - \eta|} + \frac{\nu(\eta) \cdot (\xi - \eta)}{|\xi - \eta|^2} h_2(\xi, \eta) + h_3(\xi, \eta) \right\} g(\eta) \, ds(\eta) = w(\xi)$$

for $\xi \in \Gamma$ with analytic kernels h_1, h_2 and h_3. For our purpose of exposition, it suffices to consider only the singular part, that is, the case when $h_2 = h_3 = 0$. Using the substitution $\xi = q(x)$ and $\eta = q(y)$, the integral equation over Γ can be transformed into an integral equation over Ω of the form

$$(3.94) \qquad f(x) - \int_\Omega \frac{k(x,y)F(x,y)}{|x-y|} f(y) \, ds(y) = v(x), \quad x \in \Omega,$$

with the functions f, k and v appropriately defined through g, h_1 and w and with F given as in (3.92). We write $A : C(\Omega) \to C(\Omega)$ for the weakly singular integral operator

$$(Af)(x) := \int_\Omega \frac{k(x,y)F(x,y)}{|x-y|} f(y) \, ds(y), \quad x \in \Omega,$$

occuring in (3.94). By using the quadrature rule (3.93), we arrive at an approximating quadrature operator $A_N : C(\Omega) \to C(\tilde{\Omega})$ given by

$$(3.95) \quad (A_N f)(x) := \sum_{j=1}^{N} \sum_{k=0}^{2N-1} \beta_j k(x, T_x^{-1} x_{jk}) F(x, T_x^{-1} x_{jk}) f(T_x^{-1} x_{jk}), \quad x \in \Omega.$$

We observe that the quadrature points $T_x^{-1} x_{jk}$ depend on x. Therefore, we cannot reduce the solution of the approximating equation

$$(3.96) \qquad\qquad \tilde{f}_N - A_N \tilde{f}_N = v$$

to a linear system in the usual fashion of Nyström interpolation. A possible remedy for this difficulty is to apply the projection operator Q_N a second time by solving (3.96) through the projection method, that is, the final approximating equation is of the form

$$(3.97) \qquad\qquad f_N - Q_N A_N f_N = Q_N v.$$

From the triangle inequality

$$\|Af - Q_N A_N f\|_{\infty,\Omega} \leq \|Af - Q_N Af\|_{\infty,\Omega} + \|Q_N\|_\infty \|Af - A_N f\|_{\infty,\Omega}$$

and (3.88), exponential convergence

$$\|Af - Q_N A_N f\|_{\infty,\Omega} \leq C e^{-N\sigma}$$

with positive constants C and σ depending on f can be established. Representing

$$f_N := \sum_{n=0}^{N-1} \sum_{m=-n}^{n} a_n^m Y_n^m$$

and using (3.87) and (3.95) we find that solving (3.97) is equivalent to solving the linear system

$$(3.98) \quad a_n^m - \sum_{n'=0}^{N-1} \sum_{m'=-n'}^{n'} R_{nn'}^{mm'} a_{n'}^{m'} = \frac{\pi}{N} \sum_{j=1}^{N} \sum_{k=0}^{2N-1} \alpha_j v(x_{jk}) Y_n^{-m}(x_{jk})$$

for $n = 0, \ldots, N-1$, $m = -n, \ldots, n$, where

$$R_{nn'}^{mm'} := \frac{\pi}{N} \sum_{j_1=1}^{N} \sum_{k_1=0}^{2N-1} \sum_{j_2=1}^{N} \sum_{k_2=0}^{2N-1} \alpha_{j_1} \beta_{j_2} K(x_{j_1 k_1}, x_{j_2 k_2}) Y_n^{-m}(x_{j_1 k_1}) Y_{n'}^{m'}(T_{x_{j_1 k_1}}^{-1} x_{j_2 k_2})$$

and

$$K(x,y) := k(x, T_x^{-1} y) F(x, T_x^{-1} y).$$

Since orthogonal transformations map spherical harmonics of order n into spherical harmonics of order n, we have

$$Y_{n'}^{m'}(D_T(-\theta)y) = \sum_{\mu=-n'}^{n'} Z_0(n', m', \mu, \theta) Y_{n'}^{\mu}(y)$$

with

$$Z_0(n', m', \mu, \theta) = \int_\Omega Y_{n'}^{m'}(D_T(-\theta)y)\, Y_{n'}^{-\mu}(y)\, ds(y)$$

and from (2.27) we clearly have

$$Y_{n'}^{m'}(D_P(-\varphi)y) = e^{-im'\varphi} Y_{n'}^{m'}(y).$$

From this we find that the coefficients in (3.98) can be evaluated recursively through the scheme

$$Z_1(j_1, k_1, j_2, \mu) \; := \; \sum_{k_2=0}^{2N-1} \beta_{j_2} e^{i\mu(\varphi_{k_2}-\varphi_{k_1})} K(x_{j_1 k_1}, x_{j_2 k_2}),$$

$$Z_2(j_1, k_1, n', \mu) \; := \; \sum_{j_2=1}^{N} Y_{n'}^{\mu}(x_{j_2,0}) Z_1(j_1, k_1, j_2, \mu),$$

$$Z_3(j_1, k_1, n', m') \; := \; \sum_{\mu=-n'}^{n'} Z_0(n', m', \mu, \theta_{j_1}) Z_2(j_1, k_1, n', \mu) e^{im'\varphi_{k_1}},$$

$$Z_4(j_1, m, n', m') \; := \; \sum_{k_1=0}^{2N-1} e^{-im\varphi_{k_1}} Z_3(j_1, k_1, n', m'),$$

$$R_{nn'}^{mm'} \; := \; \frac{\pi}{N} \sum_{j_1=1}^{N} \alpha_{j_1} Y_n^{-m}(x_{j_1,0}) Z_4(j_1, m, n', m'),$$

by $O(N^5)$ multiplications provided the numbers $Z_0(n', m', \mu, \theta_{j_1})$ (which do not depend on the surface) are precalculated. The latter calculations can be based on

$$Z_0(n', m', \mu, \theta) \; = \; \int_\Omega (Q_N(Y_{n'}^{m'} \circ D_T(-\theta)))(y)\, Y_{n'}^{-\mu}(y)\, ds(y)$$

$$= \; \frac{\pi}{N} \sum_{j=1}^{N} \sum_{k=0}^{2N-1} \alpha_j Y_{n'}^{m'}(D_T(-\theta)x_{jk})\, Y_{n'}^{-\mu}(x_{jk}).$$

For the details we refer to Wienert [192].

4. Ill-Posed Problems

As previously mentioned, for problems in mathematical physics Hadamard [63] postulated three requirements: a solution should exist, the solution should be unique, and the solution should depend continuously on the data. The third postulate is motivated by the fact that in all applications the data will be measured quantities. Therefore, one wants to make sure that small errors in the data will cause only small errors in the solution. A problem satisfying all three requirements is called *well-posed*. Otherwise, it is called *ill-posed*. As shown in the previous chapter, the direct obstacle scattering problem is well-posed.

For a long time, research on ill-posed problems was neglected since they were not considered relevant to the proper treatment of applied problems. However, it eventually became apparent that a growing number of important problems fail to be well-posed, for example Cauchy's problem for the Laplace equation and the initial boundary value problem for the backward heat equation. In particular, a large number of inverse problems for partial differential equations turn out to be ill-posed. Most classical problems where one assumes the partial differential equation, its domain and its initial and/or boundary data completely prescribed are well-posed in a canonical setting. Usually, such problems are referred to as *direct problems*. However, if the problem consists in determining part of the differential equation or its domain or its initial and/or boundary data then this *inverse problem* quite often will be ill-posed in any reasonable setting. In this sense, there is a close linkage and interaction between research on inverse problems and ill-posed problems.

This chapter is intended as an introduction into the basic ideas on ill-posed problems and regularization methods for their stable approximate solution. We mainly confine ourselves to linear equations of the first kind with compact operators in Hilbert spaces and base our presentation on the singular value decomposition. From the variety of regularization concepts, we will discuss only the spectral cut-off, Tikhonov regularization, the discrepancy principle and quasi-solutions. At the end of the chapter, we will include some material on nonlinear problems.

For a more comprehensive study of ill-posed problems, we refer to Baumeister [11], Groetsch [61], Kress [106], Louis [122], Morozov [134] and Tikhonov and Arsenin [174].

4.1 The Concept of Ill-Posedness

We will first make Hadamard's concept of well-posedness more precise.

Definition 4.1 *Let $A : U \subset X \to Y$ be an operator from a subset U of a normed space X into a normed space Y. The equation*

$$(4.1) \qquad\qquad A\varphi = f$$

is called well-posed *or* properly posed *if $A : U \to Y$ is bijective and the inverse operator $A^{-1} : Y \to U$ is continuous. Otherwise the equation is called* ill-posed *or* improperly posed.

According to this definition we may distinguish three types of ill-posedness. If A is not surjective, then equation (4.1) is not solvable for all $f \in Y$ (*nonexistence*). If A is not injective, then equation (4.1) may have more than one solution (*nonuniqueness*). Finally, if $A^{-1} : Y \to U$ exists but is not continuous then the solution φ of equation (4.1) does not depend continuously on the data f (*instability*). The latter case of instability is the one of primary interest in the study of ill-posed problems. We note that the three properties, in general, are not independent. For example, if $A : X \to Y$ is a bounded linear operator mapping a Banach space X bijectively onto a Banach space Y, then by the inverse mapping theorem the inverse operator $A^{-1} : Y \to X$ is bounded and therefore continuous. Note that the well-posedness of a problem is a property of the operator A together with the solution space X and the data space Y including the norms on X and Y. Therefore, if an equation is ill-posed one could try and restore stability by changing the spaces X and Y and their norms. But, in general, this approach is inadequate since the spaces X and Y including their norms are determined by practical needs. In particular, the space Y and its norm must be suitable to describe the measured data.

The typical example of an ill-posed problem is a completely continuous operator equation of the first kind. Recall that an operator $A : U \subset X \to Y$ is called compact if it maps bounded sets from U into relatively compact sets in Y and that A is called *completely continuous* if it is continuous and compact. Since linear compact operators are always continuous, for linear operators there is no need to distinguish between compactness and complete continuity.

Theorem 4.2 *Let $A : U \subset X \to Y$ be a completely continuous operator from a subspace U of a normed space X into a normed space Y. Then the equation of the first kind $A\varphi = f$ is improperly posed if U is not of finite dimension.*

Proof. Assume that $A^{-1} : Y \to U$ exists and is continuous. Then from $I = A^{-1}A$ we see that the identity operator on U is compact since the product of a continuous and a compact operator is compact. Hence U must be finite dimensional. $\square$

The ill-posed nature of an equation, of course, has consequences for its numerical treatment. We may view a numerical approximation of a given equation as the solution to perturbed data. Therefore, straightforward application of the classical methods for the approximate solution of operator equations to ill-posed problems usually will generate numerical nonsense. In terms of condition numbers, the fact that a bounded linear operator A does not have a bounded inverse means that the condition numbers of its finite dimensional approximations grow with the quality of the approximation. Hence, a careless discretization of ill-posed problems leads to a numerical behavior which at a first glance seems to be paradoxical. Namely, increasing the degree of discretization, i.e., increasing the accuracy of the approximation for the operator A will cause the approximate solution to the equation $A\varphi = f$ to become less and less reliable.

4.2 Regularization Methods

Methods for constructing a stable approximate solution of an ill-posed problem are called *regularization methods*. We shall now introduce the classical regularization concepts for linear equations of the first kind. In the sequel, we mostly will assume that the linear operator $A : X \to Y$ is injective. This is not a significant loss of generality since uniqueness for a linear equation always can be achieved by a suitable modification of the solution space X. We wish to approximate the solution φ to the equation $A\varphi = f$ from a knowledge of a perturbed right hand side f^δ with a known error level

$$(4.2) \qquad \|f^\delta - f\| \le \delta.$$

When f belongs to the range $A(X) := \{A\varphi : \varphi \in X\}$ then there exists a unique solution φ of $A\varphi = f$. For a perturbed right hand side, in general we cannot expect $f^\delta \in A(X)$. Using the erroneous data f^δ, we want to construct a reasonable approximation φ^δ to the exact solution φ of the unperturbed equation $A\varphi = f$. Of course, we want this approximation to be stable, i.e., we want φ^δ to depend continuously on the actual data f^δ. Therefore, our task requires finding an approximation of the unbounded inverse operator $A^{-1} : A(X) \to X$ by a bounded linear operator $R : Y \to X$.

Definition 4.3 *Let X and Y be normed spaces and let $A : X \to Y$ be an injective bounded linear operator. Then a family of bounded linear operators $R_\alpha : Y \to X$, $\alpha > 0$, with the property of pointwise convergence*

$$(4.3) \qquad \lim_{\alpha \to 0} R_\alpha A\varphi = \varphi$$

for all $\varphi \in X$ is called a regularization scheme *for the operator A. The parameter α is called the* regularization parameter.

Of course, (4.3) is equivalent to $R_\alpha f \to A^{-1}f$, $\alpha \to 0$, for all $f \in A(X)$. The following theorem shows that for regularization schemes for compact operators this convergence cannot be uniform.

Theorem 4.4 *Let X and Y be normed spaces, let $A : X \to Y$ be a compact linear operator, and let $\dim X = \infty$. Then for a regularization scheme the operators R_α cannot be uniformly bounded with respect to α and the operators $R_\alpha A$ cannot be norm convergent as $\alpha \to 0$.*

Proof. For the first statement, assume $\|R_\alpha\| \leq C$ for all $\alpha > 0$ and some constant C. Then from $R_\alpha f \to A^{-1} f$, $\alpha \to 0$, for all $f \in A(X)$ we deduce $\|A^{-1} f\| \leq C\|f\|$, i.e., $A^{-1} : A(X) \to X$ is bounded. By Theorem 4.2 this is a contradiction to $\dim X = \infty$.

For the second statement, assume that we have norm convergence. Then there exists $\alpha > 0$ such that $\|R_\alpha A - I\| < 1/2$. Now for all $f \in A(X)$ we can estimate

$$\|A^{-1} f\| \leq \|A^{-1} f - R_\alpha A A^{-1} f\| + \|R_\alpha f\| \leq \frac{1}{2}\|A^{-1} f\| + \|R_\alpha\|\,\|f\|,$$

whence $\|A^{-1} f\| \leq 2\|R_\alpha\|\,\|f\|$ follows. Therefore, $A^{-1} : A(X) \to X$ is bounded and we have the same contradiction as above. □

The regularization scheme approximates the solution φ of $A\varphi = f$ by the regularized solution

(4.4)
$$\varphi_\alpha^\delta := R_\alpha f^\delta.$$

Then, for the approximation error, writing

$$\varphi_\alpha^\delta - \varphi = R_\alpha f^\delta - R_\alpha f + R_\alpha A\varphi - \varphi,$$

by the triangle inequality we have the estimate

(4.5)
$$\|\varphi_\alpha^\delta - \varphi\| \leq \delta\|R_\alpha\| + \|R_\alpha A\varphi - \varphi\|.$$

This decomposition shows that the error consists of two parts: the first term reflects the influence of the incorrect data and the second term is due to the approximation error between R_α and A^{-1}. Under the assumptions of Theorem 4.4, the first term cannot be estimated uniformly with respect to α and the second term cannot be estimated uniformly with respect to φ. Typically, the first term will be increasing as $\alpha \to 0$ due to the ill-posed nature of the problem whereas the second term will be decreasing as $\alpha \to 0$ according to (4.3). Every regularization scheme requires a strategy for choosing the parameter α in dependence on the error level δ in order to achieve an acceptable total error for the regularized solution. On one hand, the accuracy of the approximation asks for a small error $\|R_\alpha A\varphi - \varphi\|$, i.e., for a small parameter α. On the other hand, the stability requires a small $\|R_\alpha\|$, i.e., a large parameter α. An optimal choice would try and make the right hand side of (4.5) minimal. The corresponding parameter effects a compromise between accuracy and stability. For a reasonable regularization strategy we expect the regularized solution to converge to the exact solution when the error level tends to zero. We express this requirement through the following definition.

Definition 4.5 *A strategy for a regularization scheme R_α, $\alpha > 0$, that is, the choice of the regularization parameter $\alpha = \alpha(\delta)$ depending on the error level δ, is called* regular *if for all $f \in A(X)$ and all $f^\delta \in Y$ with $\|f^\delta - f\| \leq \delta$ we have*

$$R_{\alpha(\delta)} f^\delta \to A^{-1} f, \quad \delta \to 0.$$

In the discussion of regularization schemes, one usually has to distinguish between an *a priori* or an *a posteriori* choice of the regularization parameter α. An a priori choice would be based on some information on smoothness properties of the exact solution which, in practical problems, in general will not be available. Therefore, a posteriori strategies based on some considerations of the data error level δ are more practical.

A natural a posteriori strategy is given by the *discrepancy* or *residue principle* indroduced by Morozov [132, 133]. Its motivation is based on the consideration that, in general, for erroneous data the residual $\|A\varphi - f\|$ should not be smaller than the accuracy of the measurements of f, i.e., the regularization parameter α should be chosen such that

$$\|AR_\alpha f^\delta - f^\delta\| = \gamma\delta$$

with some fixed parameter $\gamma \geq 1$ multiplying the error level δ. In the case of a regularization scheme R_m with a regularization parameter $m = 1, 2, 3, \ldots$ taking only discrete values, m should be chosen as the smallest integer satisfying

$$\|AR_m f^\delta - f^\delta\| \leq \gamma\delta.$$

Finally, we also need to note that quite often the only choice for selecting the regularization parameter will be *trial and error*, that is, one uses a few different parameters α and then picks the most reasonable result based on appropriate information on the expected solution.

4.3 Singular Value Decomposition

We shall now describe some regularization schemes in a Hilbert space setting. Our approach will be based on the singular value decomposition for compact operators which is a generalization of the spectral decomposition for compact self adjoint operators.

Let X be a Hilbert space and let $A : X \to X$ be a self adjoint compact operator, that is, $(A\varphi, \psi) = (\varphi, A\psi)$ for all $\varphi, \psi \in X$. Then all eigenvalues of A are real. $A \neq 0$ has at least one eigenvalue different from zero and at most a countable set of eigenvalues accumulating only at zero. All nonzero eigenvalues have finite multiplicity, that is, the corresponding eigenspaces are finite dimensional, and eigenelements corresponding to different eigenvalues are orthogonal. Assume the sequence (λ_n) of the nonzero eigenvalues is ordered such that

$$|\lambda_1| \geq |\lambda_2| \geq |\lambda_3| \geq \cdots$$

where each eigenvalue is repeated according to its multiplicity and let (φ_n) be a sequence of corresponding orthonormal eigenelements. Then for each $\varphi \in X$ we can expand

$$(4.6) \qquad \varphi = \sum_{n=1}^{\infty} (\varphi, \varphi_n)\varphi_n + Q\varphi$$

where $Q : X \to N(A)$ denotes the orthogonal projection operator of X onto the nullspace $N(A) := \{\varphi \in X : A\varphi = 0\}$ and

$$(4.7) \qquad A\varphi = \sum_{n=1}^{\infty} \lambda_n(\varphi, \varphi_n)\varphi_n.$$

For a proof of this *spectral decomposition* for self adjoint compact operators see for example [106], Theorem 15.12.

We will now describe modified forms of the expansions (4.6) and (4.7) for arbitrary compact operators in a Hilbert space. Recall that for each bounded linear operator $A : X \to Y$ between two Hilbert spaces X and Y there exists a uniquely determined bounded linear operator $A^* : Y \to X$ called the *adjoint operator* of A such that $(A\varphi, \psi) = (\varphi, A^*\psi)$ for all $\varphi \in X$ and $\psi \in Y$.

Occasionally, we will make use of the following basic connection between the nullspaces and the ranges of A and A^*. Therefore, we include the simple proof.

Theorem 4.6 *For a bounded linear operator we have*

$$A(X)^\perp = N(A^*) \quad and \quad N(A^*)^\perp = \overline{A(X)}.$$

Proof. $g \in A(X)^\perp$ means $(A\varphi, g) = 0$ for all $\varphi \in X$. This is equivalent to $(\varphi, A^*g) = 0$ for all $\varphi \in X$, which in turn is equivalent to $A^*g = 0$, that is, $g \in N(A^*)$. Hence, $A(X)^\perp = N(A^*)$. We abbreviate $U = A(X)$ and, trivially, have $\bar{U} \subset (U^\perp)^\perp$. Denote by $P : Y \to \bar{U}$ the orthogonal projection operator. Then for arbitrary $\varphi \in (U^\perp)^\perp$ we have orthogonality $P\varphi - \varphi \perp U$. But we also have $P\varphi - \varphi \perp U^\perp$ since we already know that $\bar{U} \subset (U^\perp)^\perp$. Therefore, it follows that $\varphi = P\varphi \in \bar{U}$, whence $\bar{U} = (U^\perp)^\perp$, i.e., $\overline{A(X)} = N(A^*)^\perp$. $\qquad\square$

Now let $A : X \to Y$ be a compact linear operator. Then its adjoint operator $A^* : Y \to X$ is also compact. The nonnegative square roots of the eigenvalues of the nonnegative self adjoint compact operator $A^*A : X \to X$ are called *singular values* of A.

Theorem 4.7 *Let (μ_n) denote the sequence of the nonzero singular values of the compact linear operator A (with $A \neq 0$) ordered such that*

$$\mu_1 \geq \mu_2 \geq \mu_3 \geq \cdots$$

and repeated according to their multiplicity, that is, according to the dimension of the nullspaces $N(\mu_n^2 I - A^*A)$. Then there exist orthonormal sequences (φ_n) in X and (g_n) in Y such that

$$(4.8) \qquad\qquad A\varphi_n = \mu_n g_n, \quad A^* g_n = \mu_n \varphi_n$$

for all $n \in \mathbb{N}$. For each $\varphi \in X$ we have the singular value decomposition

$$(4.9) \qquad\qquad \varphi = \sum_{n=1}^{\infty} (\varphi, \varphi_n)\varphi_n + Q\varphi$$

with the orthogonal projection operator $Q : X \to N(A)$ and

$$(4.10) \qquad\qquad A\varphi = \sum_{n=1}^{\infty} \mu_n (\varphi, \varphi_n) g_n.$$

Each system (μ_n, φ_n, g_n), $n \in \mathbb{N}$, with these properties is called a singular system *of A. When there are only finitely many singular values the series (4.9) and (4.10) degenerate into finite sums.*

Proof. Let (φ_n) denote an orthonormal sequence of the eigenelements of A^*A, that is,

$$A^* A\varphi_n = \mu_n^2 \varphi_n$$

and define a second orthonormal sequence by

$$g_n := \frac{1}{\mu_n} A\varphi_n.$$

Straightforward computations show that the system (μ_n, φ_n, g_n), $n \in \mathbb{N}$, satisfies (4.8). Application of the expansion (4.6) to the self adjoint compact operator A^*A yields

$$\varphi = \sum_{n=1}^{\infty} (\varphi, \varphi_n)\varphi_n + Q\varphi$$

for all $\varphi \in X$ where Q denotes the orthogonal projection operator from X onto $N(A^*A)$. Let $\psi \in N(A^*A)$. Then $(A\psi, A\psi) = (\psi, A^*A\psi) = 0$ and this implies that $N(A^*A) = N(A)$. Therefore, (4.9) is proven and (4.10) follows by applying A to (4.9). $\square$

Note that the singular value decomposition implies that for all $\varphi \in X$ we have

$$(4.11) \qquad\qquad \|\varphi\|^2 = \sum_{n=1}^{\infty} |(\varphi, \varphi_n)|^2 + \|Q\varphi\|^2,$$

$$(4.12) \qquad\qquad \|A\varphi\|^2 = \sum_{n=1}^{\infty} \mu_n^2 |(\varphi, \varphi_n)|^2.$$

In the following theorem, we express the solution to an equation of the first kind with a compact operator in terms of a singular system.

Theorem 4.8 (Picard) *Let $A : X \to Y$ be a compact linear operator with singular system (μ_n, φ_n, g_n). The equation of the first kind*

$$(4.13) \qquad\qquad A\varphi = f$$

is solvable if and only if f belongs to the orthogonal complement $N(A^*)^\perp$ and satisfies

$$(4.14) \qquad\qquad \sum_{n=1}^{\infty} \frac{1}{\mu_n^2} |(f, g_n)|^2 < \infty.$$

In this case a solution is given by

$$(4.15) \qquad\qquad \varphi = \sum_{n=1}^{\infty} \frac{1}{\mu_n} (f, g_n)\varphi_n.$$

Proof. The necessity of $f \in N(A^*)^\perp$ follows from Theorem 4.6. If φ is a solution of (4.13) then

$$\mu_n(\varphi, \varphi_n) = (\varphi, A^* g_n) = (A\varphi, g_n) = (f, g_n)$$

and (4.11) implies

$$\sum_{n=1}^{\infty} \frac{1}{\mu_n^2} |(f, g_n)|^2 = \sum_{n=1}^{\infty} |(\varphi, \varphi_n)|^2 \le \|\varphi\|^2,$$

whence the necessity of (4.14) follows.

Conversely, assume that $f \in N(A^*)^\perp$ and (4.14) is fulfilled. Then, by considering the partial sums of (4.14), we see that the series (4.15) converges in the Hilbert space X. We apply A to (4.15), use (4.9) with the singular system (μ_n, g_n, φ_n) of the operator A^* and observe $f \in N(A^*)^\perp$ to obtain

$$A\varphi = \sum_{n=1}^{\infty} (f, g_n) g_n = f.$$

This ends the proof. □

Picard's theorem demonstrates the ill-posed nature of the equation $A\varphi = f$. If we perturb the right hand side by setting $f^\delta = f + \delta g_n$ we obtain a perturbed solution $\varphi^\delta = \varphi + \delta\varphi_n/\mu_n$. Hence, the ratio $\|\varphi^\delta - \varphi\|/\|f^\delta - f\| = 1/\mu_n$ can be made arbitrarily large due to the fact that the singular values tend to zero. The influence of errors in the data f is obviously controlled by the rate of this convergence. In this sense, we may say that the equation is *mildly ill-posed* if the singular values decay slowly to zero and that it is *severely ill-posed* if they decay very rapidly.

Coming back to the far field mapping introduced in Section 2.5, we may consider it as a compact operator $F : L^2(\Omega_R) \to L^2(\Omega)$ transfering the restriction of a radiating solution u to the Helmholtz equation to the sphere Ω_R with radius R and center at the origin onto its far field pattern u_∞. From Theorems 2.14 and 2.15 we see that $\varphi \in L^2(\Omega_R)$ with the expansion

$$\varphi(x) = \sum_{n=0}^{\infty} \sum_{m=-n}^{n} a_n^m Y_n^m \left(\frac{x}{|x|} \right), \quad |x| = R,$$

is mapped onto

$$(F\varphi)(\hat{x}) = \frac{1}{k} \sum_{n=0}^{\infty} \sum_{m=-n}^{n} \frac{a_n^m}{i^{n+1} h_n^{(1)}(kR)} Y_n^m(\hat{x}), \quad \hat{x} \in \Omega.$$

Therefore, the singular values of F are given by

$$\mu_n = \frac{1}{kR^2 |h_n^{(1)}(kR)|}, \quad n = 0, 1, 2, \ldots,$$

and from (2.38) we have the asymptotic behavior

$$\mu_n = O \left(\frac{ekR}{2n} \right)^n, \quad n \to \infty,$$

indicating severe ill-posedness.

As already pointed out, Picard's Theorem 4.8 illustrates the fact that the ill-posedness of an equation of the first kind with a compact operator stems from the behavior of the singular values $\mu_n \to 0$, $n \to \infty$. This suggests to try to regularize the equation by damping or filtering out the influence of the factor $1/\mu_n$ in the solution formula (4.15).

Theorem 4.9 *Let $A : X \to Y$ be an injective compact linear operator with singular system (μ_n, φ_n, g_n), $n \in \mathbb{N}$, and let $q : (0, \infty) \times (0, \|A\|] \to \mathbb{R}$ be a bounded function such that for each $\alpha > 0$ there exists a positive constant $c(\alpha)$ with*

(4.16) $$|q(\alpha, \mu)| \le c(\alpha)\mu, \quad 0 < \mu \le \|A\|,$$

and

(4.17) $$\lim_{\alpha \to 0} q(\alpha, \mu) = 1, \quad 0 < \mu \le \|A\|.$$

Then the bounded linear operators $R_\alpha : Y \to X$, $\alpha > 0$, defined by

(4.18) $$R_\alpha f := \sum_{n=1}^{\infty} \frac{1}{\mu_n} q(\alpha, \mu_n) (f, g_n) \varphi_n, \quad f \in Y,$$

describe a regularization scheme with

(4.19) $$\|R_\alpha\| \le c(\alpha).$$

Proof. From (4.11) and (4.16) we have

$$\|R_\alpha f\|^2 = \sum_{n=1}^{\infty} \frac{1}{\mu_n^2} \left[q(\alpha, \mu_n)\right]^2 |(f, g_n)|^2 \le [c(\alpha)]^2 \sum_{n=1}^{\infty} |(f, g_n)|^2 \le [c(\alpha)]^2 \|f\|^2$$

for all $f \in Y$, whence the bound (4.19) follows. With the aid of

$$(R_\alpha A\varphi, \varphi_n) = \frac{1}{\mu_n} q(\alpha, \mu_n) (A\varphi, g_n) = q(\alpha, \mu_n) (\varphi, \varphi_n)$$

and the singular value decomposition for $R_\alpha A\varphi - \varphi$ we obtain

$$\|R_\alpha A\varphi - \varphi\|^2 = \sum_{n=1}^{\infty} |(R_\alpha A\varphi - \varphi, \varphi_n)|^2 = \sum_{n=1}^{\infty} [q(\alpha, \mu_n) - 1]^2 |(\varphi, \varphi_n)|^2.$$

Here we have used the fact that A is injective. Let $\varphi \in X$ with $\varphi \ne 0$ and $\varepsilon > 0$ be given and let M denote a bound for q. Then there exists $N(\varepsilon) \in \mathbb{N}$ such that

$$\sum_{n=N+1}^{\infty} |(\varphi, \varphi_n)|^2 < \frac{\varepsilon}{2(M+1)^2}.$$

By the convergence condition (4.17), there exists $\alpha_0(\varepsilon) > 0$ such that

$$[q(\alpha, \mu_n) - 1]^2 < \frac{\varepsilon}{2\|\varphi\|^2}$$

for all $n = 1, \ldots, N$ and all $0 < \alpha \le \alpha_0$. Splitting the series in two parts and using (4.11), it now follows that

$$\|R_\alpha A\varphi - \varphi\|^2 < \frac{\varepsilon}{2\|\varphi\|^2} \sum_{n=1}^{N} |(\varphi, \varphi_n)|^2 + \frac{\varepsilon}{2} \le \varepsilon$$

for all $0 < \alpha \le \alpha_0$. Thus we have established that $R_\alpha A\varphi \to \varphi$, $\alpha \to 0$, for all $\varphi \in X$ and the proof is complete. $\qquad\square$

We now describe two basic regularization schemes, namely the spectral cut-off and the Tikhonov regularization, obtained by choosing the damping or filter function q appropriately.

Theorem 4.10 *Let $A : X \to Y$ be an injective compact linear operator with singular system (μ_n, φ_n, g_n), $n \in \mathbb{N}$. Then the spectral cut-off*

(4.20)
$$R_m f := \sum_{\mu_n \ge \mu_m} \frac{1}{\mu_n} (f, g_n) \varphi_n$$

describes a regularization scheme with regularization parameter $m \to \infty$ and $\|R_m\| = 1/\mu_m$.

Proof. The function q with $q(m, \mu) = 1$ for $\mu \geq \mu_m$ and $q(m, \mu) = 0$ for $\mu < \mu_m$ satisfies the conditions (4.16) and (4.17). For the norm, by Bessel's inequality, we can estimate

$$\|R_m f\|^2 = \sum_{\mu_n \geq \mu_m} \frac{1}{\mu_n^2} |(f, g_n)|^2 \leq \frac{1}{\mu_m^2} \sum_{\mu_n \geq \mu_m} |(f, g_n)|^2 \leq \frac{1}{\mu_m^2} \|f\|^2,$$

whence $\|R_m\| \leq 1/\mu_m$. Equality follows from $R_m g_m = \varphi_m/\mu_m$. □

The regularization parameter m determines the number of terms in the sum (4.20). Accuracy of the approximation requires this number to be large and stability requires it to be small. In particular, the following discrepancy principle turns out to be a regular a posteriori strategy for determining the stopping point for the spectral cut-off.

Theorem 4.11 *Let $A : X \to Y$ be an injective compact linear operator with dense range in Y, let $f \in Y$ and let $\delta > 0$. Then there exists a smallest integer m such that*

$$\|AR_m f - f\| \leq \delta.$$

Proof. By Theorem 4.6, the dense range $\overline{A(X)} = Y$ implies that A^* is injective. Hence, the singular value decomposition (4.9) with the singular system (μ_n, g_n, φ_n) for the adjoint operator A^*, applied to an element $f \in Y$, yields

$$(4.21) \qquad f = \sum_{n=1}^{\infty} (f, g_n) g_n$$

and consequently

$$(4.22) \qquad \|(AR_m - I)f\|^2 = \sum_{\mu_n < \mu_m} |(f, g_n)|^2 \to 0, \quad m \to \infty.$$

From this we conclude that there exists a smallest integer $m = m(\delta)$ such that $\|AR_m f - f\| \leq \delta$. □

From (4.21) and (4.22), we see that

$$\|AR_m f - f\|^2 = \|f\|^2 - \sum_{\mu_n \geq \mu_m} |(f, g_n)|^2,$$

which allows a stable determination of the stopping parameter $m(\delta)$ by terminating the sum when the right hand side becomes smaller than or equal to δ^2 for the first time.

The regularity of the discrepancy principle for the spectral cut-off described through Theorem 4.11 is established in the following theorem.

Theorem 4.12 *Let $A : X \to Y$ be an injective compact linear operator with dense range in Y. Let $f \in A(X)$, $f^\delta \in Y$ satisfy $\|f^\delta - f\| \leq \delta$ with $\delta > 0$ and let $\gamma > 1$. Then there exists a smallest integer $m = m(\delta)$ such that*

$$(4.23) \qquad \|AR_{m(\delta)}f^\delta - f^\delta\| \leq \gamma\delta$$

is satisfied and

$$(4.24) \qquad R_{m(\delta)}f^\delta \to A^{-1}f, \quad \delta \to 0.$$

Proof. In view of Theorem 4.11, we only need to establish the convergence (4.24). We first note that (4.22) implies $\|I - AR_m\| = 1$ for all $m \in \mathbb{N}$. Therefore, writing

$$(AR_m f^\delta - f^\delta) - (AR_m f - f) = (AR_m - I)(f^\delta - f)$$

we have the triangle inequalities

$$(4.25) \qquad \|AR_m f - f\| \leq \delta + \|AR_m f^\delta - f^\delta\|,$$

$$(4.26) \qquad \|AR_m f^\delta - f^\delta\| \leq \delta + \|AR_m f - f\|.$$

From (4.23) and (4.25) we obtain

$$\|AR_{m(\delta)}f - f\| \leq \delta + \|AR_{m(\delta)}f^\delta - f^\delta\| \leq (1 + \gamma)\delta \to 0, \quad \delta \to 0.$$

Therefore, from the expansion (4.22), we conclude that either the number of terms in the sum (4.20) tends to infinity, $m(\delta) \to \infty$, $\delta \to 0$, or the expansion for f degenerates into a finite sum

$$f = \sum_{\mu_n \geq \mu_{m_0}} (f, g_n)g_n$$

and $m(\delta) \geq m_0$. In the first case, from $\|AR_{m(\delta)-1}f^\delta - f^\delta\| > \gamma\delta$ and the triangle inequality (4.26), we conclude

$$\gamma\delta < \delta + \|A(R_{m(\delta)-1}f - A^{-1}f)\|$$

whence

$$\delta < \frac{1}{\gamma - 1}\, \|A(R_{m(\delta)-1}f - A^{-1}f)\|$$

follows. In order to establish the convergence (4.24), in this case, from (4.5) it suffices to show that

$$\|R_m\|\, \|A(R_{m-1}A\varphi - \varphi)\| \to 0, \quad m \to \infty,$$

for all $\varphi \in X$. But the latter property is obvious from

$$\|R_m\|^2\, \|A(R_{m-1}A\varphi - \varphi)\|^2 = \frac{1}{\mu_m^2} \sum_{\mu_n < \mu_{m-1}} \mu_n^2 |(\varphi, \varphi_n)|^2 \leq \sum_{\mu_n \leq \mu_m} |(\varphi, \varphi_n)|^2.$$

In the case where f has a finite expansion then clearly

$$A^{-1}f = \sum_{\mu_n \geq \mu_{m_0}} \frac{1}{\mu_n} (f, g_n)\varphi_n = R_m f$$

for all $m \geq m_0$. Hence

$$\|AR_m f^\delta - f^\delta\| = \|(AR_m - I)(f^\delta - f)\| \leq \|f^\delta - f\| \leq \delta < \gamma\delta,$$

and therefore $m(\delta) \leq m_0$. This implies equality $m(\delta) = m_0$ since $m(\delta) \geq m_0$ as already noted above. Now observing

$$\|R_{m(\delta)} f^\delta - A^{-1}f\| = \|R_{m_0}(f^\delta - f)\| \leq \frac{\delta}{\mu_{m_0}} \to 0, \quad \delta \to 0,$$

the proof is finished. □

4.4 Tikhonov Regularization

We continue our study of regularization methods by introducing Tikhonov's [172, 173] regularization scheme first as a special case of Theorem 4.9 and then also as a penalized residual minimization.

Theorem 4.13 *Let $A : X \to Y$ be a compact linear operator. Then for each $\alpha > 0$ the operator $\alpha I + A^*A : X \to X$ is bijective and has a bounded inverse. Furthermore, if A is injective then*

(4.27)
$$R_\alpha := (\alpha I + A^*A)^{-1} A^*$$

describes a regularization scheme with $\|R_\alpha\| \leq 1/2\sqrt{\alpha}$.

Proof. From

(4.28)
$$\alpha\|\varphi\|^2 \leq (\alpha\varphi + A^*A\varphi, \varphi)$$

for all $\varphi \in X$ we conclude that for $\alpha > 0$ the operator $\alpha I + A^*A$ is injective. Let (μ_n, φ_n, g_n), $n \in \mathbb{N}$, be a singular system for A and let $Q : X \to N(A)$ denote the orthogonal projection operator. Then the operator $T : X \to X$ defined by

$$T\varphi := \sum_{n=1}^{\infty} \frac{1}{\alpha + \mu_n^2} (\varphi, \varphi_n)\varphi_n + \frac{1}{\alpha} Q(\varphi)$$

can be easily seen to be bounded and to satisfy $(\alpha I + A^*A)T = T(\alpha I + A^*A) = I$, i.e., $T = (\alpha I + A^*A)^{-1}$.

If A is injective then for the unique solution φ_α of

(4.29)
$$\alpha\varphi_\alpha + A^*A\varphi_\alpha = A^*f$$

we deduce from the above expression for $(\alpha I + A^* A)^{-1}$ and $(A^* f, \varphi_n) = \mu_n (f, g_n)$ that

$$(4.30) \qquad \varphi_\alpha = \sum_{n=1}^{\infty} \frac{\mu_n}{\alpha + \mu_n^2} (f, g_n) \varphi_n.$$

Hence, R_α can be brought into the form (4.18) with

$$q(\alpha, \mu) = \frac{\mu^2}{\alpha + \mu^2} .$$

This function q is bounded by $0 < q(\alpha, \mu) < 1$ and satisfies the conditions (4.16) and (4.17) with

$$c(\alpha) = \frac{1}{2\sqrt{\alpha}}$$

because of the arithmetic-geometric mean inequality

$$\sqrt{\alpha} \mu \le \frac{\alpha + \mu^2}{2} .$$

The proof of the theorem now follows from Theorem 4.9. $\qquad\qquad\square$

The following theorem presents another aspect of the Tikhonov regularization complementing its introduction via Theorems 4.9 and 4.13

Theorem 4.14 *Let $A : X \to Y$ be a compact linear operator and let $\alpha > 0$. Then for each $f \in Y$ there exists a unique $\varphi_\alpha \in X$ such that*

$$(4.31) \qquad \|A\varphi_\alpha - f\|^2 + \alpha \|\varphi_\alpha\|^2 = \inf_{\varphi \in X} \{ \|A\varphi - f\|^2 + \alpha \|\varphi\|^2 \}.$$

The minimizer φ_α is given by the unique solution of (4.29) and depends continuously on f.

Proof. From the equation

$$\|A\varphi - f\|^2 + \alpha \|\varphi\|^2 = \|A\varphi_\alpha - f\|^2 + \alpha \|\varphi_\alpha\|^2$$
$$+ 2 \operatorname{Re}(\varphi - \varphi_\alpha, \alpha \varphi_\alpha + A^*(A\varphi_\alpha - f)) + \|A(\varphi - \varphi_\alpha)\|^2 + \alpha \|\varphi - \varphi_\alpha\|^2,$$

which is valid for all $\varphi \in X$, we observe that the condition (4.29) is necessary and sufficient for φ_α to minimize the *Tikhonov functional* defined by (4.31). The theorem now follows from the Riesz–Fredholm theory or from the first part of the statement of the previous Theorem 4.13. $\qquad\square$

We note that Theorem 4.14 remains valid for bounded operators. This follows from the Lax–Milgram theorem since by (4.28) the operator $\alpha I + A^* A$ is strictly coercive (see [106]).

By the interpretation of the Tikhonov regularization as minimizer of the Tikhonov functional, its solution keeps the residual $\|A\varphi_\alpha - f\|^2$ small and is stabilized through the penalty term $\alpha \|\varphi_\alpha\|^2$. We can view this quadratic optimization problem as a *penalty method* for either one of the following constrained optimization problems:

a) For given $\delta > 0$, minimize the norm $\|\varphi\|$ subject to the constraint that the defect is bounded by $\|A\varphi - f\| \leq \delta$.

b) For given $\rho > 0$, minimize the defect $\|A\varphi - f\|$ subject to the constraint that the norm is bounded by $\|\varphi\| \leq \rho$.

The first interpretation leads to the *discrepancy principle* and the second to the concept of *quasi-solutions*. We begin by discussing the discrepancy principle.

Theorem 4.15 *Let $A : X \to Y$ be an injective compact linear operator with dense range in Y and let $f \in Y$ with $0 < \delta < \|f\|$. Then there exists a unique parameter α such that*

(4.32)
$$\|AR_\alpha f - f\| = \delta.$$

Proof. We have to show that the function $F : (0, \infty) \to \mathbb{R}$ defined by

$$F(\alpha) := \|AR_\alpha f - f\|^2 - \delta^2$$

has a unique zero. From (4.21) and (4.30) we find

$$F(\alpha) = \sum_{n=1}^{\infty} \frac{\alpha^2}{(\alpha + \mu_n^2)^2} \, |(f, g_n)|^2 - \delta^2.$$

Therefore, F is continuous and strictly monotonically increasing with the limits $F(\alpha) \to -\delta^2 < 0$, $\alpha \to 0$, and $F(\alpha) \to \|f\|^2 - \delta^2 > 0$, $\alpha \to \infty$. Hence, F has exactly one zero $\alpha = \alpha(\delta)$. $\qquad\square$

In general, we will have data satisfying $\|f\| > \delta$, i.e., data exceeding the error level. Then the regularization parameter satisfying (4.32) can be obtained numerically by Newton's method for solving $F(\alpha) = 0$. With the unique solution φ_α of (4.29), we can write $F(\alpha) = \|A\varphi_\alpha - f\|^2 - \delta^2$ and get

$$F'(\alpha) = 2 \operatorname{Re}\left(A \frac{d\varphi_\alpha}{d\alpha}, A\varphi_\alpha - f \right)$$

from which, again using (4.29), we deduce that

$$F'(\alpha) = -2\alpha \operatorname{Re}\left(\frac{d\varphi_\alpha}{d\alpha}, \varphi_\alpha \right).$$

Differentiating (4.29) with respect to the parameter α yields

$$\alpha \frac{d\varphi_\alpha}{d\alpha} + A^* A \frac{d\varphi_\alpha}{d\alpha} = -\varphi_\alpha$$

as an equation for $d\varphi_\alpha/d\alpha$ which has to be solved for the evaluation of $F'(\alpha)$.

The regularity of the discrepancy principle for the Tikhonov regularization is established through the following theorem.

Theorem 4.16 *Let $A : X \to Y$ be an injective compact linear operator with dense range in Y. Let $f \in A(X)$ and $f^\delta \in Y$ satisfy $\|f^\delta - f\| \leq \delta < \|f^\delta\|$ with $\delta > 0$. Then there exists a unique parameter $\alpha = \alpha(\delta)$ such that*

$$(4.33) \qquad \|AR_{\alpha(\delta)}\overset{\cdot}{f}^\delta - f^\delta\| = \delta$$

is satisfied and

$$(4.34) \qquad R_{\alpha(\delta)}f^\delta \to A^{-1}f, \quad \delta \to 0.$$

Proof. In view of Theorem 4.15, we only need to establish the convergence (4.34). Since $\varphi^\delta = R_{\alpha(\delta)}f^\delta$ minimizes the Tikhonov functional for the right hand side f^δ, we have

$$\delta^2 + \alpha\|\varphi^\delta\|^2 = \|A\varphi^\delta - f^\delta\|^2 + \alpha\|\varphi^\delta\|^2$$
$$\leq \|AA^{-1}f - f^\delta\|^2 + \alpha\|A^{-1}f\|^2 \leq \delta^2 + \alpha\|A^{-1}f\|^2,$$

whence

$$(4.35) \qquad \|\varphi^\delta\| \leq \|A^{-1}f\|$$

follows. Now let $g \in Y$ be arbitrary. Then we can estimate

$$|(A\varphi^\delta - f, g)| \leq \{\|A\varphi^\delta - f^\delta\| + \|f^\delta - f\|\}\|g\| \leq 2\delta\|g\| \to 0, \quad \delta \to 0.$$

This implies weak convergence $\varphi^\delta \rightharpoonup A^{-1}f$, $\delta \to 0$, since for the injective operator A the range $A^*(Y)$ is dense in X by Theorem 4.6 and φ^δ is bounded by (4.35). Then, again using (4.35), we obtain

$$\|\varphi^\delta - A^{-1}f\|^2 = \|\varphi^\delta\|^2 - 2\,\mathrm{Re}(\varphi^\delta, A^{-1}f) + \|A^{-1}f\|^2$$

$$\leq 2\{\|A^{-1}f\|^2 - \mathrm{Re}(\varphi^\delta, A^{-1}f)\} \to 0, \quad \delta \to 0,$$

which finishes the proof. $\qquad\qquad\qquad\qquad\qquad\qquad\qquad\qquad\qquad\qquad\square$

The principal idea underlying the concept of quasi-solutions as introduced by Ivanov [74, 75] is to stabilize an ill-posed problem by restricting the solution set to some subset $U \subset X$ exploiting suitable a priori information on the solution of $A\varphi = f$. For perturbed right hand sides, in general we cannot expect a solution in U. Therefore, instead of trying to solve the equation exactly, we minimize the residual. For simplicity, we restrict our presentation to the case where U is a ball of radius ρ for some $\rho > 0$.

Theorem 4.17 *Let $A : X \to Y$ be a compact injective linear operator and let $\rho > 0$. Then for each $f \in Y$ there exists a unique element $\varphi_0 \in X$ with $\|\varphi_0\| \leq \rho$ satisfying*

$$\|A\varphi_0 - f\| \leq \|A\varphi - f\|, \quad \|\varphi\| \leq \rho.$$

The element φ_0 is called the quasi-solution *of $A\varphi = f$ with constraint ρ.*

Proof. Note that φ_0 is a quasi-solution with constraint ρ if and only if $A\varphi_0$ is a best approximation to f with respect to the set $V := \{A\varphi : \|\varphi\| \leq \rho\}$. Since A is linear, the set V is clearly convex. In the Hilbert space Y there exists at most one best approximation to f with respect to the convex set V. Since A is injective this implies uniqueness of the quasi-solution.

To prove existence of the quasi-solution, let (φ_n) be a minimizing sequence, that is, $\|\varphi_n\| \leq \rho$ and

$$\lim_{n\to\infty} \|A\varphi_n - f\| = \inf_{\|\varphi\|\leq\rho} \|A\varphi - f\|.$$

Without loss of generality, for the bounded sequence (φ_n) we may assume weak convergence $\varphi_n \rightharpoonup \varphi_0$, $n \to \infty$, for some $\varphi_0 \in X$. Since A is compact this implies convergence $\|A\varphi_n - A\varphi_0\| \to 0$, $n \to \infty$, and therefore

$$\|A\varphi_0 - f\| = \inf_{\|\varphi\|\leq\rho} \|A\varphi - f\|.$$

From

$$\|\varphi_0\|^2 = \lim_{n\to\infty} (\varphi_n, \varphi_0) \leq \rho\|\varphi_0\|$$

we obtain $\|\varphi_0\| \leq \rho$ and the proof is complete. $\qquad\qquad\square$

The connection of the quasi-solution to Tikhonov regularization is described through the following theorem.

Theorem 4.18 *Let $A : X \to Y$ be a compact injective linear operator with dense range in Y and assume $f \notin V := \{A\varphi : \|\varphi\| \leq \rho\}$. Then the quasi-solution φ_0 assumes the constraint*

$$(4.36) \qquad\qquad \|\varphi_0\| = \rho$$

and there exists a unique parameter $\alpha > 0$ such that

$$(4.37) \qquad\qquad \alpha\varphi_0 + A^*A\varphi_0 = A^*f.$$

Proof. If φ_0 satisfies (4.36) and (4.37) then

$$\|A\varphi - f\|^2 = \|A\varphi_0 - f\|^2 + 2\alpha\,\mathrm{Re}(\varphi_0 - \varphi, \varphi_0) + \|A(\varphi - \varphi_0)\|^2 \geq \|A\varphi_0 - f\|^2$$

for all $\|\varphi\| \leq \rho$ and therefore φ_0 is a quasi-solution of $A\varphi = f$ with constraint ρ. Therefore, in view of the preceding Theorem 4.17, the proof is established by showing existence of a solution to equations (4.36) and (4.37) with $\alpha > 0$.

For this we define the function $G : (0, \infty) \to \mathbb{R}$ by

$$G(\alpha) := \|\varphi_\alpha\|^2 - \rho^2,$$

where φ_α denotes the unique solution of (4.29), and show that $G(\alpha)$ has a unique zero. From (4.30) we obtain

$$G(\alpha) = \sum_{n=1}^{\infty} \frac{\mu_n^2}{(\alpha + \mu_n^2)^2} |(f, g_n)|^2 - \rho^2.$$

Therefore, G is continuous and strictly monotonically decreasing with

$$G(\alpha) \to -\rho^2 < 0, \quad \alpha \to \infty,$$

and

$$G(\alpha) \to \sum_{n=1}^{\infty} \frac{1}{\mu_n^2} |(f, g_n)|^2 - \rho^2, \quad \alpha \to 0,$$

where the series may diverge. The proof is now completed by showing that the latter limit is positive or infinite. Assume, to the contrary, that

$$\sum_{n=1}^{\infty} \frac{1}{\mu_n^2} |(f, g_n)|^2 \le \rho^2.$$

Then by Picard's Theorem 4.8 (note that $N(A^*) = \{0\}$ by Theorem 4.6) the equation $A\varphi = f$ has the solution

$$\varphi = \sum_{n=1}^{\infty} \frac{1}{\mu_n} (f, g_n) \varphi_n$$

with

$$\|\varphi\|^2 = \sum_{n=1}^{\infty} \frac{1}{\mu_n^2} |(f, g_n)|^2 \le \rho^2,$$

which is a contradiction to $f \notin V$. $\qquad\qquad\qquad\qquad\qquad\qquad$ □

In the sense of Theorem 4.18, we may view the quasi-solution as a possibility of an a posteriori choice of the regularization parameter in the Tikhonov regularization. We also note that Theorems 4.15 to 4.18 remain valid for injective linear operators which are merely bounded (see [106]).

4.5 Nonlinear Operators

We conclude our introduction to ill-posed problems by making a few comments on nonlinear problems. We first show that the ill-posedness of a nonlinear problem is inherited by its linearization. This implies that whenever we try to approximately solve an ill-posed nonlinear equation by linearization, for example by a Newton method, we obtain ill-posed linear equations for which a regularization must be enforced, for example by one of the methods from the previous sections.

Theorem 4.19 *Let $A : U \subset X \to Y$ be a completely continuous operator from an open subset U of a normed space X into a Banach space Y and assume A to be Fréchet differentiable at $\psi \in U$. Then the derivative A'_ψ is compact.*

Proof. We shall use the fact that a subset V of a Banach space is relatively compact if and only if it is totally bounded, i.e., for every $\varepsilon > 0$ there exists a finite system of elements $\varphi_1, \ldots, \varphi_n$ in V such that each element $\varphi \in V$ has a distance smaller than ε from at least one of the $\varphi_1, \ldots, \varphi_n$. We have to show that

$$V := \{A'_\psi(\varphi) : \varphi \in X, \|\varphi\| \le 1\}$$

is relatively compact. Given $\varepsilon > 0$, by the definition of the Fréchet derivative there exists $\delta > 0$ such that for all $\|\varphi\| \le \delta$ we have $\psi + \varphi \in U$ and

(4.38)
$$\|A(\psi + \varphi) - A(\psi) - A'_\psi(\varphi)\| \le \frac{\varepsilon}{3} \|\varphi\|.$$

Since A is compact, the set

$$\{A(\psi + \delta\varphi) : \varphi \in X, \|\varphi\| \le 1\}$$

is relatively compact and hence totally bounded, i.e., there are finitely many elements $\varphi_1, \ldots, \varphi_n \in X$ with norm less than or equal to one such that for each $\varphi \in X$ with $\|\varphi\| \le 1$ there exists $j = j(\varphi)$ such that

(4.39)
$$\|A(\psi + \delta\varphi) - A(\psi + \delta\varphi_j)\| < \frac{\varepsilon\delta}{3}.$$

By the triangle inequality, using (4.38) and (4.39), we now have

$$\delta\|A'_\psi(\varphi) - A'_\psi(\varphi_j)\| \le \|A(\psi + \delta\varphi) - A(\psi + \delta\varphi_j)\|$$

$$\|A(\psi + \delta\varphi) - A(\psi) - A'_\psi(\delta\varphi)\| + \|A(\psi + \delta\varphi_j) - A(\psi) - A'_\psi(\delta\varphi_j)\| < \delta\varepsilon$$

and therefore V is totally bounded. This ends the proof. $\square$

In the following, we will illustrate how the classical concepts of Tikhonov regularization and quasi-solutions can be directly applied to ill-posed nonlinear problems. We begin with the nonlinear counterpart of Theorem 4.14.

Theorem 4.20 *Let $A : U \subset X \to Y$ be a weakly sequentially closed operator from a subset U of a Hilbert space X into a Hilbert space Y, i.e., for any sequence (φ_n) from U weak convergence $\varphi_n \rightharpoonup \varphi \in X$ and $A\varphi_n \rightharpoonup g \in Y$ implies $\varphi \in U$ and $A\varphi = g$. Let $\alpha > 0$. Then for each $f \in Y$ there exists $\varphi_\alpha \in U$ such that*

(4.40)
$$\|A\varphi_\alpha - f\|^2 + \alpha\|\varphi_\alpha\|^2 = \inf_{\varphi \in U}\{\|A\varphi - f\|^2 + \alpha\|\varphi\|^2\}.$$

Proof. We abbreviate the Tikhonov functional by

$$\mu(\varphi, \alpha) := \|A\varphi - f\|^2 + \alpha\|\varphi\|^2$$

and set

$$m(\alpha) := \inf_{\varphi \in U} \mu(\varphi, \alpha).$$

Let (φ_n) be a minimizing sequence in U, i.e.,

$$\lim_{n \to \infty} \mu(\varphi_n, \alpha) = m(\alpha).$$

Since $\alpha > 0$, the sequences (φ_n) and $(A\varphi_n)$ are bounded. Hence, by selecting subsequences and relabeling, we can assume weak convergence $\varphi_n \rightharpoonup \varphi_\alpha$ as $n \to \infty$ for some $\varphi_\alpha \in X$ and $A\varphi_n \rightharpoonup g$ as $n \to \infty$ for some $g \in Y$. Since A is assumed to be weakly sequentially closed φ_α belongs to U and we have $g = A\varphi_\alpha$. This now implies

$$\|A\varphi_n - A\varphi_\alpha\|^2 + \alpha\|\varphi_n - \varphi_\alpha\|^2 \to m(\alpha) - \mu(\varphi_\alpha, \alpha), \quad n \to \infty,$$

whence $m(\alpha) \geq \mu(\varphi_\alpha, \alpha)$ follows. Now observing that trivially $m(\alpha) \leq \mu(\varphi_\alpha, \alpha)$, we have shown that $m(\alpha) = \mu(\varphi_\alpha, \alpha)$ and the proof is complete.   □

For an analysis of stability, regularity and convergence of Tikhonov regularization for nonlinear operators we refer to Colonius and Kunisch [23, 24], Engl, Kunisch and Neubauer [53], Kravaris and Steinfeld [100] and Seidman and Vogel [167].

We conclude with the following result on quasi-solutions in the nonlinear case.

Theorem 4.21 *Let $A : U \subset X \to Y$ be a continuous operator from a compact subset U of a Hilbert space X into a Hilbert space Y. Then for each $f \in Y$ there exists a $\varphi_0 \in U$ such that*

(4.41) $$\|A\varphi_0 - f\|^2 = \inf_{\varphi \in U}\{\|A\varphi - f\|^2\}.$$

The element φ_0 is called the quasi-solution *of $A\varphi = f$ with respect to U.*

Proof. This is an immediate consequence of the compactness of U and the continuity of A and $\|\cdot\|$.   □

5. Inverse Acoustic Obstacle Scattering

With the analysis of the preceding chapters, we now are well prepared for studying inverse acoustic obstacle scattering problems. We recall that the direct scattering problem is, given information on the boundary of the scatterer and the nature of the boundary condition, to find the scattered wave and in particular its behavior at large distances from the scatterer, i.e., its far field. The inverse problem starts from this answer to the direct problem, i.e., a knowledge of the far field pattern, and asks for the nature of the scatterer. Of course, there is a large variety of possible inverse problems, for example, if the boundary condition is known, find the shape of the scatterer, or, if the shape is known, find the boundary condition, or, if the shape and the type of the boundary condition are known for a penetrable scatterer, find the space dependent coefficients in the transmission or resistive boundary condition, etc. Here, following the main guideline of our book, we will concentrate on one model problem for which we will develop ideas which in general can also be used to study a wider class of related problems. The inverse problem we consider is, given the far field pattern for one or several incident plane waves and knowing that the scatterer is sound-soft, to determine the shape of the scatterer. We want to discuss this inverse problem for frequencies in the *resonance region*, that is, for scatterers D and wave numbers k such that the wavelengths $2\pi/k$ is less than or of a comparable size to the diameter of the scatterer. This inverse problem turns out to be nonlinear and improperly posed. Although both of these properties make the inverse problem hard to solve, it is the latter which presents the more challenging difficulties. The inverse obstacle problem is improperly posed since, as we already know, the determination of the scattered wave u^s from a given far field pattern u_∞ is improperly posed. It is nonlinear since, given the incident wave u^i and the scattered wave u^s, the problem of finding the boundary of the scatterer as the location of the zeros of the total wave $u^i + u^s$ is nonlinear.

We begin this chapter with results on uniqueness for the inverse obstacle problem, that is, we investigate the question whether knowing the far field pattern provides enough information to completely determine the boundary of the scatterer. In the section on uniqueness, we also include an explicitly solvable problem in inverse obstacle scattering known as *Karp's theorem*.

We then proceed to briefly describe a linearization method based on the physical optics approximation and indicate its limitations. After that, we shall provide a detailed analysis of the dependence of the far field operator on variations of the boundary. Here, by the far field operator we mean the mapping which for a given incident wave maps the boundary onto the far field of the scattered wave. In particular, we will establish continuity and differentiability of this operator by using both weak solution and boundary integral equation techniques. This provides the necessary prerequisites for the solution of the inverse problem by what we call direct methods, that is, by straightforwardly solving a nonlinear operator equation with the above far field operator.

A common feature of these direct methods is that their iterative numerical implementation requires the numerical solution of the direct scattering problem for different domains at each iteration step. In contrast to this, the last two sections of this chapter introduce the main ideas of two methods which avoid this problem. We will describe the theoretical foundation of both methods and outline their numerical implementation. In numerical tests, both methods have been shown to yield satisfactory reconstructions. However, there is still work to be done on the improvement of their numerical performance, particularly for three-dimensional problems. In this sense, we provide only a state of the art survey on inverse obstacle scattering in the resonance region rather than an exposition on a subject which has already been brought to completion.

5.1 Uniqueness

In this section, we investigate under what conditions a sound-soft obstacle is uniquely determined by a knowledge of its far field patterns for incident plane waves. We note that by analyticity the far field pattern is completely determined on the whole unit sphere by only knowing it on some surface patch. We first give a uniqueness result based on the ideas of Schiffer (see [116]).

Theorem 5.1 *Assume that D_1 and D_2 are two sound-soft scatterers such that the far field patterns coincide for an infinite number of incident plane waves with distinct directions and one fixed wave number. Then $D_1 = D_2$.*

Proof. Assume that $D_1 \neq D_2$. Since by Theorem 2.13 the far field pattern uniquely determines the scattered field, for each incident wave $u^i(x) = e^{ik\,x\cdot d}$ the scattered wave u^s for both obstacles coincides in the unbounded component G of the complement of $D_1 \cup D_2$ and the total wave vanishes on ∂G. Without loss of generality, we can assume that $D^* := (\mathbb{R}^3 \setminus G) \setminus \bar{D}_2$ is nonempty. Then u^s is defined in D^* since it describes the scattered wave for D_2, that is, $u = u^i + u^s$ satisfies the Helmholtz equation in D^* and homogeneous boundary conditions $u = 0$ on ∂D^*. Hence, u is a Dirichlet eigenfunction for the negative Laplacian in the domain D^* with eigenvalue k^2. From Lemma 3.8 and the approximation technique used in its proof, we know that u belongs to the Sobolev space $H_0^1(D^*)$ without needing any regularity requirements on D^* (besides the assumption that the solution to the scattering problem exists, that is, the scattered wave is continuous up to the boundary). The proof of our theorem is now completed by showing that the total fields for distinct incoming plane waves are linearly independent and that for a fixed wave number k there exist only finitely many linearly independent Dirichlet eigenfunctions in $H_0^1(D^*)$.

Recall that we indicate the dependence of the scattered wave and the total wave on the incident direction by writing $u^s(x; d)$ and $u(x; d)$. Assume that

$$(5.1) \qquad \sum_{n=1}^{N} c_n u(\cdot; d_n) = 0$$

in D^* for some constants c_n and N distinct incident directions d_n, $n = 1, \ldots, N$. Then, by analyticity (Theorem 2.2), equation (5.1) is also satisfied in the exterior of some sphere containing D_1 and D_2. Writing

$$u(x; d_n) = e^{ik\,x\cdot d_n} + u^s(x; d_n)$$

and using the asymptotic behavior $u^s(x; d_n) = O(1/|x|)$, from (5.1) we obtain

$$(5.2) \qquad \frac{1}{R^2} \sum_{n=1}^{N} c_n \int_{|x|=R} e^{ik\,x\cdot(d_n-d_m)} ds(x) = O\left(\frac{1}{R}\right), \qquad R \to \infty,$$

for $m = 1, \ldots, N$. Now we apply the Funk–Hecke theorem (2.44), that is,

$$\int_{|x|=R} e^{ik\,x\cdot(d_n-d_m)} ds(x) = \frac{4\pi R \sin(kR|d_n - d_m|)}{k|d_n - d_m|}, \qquad n \neq m,$$

to see that

$$\frac{1}{R^2} \sum_{n=1}^{N} c_n \int_{|x|=R} e^{ik\,x\cdot(d_n-d_m)} ds(x) = 4\pi c_m + O\left(\frac{1}{R}\right), \qquad R \to \infty.$$

Hence, passing to the limit $R \to \infty$ in (5.2) yields $c_m = 0$ for $m = 1, \ldots, N$, i.e., the functions $u(\cdot; d_n)$, $n = 1, \ldots, N$, are linearly independent.

We now show that there are only finitely many linearly independent eigenfunctions u_n possible. Assume to the contrary that we have infinitely many. Then, by the Gram–Schmidt orthogonalization procedure, we may assume that

$$\int_{D^*} u_n \bar{u}_m \, dx = \delta_{nm}$$

where δ_{nm} denotes the Kronecker delta symbol. From Green's theorem (2.2) we observe that

$$\int_{D^*} |\operatorname{grad} u_n|^2 dx = k^2 \int_{D^*} |u_n|^2 dx = k^2,$$

that is, the sequence (u_n) is bounded with respect to the norm in $H_0^1(D^*)$. By the Rellich selection theorem, that is by the compactness of the imbedding of the Sobolev space $H_0^1(D^*)$ into $L^2(D^*)$, we can choose a convergent subsequence of (u_n) with respect to the norm in $L^2(D^*)$. But this is a contradiction to $\|u_n - u_m\|_{L^2(D^*)}^2 = 2$ for all $n \neq m$ for the orthonormal sequence (u_n). $\qquad \square$

According to the following theorem (Colton and Sleeman [48]), the scatterer is uniquely determined by the far field pattern of a finite number of incident plane waves provided a priori information on the size of the obstacle is available. For $n = 0, 1, \ldots$, we denote the positive zeros of the spherical Bessel functions j_n by t_{nl}, $l = 0, 1, \ldots$, i.e., $j_n(t_{nl}) = 0$.

Theorem 5.2 *Let D_1 and D_2 be two scatterers which are contained in a ball of radius R, let*

$$N := \sum_{t_{nl} < kR} 2n + 1,$$

and assume that the far field patterns coincide for $N + 1$ incident plane waves with distinct directions and one fixed wave number. Then $D_1 = D_2$.

Proof. We proceed as in the previous proof, recalling the definition of D^*. As a consequence of the Courant maximum-minimum principle for compact symmetric operators, the eigenvalues of the negative Laplacian under Dirichlet boundary conditions have the following strong monotonicity property (see [119], Theorem 4.7): the n–th eigenvalue for a ball B containing the domains D_1 and D_2 is always smaller than the n–th eigenvalue for the subdomain $D^* \subset B$ where the eigenvalues are arranged according to increasing magnitude and taken with their respective multiplicity. Hence, if $\lambda_1 \leq \lambda_2 \leq \cdots \leq \lambda_m = k^2$ are the eigenvalues of D^* that are less than or equal to k^2 and $\mu_1 \leq \mu_2 \leq \cdots \leq \mu_m$ are the first m eigenvalues of the ball of radius R, then $\mu_m < \lambda_m = k^2$. In particular, the multiplicity M of λ_m is less than or equal to the sum of the multiplicities of the eigenvalues for the ball which are less than k^2. However, from the discussion of the example after Theorem 3.17, we know that the eigenfunctions for a ball of radius R are the spherical wave functions $j_n(k|x|) Y_n(\hat{x})$ with the eigenvalues given in terms of the zeros of the spherical Bessel functions by $\mu_{nl} = t_{nl}^2/R^2$. By Theorem 2.6, the multiplicity of the eigenvalues μ_{nl} is $2n + 1$ whence $M \leq N$ follows with N as defined in the theorem. Since D^* is nonempty, this leads to a contradiction because the $N + 1$ different incident waves yield $N + 1$ linearly independent eigenfunctions with eigenvalue k^2 for D^*. Hence, we can now conclude that $D_1 = D_2$.   □

Corollary 5.3 *Let D_1 and D_2 be two scatterers which are contained in a ball of radius R such that $kR < \pi$ and assume that the far field patterns coincide for one incident plane wave with wave number k. Then $D_1 = D_2$.*

Proof. From (2.35) and Rolle's theorem, we see that between two zeros of j_n there lies a zero of j_{n-1}. Since $j_n(0) = 0$ for $n = 1, 2, \ldots$, this implies that the sequence (t_{n0}) is strictly monotonically increasing and therefore the smallest positive zero of the spherical Bessel functions is given by the smallest zero of j_0, that is, $t_{00} = \pi$ since $j_0(t) = \sin t/t$.   □

Essentially the same arguments show that the scatterer is uniquely determined by the far field patterns for an infinite number of incident plane waves with distinct wave numbers that do not accumulate at infinity and one fixed incident direction. There is also an analogue of Theorem 5.2 for a finite number of wave numbers with one fixed incident direction.

As an interesting open problem, we wish to point out that it is not known if one incoming plane wave for one single direction and one single wave number completely determines the scatterer (without any additional a priori information).

Difficulties arise in attempting to generalize Schiffer's approach to the case of the sound-hard obstacle, i.e., the Neumann boundary condition. This is due to the fact that the validity of the Rellich selection theorem in the Sobolev space $H^1(D^*)$, i.e., without homogeneous Dirichlet boundary values, requires the boundary to be sufficiently smooth. Hence, using Schiffer's method, we cannot exclude the existence of two scatterers D_1 and D_2 for which the boundary of $D_1 \setminus (\bar{D}_1 \cap \bar{D}_2)$ is not sufficiently smooth, for example if it has cusps.

Isakov [72] has obtained uniqueness results on inverse scattering for penetrable obstacles by using techniques different from Schiffer's method. In the following we use a simplified version of Isakov's approach (due to Kirsch and Kress [95]) to prove a uniqueness result for impenetrable sound-hard obstacles.

We begin with a few comments on the *interior Dirichlet problem*. The classical approach for solving the interior Dirichlet problem is to seek the solution in the form of a double-layer potential

$$u(x) = \int_{\partial D} \frac{\partial \Phi(x, y)}{\partial \nu(y)} \, \varphi(y) \, ds(y), \quad x \in D,$$

with a continuous density φ. Then, given a continuous function f on ∂D, by the jump relations of Theorem 3.1 the double-layer potential u satisfies the boundary condition $u = f$ on ∂D if the density solves the integral equation

$$\varphi - K\varphi = -2f$$

with the double-layer integral operator K defined by (3.9). If we assume that k^2 is not a Dirichlet eigenvalue for D, i.e., the homogeneous Dirichlet problem in D has only the trivial solution, then with the aid of the jump relations it can be seen that $I - K$ has a trivial null space in $C(\partial D)$ (for details see [32]). Hence, by the Riesz–Fredholm theory $I - K$ has a bounded inverse $(I - K)^{-1}$ from $C(\partial D)$ into $C(\partial D)$. This implies solvability and well-posedness of the interior Dirichlet problem. In particular, we have the following theorem.

Theorem 5.4 *Assume that k^2 is not a Dirichlet eigenvalue for the bounded domain D. Let (u_n) be a sequence of $C^2(D) \cap C(\bar{D})$ solutions to the Helmholtz equation in D such that the boundary data $f_n = u_n$ on ∂D are weakly convergent in $L^2(\partial D)$. Then the sequence (u_n) converges uniformly (together with all its derivatives) on compact subsets of D to a solution u of the Helmholtz equation.*

Proof. Following the argument used in Theorem 3.20, the Fredholm alternative can be employed to show that $(I - K)^{-1}$ is bounded from $L^2(\partial D)$ into $L^2(\partial D)$. Hence, the sequence $\varphi_n := 2(K - I)^{-1}f_n$ converges weakly to $\varphi := 2(K - I)^{-1}f$ as $n \to \infty$ provided (f_n) converges weakly towards f. Substituting this into the double-layer potential, we see that the sequence (u_n) converges pointwise in D to the double-layer potential u with density φ. Applying the Schwarz inequality to the double-layer potential, we see that

$$|u_n(x_1) - u_n(x_2)| \leq |\partial D|^{1/2} \sup_{y \in \partial D} \left| \frac{\partial \Phi(x_1, y)}{\partial \nu(y)} - \frac{\partial \Phi(x_2, y)}{\partial \nu(y)} \right| \|2(K - I)^{-1}f_n\|_{L^2(\partial D)}$$

for all $x_1, x_2 \in D$. From this we deduce that the u_n are equicontinuous on compact subsets of D since weakly convergent sequences are bounded. This, together with the pointwise convergence, implies uniform convergence of the sequence (u_n) on compact subsets of D. Convergence of the derivatives follows in an analogous manner.   □

We will use Theorem 5.4 not only for our uniqueness result in sound-hard obstacle scattering but also later in our analysis of the method of Colton and Monk for approximately solving the inverse problem. In order to prove our uniqueness result, we need the following theorem.

Theorem 5.5 *Assume that k^2 is not a Dirichlet eigenvalue for the bounded domain D and $\mathbb{R}^3 \setminus \bar{D}$ is connected. Let $u^i(x; d) = e^{ik\,x\cdot d}$. Then the restriction of the set of plane waves $\{u^i(\cdot\,; d) : d \in \Omega\}$ to ∂D is complete in $L^2(\partial D)$.*

Proof. Let $\varphi \in L^2(\partial D)$ satisfy

$$\int_{\partial D} \varphi(y)\, e^{-ik\,y\cdot d} ds(y) = 0$$

for all $d \in \Omega$. Then, by (2.14) the single-layer potential

$$u(x) := \int_{\partial D} \varphi(y)\Phi(x, y)\, ds(y), \quad x \in \mathbb{R}^3 \setminus \partial D,$$

has vanishing far field pattern $u_\infty = 0$. Therefore, by Theorem 2.13 we have $u = 0$ in $\mathbb{R}^3 \setminus \bar{D}$. The L^2 jump relation (3.20) now implies $\varphi - K'\varphi = 0$ on ∂D. Arguing as in the proof of Theorem 3.20, it can be seen that the nullspaces of $I - K'$ in $C(\partial D)$ and $L^2(\partial D)$ coincide. Hence, $\varphi \in C(\partial D)$ and the continuity of the single-layer potential with continuous densities implies that u solves the homogeneous Dirichlet problem in D. Thus, by our assumption on D, we conclude that $u = 0$ in D and the jump relation (3.2) finally implies that $\varphi = 0$.   □

We are now ready to establish a uniqueness result for the inverse Neumann problem.

Theorem 5.6 *Assume that D_1 and D_2 are two sound-hard scatterers such that for a fixed wave number the far field patterns for both scatterers coincide for all incident directions. Then $D_1 = D_2$.*

Proof. As in the proof of Theorem 5.1, we first conclude that by Theorem 2.13 the scattered waves $u^s(\,\cdot\,; d)$ for the incident fields $u^i(x; d) = e^{ik\,x\cdot d}$ coincide in the unbounded component G of the complement of $\bar{D}_1 \cup \bar{D}_2$. Choose $x_0 \in G$ and consider the two exterior Neumann problems for radiating solutions to the Helmholtz equation

$$(5.3) \qquad \Delta w_j^s + k^2 w_j^s = 0 \quad \text{in } \mathbb{R}^3 \setminus \bar{D}_j, \quad j = 1, 2,$$

with boundary condition

$$(5.4) \qquad \frac{\partial w_j^s}{\partial \nu} + \frac{\Phi(\,\cdot\,, x_0)}{\partial \nu} = 0 \quad \text{on } \partial D_j, \quad j = 1, 2.$$

We will show that

$$(5.5) \qquad w_1^s = w_2^s \quad \text{in } G.$$

To this end, we choose a bounded C^2 domain B such that $\mathbb{R}^3 \setminus B$ is connected, $\bar{D}_1 \cup \bar{D}_2 \subset B$, $x_0 \notin \bar{B}$ and k^2 is not a Dirichlet eigenvalue for B. The latter is possible because of the strong monotonicity property of the eigenvalues which we have used in the proof of Theorem 5.2. Then, by Theorem 5.5, there exists a sequence (v_n) in $\text{span}\{u^i(\,\cdot\,; d) : d \in \Omega\}$ such that

$$\|v_n - \Phi(\,\cdot\,, x_0)\|_{L^2(\partial B)} \to 0, \quad n \to \infty.$$

By Theorem 5.4, this implies that

$$(5.6) \qquad \text{grad } v_n \to \text{grad } \Phi(\,\cdot\,, x_0), \quad n \to \infty,$$

uniformly on $\bar{D}_1 \cup \bar{D}_2$. Since the v_n are linear combinations of plane waves, the corresponding scattered waves $v_{n,1}^s$ and $v_{n,2}^s$ for the obstacles D_1 and D_2 coincide in G, that is, for $v_n^s = v_{n,1}^s = v_{n,2}^s$ we have

$$(5.7) \qquad \frac{\partial v_n^s}{\partial \nu} + \frac{\partial v_n}{\partial \nu} = 0 \quad \text{on } \partial D_j \cap \partial G, \quad j = 1, 2.$$

The well-posedness of the exterior Neumann problem (Theorem 3.10), the boundary conditions (5.4) and (5.7) and the convergence (5.6) now imply that

$$v_n^s \to w_j^s, \quad n \to \infty,$$

uniformly on compact subsets of G for $j = 1, 2$, whence (5.5) follows.

Now assume that $D_1 \neq D_2$. Then, without loss of generality, there exists $x^* \in \partial G$ such that $x^* \in \partial D_1$ and $x^* \notin \bar{D}_2$. We can choose $h > 0$ such that the sequence

$$x_n := x^* + \frac{h}{n}\, \nu(x^*), \quad n = 1, 2, \dots,$$

is contained in G and consider the solutions $w_{n,j}^s$ to the exterior Neumann problems (5.3), (5.4) with x_0 replaced by x_n. By (5.5) we have $w_{n,1}^s = w_{n,2}^s$ in G. Considering $w_n^s = w_{n,2}^s$ as the scattered wave corresponding to the obstacle D_2, we observe that the Neumann data

$$\frac{\partial w_n^s}{\partial \nu} = -\frac{\Phi(\cdot, x_n)}{\partial \nu} \quad \text{on } \partial D_2$$

are uniformly bounded with respect to the maximum norm on ∂D_2. Therefore, by the well-posedness of the exterior Neumann problem (Theorem 3.10) we have that grad w_n^s is uniformly bounded with respect to the maximum norm on closed subsets of $\mathbb{R}^3 \setminus \bar{D}_2$. In particular, this implies that

$$\left| \frac{\partial w_n^s}{\partial \nu} (x^*) \right| \leq C$$

for all n and some positive constant C. On the other hand, considering $w_n^s = w_{n,1}^s$ as the scattered wave corresponding to the obstacle D_1, from the boundary condition we have

$$\left| \frac{\partial w_n^s}{\partial \nu} (x^*) \right| = \left| \frac{\partial \Phi(x^*, x_n)}{\partial \nu} (x^*) \right| = \frac{|1 - ik|x^* - x_n||}{4\pi |x^* - x_n|^2} \to \infty, \quad n \to \infty.$$

This is a contradiction. Therefore $D_1 = D_2$ and the proof is complete. $\square$

Closely related to the uniqueness question is the following example which gives an explicit solution to the inverse obstacle problem. If the sound-soft scatterer D is a ball centered at the origin, it is obvious from symmetry that the far field pattern only depends on the angle between the incident and the observation direction (see also example (3.30)). Hence, we have

$$(5.8) \qquad u_\infty(\hat{x}; d) = u_\infty(Q\hat{x}; Qd)$$

for all $\hat{x}, d \in \Omega$ and all rotations, that is, all real orthogonal matrices Q with $\det Q = 1$. *Karp's theorem* [82] says that the converse of this statement is also true. We shall now show how the approach of Colton and Kirsch [27] to proving this result can be considerably simplified. For this, in view of the Funk–Hecke formula (2.44), we consider the superposition of incident plane waves given by

$$(5.9) \qquad v^i(x) = \int_\Omega e^{ik\, x \cdot d} ds(d) = \frac{4\pi \sin k|x|}{k|x|}.$$

Then, by Lemma 3.16, the corresponding scattered wave v^s has the far field pattern

$$v_\infty(\hat{x}) = \int_\Omega u_\infty(\hat{x}; d)\, ds(d),$$

and the condition (5.8) implies that $v_\infty(\hat{x}) = c$ for some constant c. Hence,

$$(5.10) \qquad v^s(x) = c\, \frac{e^{ik|x|}}{|x|}$$

follows. From (5.9) and (5.10) together with the boundary condition $v^i + v^s = 0$ on ∂D we now have

$$\sin k|x| + \frac{kc}{4\pi} e^{ik|x|} = 0$$

for all $x \in \partial D$. From this we conclude that $|x| = $ constant for all $x \in \partial D$, i.e., D is a ball with center at the origin.

5.2 Physical Optics Approximation

From the boundary integral equation approach to the solution of the direct scattering problem, it is obvious that the far field pattern depends nonlinearly on the boundary of the scatterer. Therefore, the inverse problem to determine the boundary of the scatterer from a knowledge of the far field pattern is a nonlinear problem. We begin our discussion on the solution of the inverse problem by presenting a linearized method based on the Kirchhoff or physical optics approximation.

In the physical optics approximation for a convex scatterer D for large wave numbers k, by (3.34) the far field pattern is given by

$$u_\infty(\hat{x}; d) = -\frac{ik}{2\pi} \int_{\partial D_-} \nu(y) \cdot d \, e^{ik(d-\hat{x})\cdot y} \, ds(y), \quad \hat{x} \in \Omega,$$

where $\partial D_- := \{y \in \partial D : \nu(y) \cdot d < 0\}$ denotes the part of the boundary which is illuminated by the plane wave with incident direction d. In particular, for $\hat{x} = -d$, i.e., for the far field in the *back scattering* direction we have

$$u_\infty(-d; d) = -\frac{1}{4\pi} \int_{\nu(y)\cdot d<0} \frac{\partial}{\partial \nu(y)} e^{2ik \, d\cdot y} \, ds(y).$$

Analogously, replacing d by $-d$, we have

$$u_\infty(d; -d) = -\frac{1}{4\pi} \int_{\nu(y)\cdot d>0} \frac{\partial}{\partial \nu(y)} e^{-2ik \, d\cdot y} \, ds(y).$$

Combining the last two equations we find

$$u_\infty(-d; d) + \overline{u_\infty(d; -d)} = -\frac{1}{4\pi} \int_{\partial D} \frac{\partial}{\partial \nu(y)} e^{2ik \, d\cdot y} \, ds(y)$$

whence by Green's theorem

$$u_\infty(-d; d) + \overline{u_\infty(d; -d)} = -\frac{1}{4\pi} \int_D \Delta e^{2ik \, d\cdot y} \, dy = \frac{k^2}{\pi} \int_D e^{2ik \, d\cdot y} \, dy$$

follows. Denoting by χ the characteristic function of the domain D, we rewrite this equation in the form

(5.11) $$\int_{\mathbb{R}^3} \chi(y) e^{2ik \, d\cdot y} \, dy = \frac{\pi}{k^2} \left\{ u_\infty(-d; d) + \overline{u_\infty(d; -d)} \right\}$$

which is known as the *Bojarski identity* [15, 16]. Hence, in the physical optics approximation the Fourier transform of the characteristic function of the scatterer, in principle, can be completely obtained from measurements of the back scattering far field for all incident directions $d \in \Omega$ and all wave numbers $k > 0$. Then, by inverting the Fourier transform (which is a bounded operator from $L^2(\mathbb{R}^3)$ onto $L^2(\mathbb{R}^3)$ with bounded inverse) one can determine χ and therefore D. Thus, the physical optics approximation leads to a linearization of the inverse problem. For details on the implementation, we refer to Bleistein [13], Langenberg [114] and Ramm [156].

However, there are several drawbacks to this procedure. Firstly, we need the far field data for all wave numbers. But the physical optics approximation is valid only for large wave numbers. Therefore, in practice, the Fourier transform of χ is available only for wavenumbers $k \geq k_0$ for some $k_0 > 1$. This means that we have to invert a Fourier transform with incomplete data. This may cause uniqueness ambiguities and it leads to severe ill-posedness of the inversion as is known from corresponding situations in the inversion of the Radon transform in computerized tomography (see Natterer [145]). Thus, the ill-posedness which seemed to have disappeared through the inversion of the Fourier transform is back on stage. Secondly, and particularly important in the context of the scope of this book, the physical optics approximation will not work at all in situations where far field data are available only for frequencies in the resonance region. Therefore, after this brief mentioning of the physical optics approximation we turn our attention to a solution of the full nonlinear inverse scattering problem.

5.3 The Direct Approach to the Inverse Problem

The solution to the direct scattering problem with a fixed incident plane wave u^i defines an operator

$$F : \partial D \mapsto u_\infty$$

which maps the boundary ∂D of the scatterer D onto the far field pattern u_∞ of the scattered wave. In terms of this operator, given a (measured) far field pattern u_∞, the inverse problem just consists in solving the nonlinear and ill-posed equation

$$(5.12) \qquad F(\partial D) = u_\infty$$

for the unknown surface ∂D. Hence, it is quite natural to try some of the regularization methods mentioned at the end of the previous chapter. For this we need to establish some properties of the operator F.

So far, we have not defined the solution and data space for the operator F. Having in mind that for ill-posed problems the norm in the data space has to be suitable for describing the measurement errors, our natural choice is the

Hilbert space $L^2(\Omega)$ of square integrable functions on the unit sphere. For the time being, for the solution space we will be rather vague and think of it as a class of admissable surfaces described by some suitable parametrization and equipped with an appropriate norm.

The main question we wish to address is that of continuity and compactness of the operator F. For this, we have to investigate the dependence of the solution to the direct problem on the boundary surface. In principle, this analysis can be based either on the boundary integral equation method described in Section 3.2 or on weak solution methods. In both approaches, in order to compare the solution operators for different domains, we have to transform the boundary value problem for the variable domain into one for a fixed reference domain. Since both methods have their merits, we will present both. We postpone the presentation of the continuous dependence results via integral equation methods until the end of this section and first follow Pironneau [153] in using a weak solution approach. In doing this, we expect the reader to be familiar with the basic theory of Sobolev spaces which we will not explain in detail. For an introduction, we refer to Adams [1], Gilbarg and Trudinger [56] and Treves [177].

Theorem 5.7 *For a fixed incident wave u^i, the operator $F : \partial D \mapsto u_\infty$ which maps the boundary ∂D onto the far field pattern u_∞ of the scattered wave u^s is completely continuous from C^1 into $L^2(\Omega)$.*

Proof. Due to the unbounded domain and the radiation condition, we have to couple the weak solution technique either with a boundary integral equation method or a spectral method. Our proof consists of two parts. First, for a fixed domain, we will establish that the linear operator which maps the incident field u^i onto the normal derivative $\partial u^s/\partial \nu$ of the scattered field is bounded from the Sobolev space $H^{1/2}(\Omega_R)$ into its dual space $H^{-1/2}(\Omega_R)$. Here, Ω_R is a sphere of radius R centered at the origin where R is chosen large enough such that D is contained in the interior of Ω_R. In the second step, we will show that this mapping depends continuously on the boundary ∂D whence the statement of the theorem follows by using Theorem 2.5.

We denote $D_R := \{x \in \mathbb{R}^3 \setminus \bar{D} : |x| < R\}$ and introduce the Sobolev space $\tilde{H}_0^1(D_R) := \{v \in H^1(D_R) : v = 0 \text{ on } \partial D\}$ where the boundary condition $v = 0$ on ∂D has to be understood in the sense of the trace operator. As in the proof of Theorem 5.1, from Lemma 3.8 we see that the solution u to the direct scattering problem belongs to $\tilde{H}_0^1(D_R)$. Then, by Green's theorem (2.2), we see that u satisfies

$$(5.13) \qquad \int_{D_R} \{\text{grad } u \cdot \text{grad } \bar{v} - k^2 u \bar{v}\} dx = \int_{\Omega_R} \frac{\partial u}{\partial \nu} \bar{v} \, ds$$

for all $v \in \tilde{H}_0^1(D_R)$ where ν denotes the exterior unit normal to Ω_R.

We denote by A the Dirichlet to Neumann map for radiating solutions w to the Helmholtz equation in the exterior of Ω_R which was introduced in

Theorem 3.11. It transforms the boundary values into the normal derivative on the boundary

$$A : w \mapsto \frac{\partial w}{\partial \nu} \quad \text{on } \Omega_R.$$

By considering the boundary integral equation method in Sobolev spaces on the boundary (see Theorem 3.6), analogous to Theorem 3.11 it can be shown that A is bounded from $H^{1/2}(\Omega_R)$ into $H^{-1/2}(\Omega_R)$ and has a bounded inverse. However, for the simple shape of the sphere this result and further properties of A can also be established by expansion of w with respect to spherical wave functions as we will indicate briefly.

From the expansion (2.48) of radiating solutions to the Helmholtz equation with respect to spherical wave functions, we see that A maps

$$w = \sum_{n=0}^{\infty} \sum_{m=-n}^{n} a_n^m Y_n^m$$

with coefficients a_n^m onto

$$Aw = \sum_{n=0}^{\infty} \gamma_n \sum_{m=-n}^{n} a_n^m Y_n^m$$

where

(5.14) $$\gamma_n := \frac{k h_n^{(1)\prime}(kR)}{h_n^{(1)}(kR)}, \quad n = 0, 1, \dots.$$

The spherical Hankel functions and their derivatives do not have real zeros since otherwise the Wronskian (2.36) would vanish. From this we observe that A is bijective. In view of the differentiation formula (2.34) and the asymptotic formula (2.38) for the spherical Hankel functions, we see that

$$c_1(n+1) \leq |\gamma_n| \leq c_2(n+1)$$

for all n and some constants $0 < c_1 < c_2$. From this the boundedness of $A : H^{1/2}(\Omega_R) \to H^{-1/2}(\Omega_R)$ and $A^{-1} : H^{-1/2}(\Omega_R) \to H^{1/2}(\Omega_R)$ is obvious since for $p \in \mathbb{R}$ the norm on $H^p(\Omega_R)$ can be described in terms of the Fourier coefficients by

$$\|w\|_p^2 = \sum_{n=0}^{\infty} (n+1)^{2p} \sum_{m=-n}^{n} |a_n^m|^2 .$$

For the limiting operator $A_0 : H^{1/2}(\Omega_R) \to H^{-1/2}(\Omega_R)$ given by

$$A_0 w = -\sum_{n=0}^{\infty} \frac{n+1}{R} \sum_{m=-n}^{n} a_n^m Y_n^m,$$

we clearly have

$$-\int_{\Omega_R} A_0^{-1} w \, \bar{w} \, ds = \sum_{n=0}^{\infty} \frac{R}{n+1} \sum_{m=-n}^{n} |a_n^m|^2$$

with the integral to be understood as the duality pairing between $H^{1/2}(\Omega_R)$ and $H^{-1/2}(\Omega_R)$. Hence,

$$-\int_{\Omega_R} A_0^{-1} w \, \bar{w} \, ds \geq c\|w\|^2_{H^{-1/2}(\Omega_R)}$$

for some constant $c > 0$, that is, the operator $-A_0^{-1}$ is strictly coercive. Finally, from the power series expansions (2.31) and (2.32) for the spherical Hankel functions, for fixed k we derive

$$\gamma_n = -\frac{n+1}{R} \left\{1 + O\left(\frac{1}{n}\right)\right\}, \quad n \to \infty.$$

This implies that $A^{-1} - A_0^{-1}$ is compact from $H^{-1/2}(\Omega_R)$ into $H^{1/2}(\Omega_R)$ since it is bounded from $H^{-1/2}(\Omega_R)$ into $H^{3/2}(\Omega_R)$ and the imbedding from $H^{3/2}(\Omega_R)$ into $H^{1/2}(\Omega_R)$ is compact.

From (5.13) it can now be deduced that if u is the solution to the scattering problem, then $u \in \tilde{H}^1_0(D_R)$ and $g = \partial u^s/\partial \nu \in H^{-1/2}(\Omega_R)$ satisfy the sesquilinear equation

$$
\begin{aligned}
(5.15) \quad & \int_{D_R} \{\text{grad } u \cdot \text{grad } \bar{v} - k^2 u \bar{v}\} dx - \int_{\Omega_R} g\bar{v} \, ds - \int_{\Omega_R} (A^{-1}g - u)\bar{h} \, ds \\
& \qquad\qquad = \int_{\Omega_R} \left\{\frac{\partial u^i}{\partial \nu} \bar{v} + u^i \bar{h}\right\} ds
\end{aligned}
$$

for all $v \in \tilde{H}^1_0(D_R)$ and all $h \in H^{-1/2}(\Omega_R)$. The sesquilinear form S on $\tilde{H}^1_0(D_R) \times H^{-1/2}(\Omega_R)$ defined by the left hand side of (5.15) can be written as $S = S_0 + S_1$ where

$$S_0(u, g; v, h) := \int_{D_R} \text{grad } u \cdot \text{grad } \bar{v} \, dx - \int_{\Omega_R} \{g\bar{v} - u\bar{h}\} ds - \int_{\Omega_R} A_0^{-1}g \, \bar{h} \, ds$$

and

$$S_1(u, g; v, h) := -k^2 \int_{D_R} u\bar{v} \, dx - \int_{\Omega_R} (A^{-1} - A_0^{-1})g \, \bar{h} \, ds.$$

Clearly S_0 is bounded, and since

$$S_0(u, g; u, g) = \int_{D_R} |\text{grad } u|^2 dx - \int_{\Omega_R} A_0^{-1}g \, \bar{g} \, ds + 2i \, \text{Im} \int_{\Omega_R} u\bar{g} \, ds,$$

by the strict coerciveness of $-A_0^{-1}$ and Friedrich's inequality we have

$$\text{Re } S_0(u, g; u, g) \geq c\left\{\|u\|^2_{\tilde{H}^1_0(D_R)} + \|g\|^2_{H^{-1/2}(\Omega_R)}\right\}$$

for all $u \in \tilde{H}^1_0(D_R)$ and $g \in H^{-1/2}(\Omega_R)$ for some constant $c > 0$, that is, S_0 is strictly coercive. By the compactness of $A^{-1} - A_0^{-1}$ and the Rellich selection theorem, that is the compact imbedding of $\tilde{H}_0^1(D_R)$ into $L^2(D_R)$, the term S_1 is compact. Through the Riesz representation theorem we can write

$$(5.16) \qquad S(u, g; v, h) = \left(L \begin{pmatrix} u \\ g \end{pmatrix}, \begin{pmatrix} v \\ h \end{pmatrix} \right)$$

with a bounded linear operator L mapping $\tilde{H}_0^1(D_R) \times H^{-1/2}(\Omega_R)$ into itself. Corresponding to $S = S_0 + S_1$, we have $L = L_0 + L_1$ where L_0 is strictly coercive and L_1 is compact. Hence, by the Lax–Milgram theorem (see [106]) and the Riesz–Fredholm theory for compact operators, in order to establish unique solvability of the sesquilinear equation (5.15) and continuous dependence of the solution on the right hand side, that is the existence of a bounded inverse L^{-1} to L, it suffices to prove uniqueness for the homogeneous form of (5.15).

This can be shown as a consequence of the uniqueness for the direct scattering problem. Assume that $u \in \tilde{H}_0^1(D_R)$ and $g \in H^{-1/2}(\Omega_R)$ satisfy

$$S(u, g; v, h) = 0$$

for all $v \in \tilde{H}_0^1(D_R)$ and $h \in H^{-1/2}(\Omega_R)$. Then

$$(5.17) \qquad \int_{D_R} \{\mathrm{grad}\, u \cdot \mathrm{grad}\, \bar{v} - k^2 u \bar{v}\} dx - \int_{\Omega_R} g \bar{v}\, ds = 0$$

for all $v \in \tilde{H}_0^1(D_R)$ and

$$(5.18) \qquad \int_{\Omega_R} (A^{-1} g - u) \bar{h}\, ds = 0$$

for all $h \in H^{-1/2}(\Omega_R)$. Equation (5.18) implies $g = Au$. Substituting $v = u$ into (5.17) we obtain

$$\mathrm{Im} \int_{\Omega_R} Au\, \bar{u}\, ds = 0,$$

that is,

$$\sum_{n=0}^{\infty} \mathrm{Im}\, \gamma_n \sum_{m=-n}^{n} |a_n^m|^2 = 0$$

for

$$u = \sum_{n=0}^{\infty} \sum_{m=-n}^{n} a_n^m Y_n^m \quad \text{on } \Omega_R.$$

From this we deduce $u = 0$ on Ω_R and $g = 0$ since from (5.14) and the Wronskian (2.36) we have that

$$\mathrm{Im}\, \gamma_n = \frac{1}{kR^2 \left| h_n^{(1)}(kR) \right|^2} > 0$$

for all n. Now (5.17) reads

$$\int_{D_R} \{\mathrm{grad}\, u \cdot \mathrm{grad}\, \bar{v} - k^2 u \bar{v}\} dx = 0,$$

i.e., u is a weak solution to the Helmholtz equation in D_R satisfying weakly a homogeneous Dirichlet condition on ∂D and on Ω_R and in addition a homogeneous Neumann condition on Ω_R. By the classical regularity properties of weak solutions to elliptic boundary value problems (see [56]), it follows that u also is a classical solution. Then the vanishing Cauchy data on Ω_R and Green's representation Theorem 2.1 imply that u can be extended to all of the exterior domain $\mathbb{R}^3 \setminus \bar{D}$ as a radiating solution to the Helmholtz equation. Then, by the uniqueness of the exterior Dirichlet problem (Theorem 3.7), from $u = 0$ on ∂D we get $u = 0$ in D_R. This completes the uniqueness proof. In particular, we now know that the solution operator

$$(5.19) \qquad u^i \mapsto g = \frac{\partial u^s}{\partial \nu}$$

is continuous from $H^{1/2}(\Omega_R)$ into $H^{-1/2}(\Omega_R)$.

We now wish to study the dependence of the sesquilinear form S on the shape of the domain D. For this we map D_R onto a fixed reference domain B_R. For the sake of simplicity, we restrict ourselves to the case of domains D which are starlike with respect to the origin. Assume that ∂D is represented in the parametric form

$$x = r(\hat{x})\,\hat{x}, \quad \hat{x} \in \Omega,$$

with a positive function $r \in C^2(\Omega)$. Without loss of generality, we may assume that $r(\hat{x}) > 1$ for all $\hat{x} \in \Omega$. After setting $R = 2\|r\|_\infty$ and $\hat{y} = y/|y|$, the mapping

$$x(y) := \begin{cases} y + (r(\hat{y}) - 1)\left(\dfrac{R - |y|}{R - 1}\right)^2 \hat{y}, & 1 \leq |y| \leq R, \\[2em] y, & R \leq |y| < \infty, \end{cases}$$

is a diffeomorphism from the closed exterior of the unit sphere onto $\mathbb{R}^3 \setminus D$ such that $B_R := \{y \in \mathbb{R}^3 : 1 < |y| < R\}$ is mapped onto D_R. With the inverse map $y = y(x)$ we substitute to obtain

$$\int_{D_R} \{\operatorname{grad} u \cdot \operatorname{grad} \bar{v} - k^2 u \bar{v}\}\,dx = T(\tilde{u}, \tilde{v})$$

where $\tilde{u}(y) = u(x(y))$, $\tilde{v}(y) = v(x(y))$, and the sesquilinear form T on $\tilde{H}_0^1(B_R)$ is defined by

$$(5.20) \qquad T(u, v) := \int_{B_R} \left\{ \sum_{i,j=1}^{3} a_{ij} \frac{\partial u}{\partial y_i} \frac{\partial \bar{v}}{\partial y_j} + b\, u\bar{v} \right\} dy$$

for $u, v \in \tilde{H}_0^1(B_R)$. The coefficients in (5.20) are given by

$$a_{ij} := \sum_{m=1}^{3} \frac{\partial y_i}{\partial x_m} \frac{\partial y_j}{\partial x_m} J \quad \text{and} \quad b := -k^2 J$$

where J denotes the Jacobian of the substitution. After this transformation we can consider the sesquilinear equation on $\tilde{H}_0^1(B_R) \times H^{-1/2}(\Omega_R)$ with the fixed domain B_R independent of the boundary. As easily seen, the coefficients a_{ij} and b depend continuously on r in the C^1 norm, i.e., there exists a constant C such that

$$\|a_{ij}(r+q) - a_{ij}(r)\|_\infty \le C\|q\|_{C^1(\Omega)}, \quad \|b(r+q) - b(r)\|_\infty \le C\|q\|_{C^1(\Omega)}$$

for all sufficiently small q. Hence, from (5.20) we deduce that after the transformation the sesquilinear form S depends continuously on r with respect to the C^1 norm. Using (5.16), in terms of the operator L this can be written as

$$(5.21) \qquad \|L_{r+q} - L_r\| \le C_1(r)\|q\|_{C^1(\Omega)}$$

for all sufficiently small $q \in C^2(\Omega)$ and some constant $C_1(r)$ depending on r. Therefore, a perturbation argument based on the Neumann series shows that the inverse operator also satisfies an estimate of the form

$$(5.22) \qquad \|L_{r+q}^{-1} - L_r^{-1}\| \le C_2(r)\|q\|_{C^1(\Omega)}$$

for some constant $C_2(r)$. This inequality carries over to the solution operator introduced through (5.19). Hence,

$$\left\| \frac{\partial u_{r+q}^s}{\partial \nu} - \frac{\partial u_r^s}{\partial \nu} \right\|_{H^{-1/2}(\Omega_R)} \le C_3(r)\|q\|_{C^1(\Omega)}$$

follows for some constant $C_3(r)$. Thus, the mappings $\partial D \mapsto \partial u^s/\partial \nu$ and $\partial D \mapsto u^s$ are continuous from $C^1(\Omega)$ into $H^{-1/2}(\Omega_R)$ and $H^{1/2}(\Omega_R)$, respectively. The theorem now follows from the analyticity of the kernel in the integral representation of the far field pattern in Theorem 2.5 applied in the exterior of Ω_R. □

We actually can pursue the ideas of the preceding proof one step further and show that $F : \partial D \mapsto u_\infty$ has a Fréchet derivative. To this end, we observe that the coefficients in the sesquilinear form (5.20) are of the form

$$a_{ij} = a_{ij}(r(\hat{y}), r_1(\hat{y}), r_2(\hat{y}), r_3(\hat{y}), y), \quad b = b(r(\hat{y}), r_1(\hat{y}), r_2(\hat{y}), r_3(\hat{y}), y),$$

where r_1, r_2, r_3 denote the cartesian components of the surface gradient $\operatorname{Grad} r$ on Ω. In particular, the coefficients are twice continuously differentiable with respect to r, r_1, r_2, r_3. By Taylor's formula, applied to a_{ij} and b, it follows that

$$\sup_{\|u\|,\|v\|\le 1} \|T_{r+q}(u,v) - T_r(u,v) - (T_r'q)(u,v)\| \le c_1(r)\|q\|_{C^1(\Omega)}^2$$

in a $C^1(\Omega)$ neighborhood of r for some constant $c_1(r)$ where

$$(T_r'q)(u,v) := \int_{B_R} \left\{ \sum_{i,j=1}^3 \frac{\partial a_{ij}}{\partial r} \frac{\partial u}{\partial y_i} \frac{\partial \bar{v}}{\partial y_j} + \frac{\partial b}{\partial r} u\bar{v} \right\} q \, dy$$

$$+ \int_{B_R} \sum_{m=1}^3 \left\{ \sum_{i,j=1}^3 \frac{\partial a_{ij}}{\partial r_m} \frac{\partial u}{\partial y_i} \frac{\partial \bar{v}}{\partial y_j} + \frac{\partial b}{\partial r_m} u\bar{v} \right\} q_m \, dy,$$

i.e., the mapping $r \mapsto T_r$ is Fréchet differentiable. By the Riesz representation theorem, this implies that there exists a bounded linear operator L'_r from $C^1(\Omega)$ into the space of bounded linear operators $\mathcal{L}\left(\tilde{H}^1_0(D_R) \times H^{-1/2}(\Omega_R)\right)$ from $\tilde{H}^1_0(D_R) \times H^{-1/2}(\Omega_R)$ into itself such that

$$(5.23) \qquad \|L_{r+q} - L_r - L'_r q\| \leq c_1(r)\|q\|^2_{C^1(\Omega)},$$

that is, the mapping $r \mapsto L_r$ from $C^1(\Omega)$ into $\mathcal{L}\left(\tilde{H}^1_0(D_R) \times H^{-1/2}(\Omega_R)\right)$ is also Fréchet differentiable. Then from

$$L_r\{L^{-1}_{r+q} - L^{-1}_r + L^{-1}_r L'_r q L^{-1}_r\}L_r = (L_{r+q} - L_r)L^{-1}_{r+q}(L_{r+q} - L_r) - (L_{r+q} - L_r - L'_r q)$$

and (5.21)–(5.23) we see that

$$\|L^{-1}_{r+q} - L^{-1}_r + L^{-1}_r L'_r q L^{-1}_r\| \leq c_2(r)\|q\|^2_{C^1(\Omega)}$$

for some constant $c_2(r)$, that is, the mapping $r \mapsto L^{-1}_r$ is also Fréchet differentiable with the derivative given by $q \mapsto -L^{-1}_r L'_r q L^{-1}_r$. As in the previous proof, this now implies that the mapping $\partial D \mapsto u^s$ is Fréchet differentiable from C^1 into $H^{1/2}(\Omega_R)$. Since $u^s \mapsto u_\infty$ is linear and bounded from $H^{1/2}(\Omega_R)$ into $L^2(\Omega)$, we have established the following result.

Theorem 5.8 *The mapping $F : \partial D \mapsto u_\infty$ is Fréchet differentiable from C^1 into $L^2(\Omega)$.*

Kirsch [91] has recently verified that evaluating the derivative $F'_r q$ basically amounts to finding the solution v of the exterior Dirichlet problem for the domain D described by r with boundary values given by

$$v = -q \frac{\partial u^s}{\partial \nu} \quad \text{on } \partial D.$$

Theorem 5.8, in principle, opens up the possibility to employ Newton's method or variants of it for the approximate solution of (5.12). In view of Theorem 4.19, a regularization has to be incorporated since by Theorem 5.7 the operator F is completely continuous. In the literature, Newton's method for the approximate solution of the inverse obstacle scattering problem has been tested only in the two-dimensional case by Roger [162] and by Murch, Tan and Wall [143]. Hence, a definitive value judgement on this approach at present is premature. The method of quasi-solutions, as described in Theorem 4.21, has been investigated by Angell, Colton and Kirsch [2]. For related numerical work we refer to [86]. A Tikhonov type regularization similar to the one described in Theorem 4.20 has been employed by Kristensson and Vogel [112]. However, it is not our intention to give a detailed description of the various numerical implementations of Newton type methods, quasi-solutions or Tikhonov regularization methods for solving the inverse obstacle problem.

A common feature of the above methods is that they are of an iterative nature. They require the evulation of the operator F, i.e., the solution of the direct scattering problem for different domains at each iteration step in the Newton method or in the iterative solution of the nonlinear optimization problem involved in the quasi-solution or Tikhonov regularization. Hence, these direct methods need an efficient forward solver and tend to be computationally costly.

Since there are no explicit solutions of the direct scattering problem available, numerical tests of approximate methods for the inverse problem usually rely on synthetic far field data obtained through the numerical solution of the forward scattering problem. Here we take the opportunity to put up a warning sign against *inverse crimes*. In order to avoid trivial inversion of finite dimensional problems, for reliably testing the performance of an approximation method for the inverse problem it is crucial that the synthetic data be obtained by a forward solver which has no connection to the inverse solver under consideration. Unfortunately, not all of the numerical reconstructions which have appeared in the literature meet with this obvious requirement. To be more precise about our objections, consider a m–parameter family G_m of boundary surfaces and use a numerical method $\mathcal{M}$ for the solution of the direct problem to obtain a number n of evaluations of the far field pattern u_∞, for example, point evaluations or Fourier coefficients. This obviously may be considered as defining some function $g : \mathbb{R}^m \to \mathbb{C}^n$. Now use the method $\mathcal{M}$ to create the synthetic data for a boundary surface $\partial D \in G_m$, that is, evaluate g for a certain parameter $a_0 \in \mathbb{R}^m$. Then, for example in Newton's method, incorporating the same method $\mathcal{M}$ in the inverse solver now just means nothing else than solving the finite dimensional nonlinear problem $g(a) = g(a_0)$. Hence, it is no surprise, in particular if m and n are not too large, that the surface ∂D is recovered pretty well.

The last two sections of this chapter will be devoted to a presentation of two inverse methods which avoid the solution of the direct scattering problem. Both stabilize the inverse problem by reformulating it as a nonlinear optimization problem. Of course, the numerical solution of these optimization problems again relies on iteration techniques. However, the actual performance of these iterations is less costly due to a simpler structure of the cost functional which does not need the solution of the direct problem.

In the convergence analysis for these methods, we will need the following continuous dependence result due to Angell, Colton and Kirsch [2] which is slightly stronger than that of Theorem 5.7. It cannot be obtained through weak solution methods since these would only give a result with respect to the $H^{1/2}$ norm instead of the L^2 norm on the boundary.

We consider the set of all surfaces Λ which are starlike with respect to the origin and described by

$$x = r(\hat{x})\,\hat{x}$$

where r denotes a positive function from the Hölder space $C^{1,\alpha}(\Omega)$ with $0 < \alpha < 1$. For a sequence of such surfaces, by convergence $\Lambda_n \to \Lambda$, $n \to \infty$, we mean the convergence $\|r_n - r\|_{1,\alpha} \to 0$, $n \to \infty$, of the representing functions in the $C^{1,\alpha}$ Hölder norm on Ω. We say that a sequence of functions f_n from $L^2(\Lambda_n)$ is L^2 convergent to a function f in $L^2(\Lambda)$ if

$$\lim_{n\to\infty} \int_\Omega |f_n(r_n(\hat{x})\,\hat{x}) - f(r(\hat{x})\,\hat{x})|^2 ds(\hat{x}) = 0.$$

We can now state the following convergence theorem. (In the theorem we can replace C^2 surfaces by $C^{1,\alpha}$ surfaces. However, for consistency with the rest of our book, we have used the more restrictive assumption of C^2 surfaces.)

Theorem 5.9 *Let (Λ_n) be a sequence of starlike C^2 surfaces which converges with respect to the $C^{1,\alpha}$ norm to a C^2 surface Λ as $n \to \infty$ and let u_n and u be radiating solutions to the Helmholtz equation in the exterior of Λ_n and Λ, respectively. Assume that the continuous boundary values of u_n on Λ_n are L^2 convergent to the boundary values of u on Λ. Then the sequence (u_n), together with all its derivatives, converges to u uniformly on compact subsets of the open exterior of Λ.*

Proof. For the solution to the exterior Dirichlet problem, we refer back to the combined double- and single-layer potential approach (3.25), that is, we represent u in the form

$$u(x) = \int_\Lambda \left\{ \frac{\partial \Phi(x,y)}{\partial \nu(y)} - i\Phi(x,y) \right\} \varphi(y)\, ds(y)$$

with a density $\varphi \in C(\Lambda)$ and, analogously, we write u_n as a combined potential with density $\varphi_n \in C(\Lambda_n)$. For a discussion of the convergence of the unique solutions of the boundary integral equations corresponding to (3.26), we need to transform the equations onto a fixed reference surface. For this we substitute

$$x = x_r = r(\hat{x})\,\hat{x}$$

and, for the density φ representing u, we obtain an integral equation over the unit sphere of the form

(5.24) $$\psi(\hat{x}) + \int_\Omega H_r(\hat{x},\hat{y})\psi(\hat{y})\, ds(\hat{y}) = f(\hat{x}), \quad \hat{x} \in \Omega,$$

where we have set $\psi(\hat{x}) := \varphi(r(\hat{x})\,\hat{x})$ and $f(\hat{x}) := 2u(r(\hat{x})\,\hat{x})$ and the kernel is given by

(5.25) $$H_r(\hat{x},\hat{y}) := 2 \left\{ \frac{\partial \Phi(x_r,y_r)}{\partial \nu(y_r)} - i\Phi(x_r,y_r) \right\} J_r(\hat{y}).$$

Here, J_r is the Jacobian of our transformation for which straightforward calculation yields

(5.26) $$J_r(\hat{x}) = r(\hat{x})\sqrt{|r(\hat{x})|^2 + |\operatorname{Grad} r(\hat{x})|^2}$$

in terms of the surface gradient $\operatorname{Grad} r$. The corresponding integral equations for the densities φ_n representing u_n are obtained by replacing r by r_n. For the moment, we postpone the more technical parts of our analysis and make use of the results of Lemma 5.12. Denoting the weakly singular integral operator in (5.24) by $A_r : C(\Omega) \to C(\Omega)$, from (5.34) there exist constants $0 < \delta, \alpha^* < 1$ and $M > 0$ such that

$$|(A_{r_n}\psi)(\hat{x}) - (A_r\psi)(\hat{x})| \leq M\|r_n - r\|_{1,\alpha}^{\delta}\,\|\psi\|_{\infty} \int_{\Omega} \frac{ds(\hat{y})}{|\hat{x} - \hat{y}|^{2-\alpha^*}}$$

for all $\psi \in C(\Omega)$ and all $\hat{x} \in \Omega$. From this it follows that

$$\|A_{r_n} - A_r\|_{\infty} \leq C\|r_n - r\|_{1,\alpha}^{\delta}$$

for some constant C since for $\alpha^* > 0$ the integral on the right hand side exists and does not depend on $\hat{x}$. The same inequality can also be established for the adjoint of A_r where the kernel variables are interchanged. Therefore, by Lax's Theorem 3.5 it follows that

$$(5.27) \qquad \|A_{r_n} - A_r\|_{L^2(\Omega)} \leq C\|r_n - r\|_{1,\alpha}^{\delta}.$$

A Neumann series argument now shows that $\|(I + A_{r_n})^{-1}\|_{L^2(\Omega)}$ is uniformly bounded. From

$$(I + A_{r_n})(\psi_{r_n} - \psi_r) = f_{r_n} - f_r - (A_{r_n} - A_r)\psi_r$$

we then derive

$$\|\psi_{r_n} - \psi_r\|_{L^2(\Omega)} \leq \tilde{C}\|f_{r_n} - f_r\|_{L^2(\Omega)} + \|A_{r_n} - A_r\|_{L^2(\Omega)}\|\psi_r\|_{L^2(\Omega)}$$

for some constant $\tilde{C}$ whence L^2 convergence of the densities $\varphi_n \to \varphi$, $n \to \infty$, follows. The convergence of (u_n) on compact subsets of the exterior of Λ is now obtained by substituting the densities into the combined double- and single-layer potentials for u_n and u and then using the Schwarz inequality. $\qquad \square$

We now establish the technical results needed for the preceding proof.

Lemma 5.10 *The inequality*

$$(5.28) \qquad |\nu(x_r) \cdot (x_r - y_r)| \leq \frac{C\|r\|_{1,\alpha}^2}{\min\limits_{\hat{z} \in \Omega} |r(\hat{z})|}\, |\hat{x} - \hat{y}|^{1+\alpha}$$

is valid for all $\hat{x}, \hat{y} \in \Omega$ and some constant C.

Proof. Straightforward calculation shows that the normal vector is given by

$$(5.29) \qquad \nu(\hat{x}) = \frac{r(\hat{x})\,\hat{x} - \mathrm{Grad}\, r(\hat{x})}{\sqrt{|r(\hat{x})|^2 + |\,\mathrm{Grad}\, r(\hat{x})|^2}}\,.$$

Using $\mathrm{Grad}\, r(\hat{x}) \cdot \hat{x} = 0$, we can write

$$\{x_r - \mathrm{Grad}\, r(\hat{x})\} \cdot \{x_r - y_r\} = a_1 + a_2 + a_3$$

where
$$a_1 := \operatorname{Grad} r(\hat{x}) \cdot \{\hat{y} - \hat{x}\} \, \{r(\hat{y}) - r(\hat{x})\},$$
$$a_2 := r(\hat{x})\{r(\hat{x}) - r(\hat{y}) + \operatorname{Grad} r(\hat{x}) \cdot (\hat{y} - \hat{x})\},$$
$$a_3 := r(\hat{x})r(\hat{y}) \, \hat{x} \cdot \{\hat{x} - \hat{y}\}.$$

By writing
$$r(\hat{y}) - r(\hat{x}) = \int_\Gamma \frac{\partial r}{\partial s}(\hat{z}) \, ds(\hat{z})$$

where Γ denotes the shorter great circle arc on Ω connecting $\hat{x}$ and $\hat{y}$, we can estimate
$$|r(\hat{y}) - r(\hat{x})| = \left| \int_\Gamma \frac{\partial r}{\partial s}(\hat{z}) \, ds(\hat{z}) \right| \leq \| \operatorname{Grad} \dot{r} \|_\infty \theta$$

where θ is the angle between $\hat{x}$ and $\hat{y}$. With the aid of the elementary inequality $2\theta \leq \pi|\hat{y} - \hat{x}|$ we now have

(5.30)
$$|r(\hat{y}) - r(\hat{x})| \leq \frac{\pi}{2} \| \operatorname{Grad} r \|_\infty |\hat{y} - \hat{x}|$$

and consequently
$$|a_1| \leq \frac{\pi}{2} \| \operatorname{Grad} r \|_\infty^2 |\hat{y} - \hat{x}|^2 \leq \frac{\pi}{2} \|r\|_{1,\alpha}^2 |\hat{y} - \hat{x}|^2.$$

We can also write
$$r(\hat{y}) - r(\hat{x}) - \operatorname{Grad} r(\hat{x}) \cdot (\hat{y} - \hat{x}) = \int_\Gamma \left\{ \frac{\partial r}{\partial s}(\hat{z}) - \frac{\partial r}{\partial s}(\hat{x}) \right\} ds(\hat{z}) + (\theta - \sin\theta)\frac{\partial r}{\partial s}(\hat{x}).$$

From this, using the definition of the Hölder norm and $|\theta - \sin\theta| \leq \theta^2/2$ which follows from Taylor's formula, we can estimate
$$|r(\hat{y}) - r(\hat{x}) - \operatorname{Grad} r(\hat{x}) \cdot (\hat{y} - \hat{x})| \leq c\|r\|_{1,\alpha}|\hat{y} - \hat{x}|^{1+\alpha}$$

for some constant c, whence
$$|a_2| \leq c\|r\|_{1,\alpha}^2 |\hat{y} - \hat{x}|^{1+\alpha}$$

follows. Finally, since on the unit sphere $2\,\hat{x} \cdot (\hat{x} - \hat{y}) = |\hat{x} - \hat{y}|^2$, we have
$$|a_3| \leq \frac{1}{2} \|r\|_\infty^2 |\hat{y} - \hat{x}|^2 \leq \frac{1}{2} \|r\|_{1,\alpha}^2 |\hat{y} - \hat{x}|^2.$$

We can now sum our three inequalities to obtain the estimate (5.28). $\qquad\square$

Lemma 5.11 *The inequalities*

(5.31)
$$\frac{1}{2} \min_{\hat{z} \in \Omega} |r(\hat{z})| \, |\hat{x} - \hat{y}| \leq |x_r - y_r| \leq \left(\frac{\pi}{2} + 1\right) \|r\|_{1,\alpha}|\hat{x} - \hat{y}|$$

are valid for all $\hat{x}, \hat{y} \in \Omega$.

Proof. The first inequality in (5.31) follows from

$$|r(\hat{x})(\hat{x} - \hat{y})| \le |x_r - y_r| + |\{r(\hat{y}) - r(\hat{x})\}\hat{y}|$$

and the triangle inequality $|r(\hat{y}) - r(\hat{x})| \le |x_r - y_r|$. The second inequality follows with (5.30) from

$$|x_r - y_r| \le |\{r(\hat{x}) - r(\hat{y})\}\hat{x}| + |r(\hat{y})\{\hat{x} - \hat{y}\}|$$

and the proof is finished. $\square$

Lemma 5.12 *Let U be a bounded subset of $C^{1,\alpha}(\Omega)$ satisfying*

(5.32) $$\min_{\hat{z} \in \Omega} |r(\hat{z})| \ge a$$

for all $r \in U$ and some positive constant a. Then there exist constants $M > 0$ and $0 < \delta, \alpha^* < 1$ such that

(5.33) $$|H_r(\hat{x}, \hat{y})| \le \frac{M}{|\hat{x} - \hat{y}|^{2-\alpha^*}}$$

and

(5.34) $$|H_r(\hat{x}, \hat{y}) - H_q(\hat{x}, \hat{y})| \le \frac{M\|r - q\|_{1,\alpha}^{\delta}}{|\hat{x} - \hat{y}|^{2-\alpha^*}}$$

holds for all $\hat{x}, \hat{y} \in \Omega$ and all $r, q \in U$.

Proof. Since U is bounded in $C^{1,\alpha}(\Omega)$, there exists a constant γ such that

(5.35) $$\|r\|_{1,\alpha} \le \gamma$$

for all $r \in U$. In the following inequalities, by c we will denote a generic constant depending on a and γ with different values in each inequality. The kernel H_r is of the form

$$H_r(\hat{x}, \hat{y}) = \left\{ \nu(y_r) \cdot (x_r - y_r) \frac{h_1(|x_r - y_r|)}{|x_r - y_r|^3} + \frac{h_2(|x_r - y_r|)}{|x_r - y_r|} \right\} J_r(\hat{y})$$

where h_1 and h_2 are C^1 functions. Hence, (5.33) follows immediately from combining (5.26), (5.28) and (5.31) with the conditions (5.32) and (5.35).

From (5.31) we have

(5.36) $$||x_r - y_r| - |x_q - y_q|| \le (\pi + 2)\gamma|\hat{x} - \hat{y}|,$$

and we also have

(5.37) $$||x_r - y_r| - |x_q - y_q|| \le |r(\hat{x}) - q(\hat{x})| + |r(\hat{y}) - q(\hat{y})| \le 2\|r - q\|_{1,\alpha}.$$

For arbitrary δ with $0 < \delta < 1$ we can combine the two inequalities (5.36) and (5.37) to get

$$(5.38) \qquad |\,|x_r - y_r| - |x_q - y_q|\,| \leq c\|r - q\|_{1,\alpha}^{\delta}|\hat{x} - \hat{y}|^{1-\delta}.$$

By Taylor's formula for a C^1 function g, we have

$$\left| \frac{g(t)}{t^m} - \frac{g(s)}{s^m} \right| \leq c \, \frac{|t - s|}{\min\{t^{m+1}, s^{m+1}\}}$$

for all t, s with $0 < t, s \leq 6\gamma$ and for $m = 1, 3$. Hence, using (5.31), (5.32) and (5.38), we obtain

$$(5.39) \qquad \left| \frac{g(|x_r - y_r|)}{|x_r - y_r|^m} - \frac{g(|x_q - y_q|)}{|x_q - y_q|^m} \right| \leq c \, \frac{\|r - q\|_{1,\alpha}^{\delta}}{|\hat{x} - \hat{y}|^{m+\delta}}$$

for all $\hat{x}, \hat{y} \in \Omega$ and all $r, q \in U$. With the aid of Lemma 5.10 and the conditions (5.32) and (5.35), we can estimate

$$(5.40) \qquad |\nu(y_r) \cdot \{x_r - y_r\} - \nu(y_q) \cdot \{x_q - y_q\}| \leq c|\hat{x} - \hat{y}|^{1+\alpha}.$$

From (5.29), by Taylor's formula and the condition (5.32), we have

$$|\nu(y_r) - \nu(y_q)| \leq c\|r - q\|_{1,\alpha}.$$

Hence, writing

$$\nu(y_r) \cdot \{x_r - y_r\} - \nu(y_q) \cdot \{x_q - y_q\}$$
$$= \{\nu(y_r) - \nu(y_q)\} \cdot \{x_r - y_r\} - \nu(y_q) \cdot \{(x_q - x_r) - (y_q - y_r)\}$$

and, for the second term, proceeding as in (5.37) we obtain

$$(5.41) \qquad |\nu(y_r) \cdot \{x_r - y_r\} - \nu(y_q) \cdot \{x_q - y_q\}| \leq c\|r - q\|_{1,\alpha}.$$

For arbitrary δ with $0 < \delta < 1$ we combine the two inequalities (5.40) and (5.41) to find

$$(5.42) \quad |\nu(y_r) \cdot \{x_r - y_r\} - \nu(y_q) \cdot \{x_q - y_q\}| \leq c|\hat{y} - \hat{x}|^{(1+\alpha)(1-\delta)}\|r - q\|_{1,\alpha}^{\delta}.$$

Finally, from (5.26) and Taylor's formula, we have

$$(5.43) \qquad |J_r(\hat{y}) - J_q(\hat{y})| \leq c\|r - q\|_{1,\alpha}.$$

We now choose

$$\delta = \frac{\alpha}{2(1 + \alpha)}$$

and combine the inequalities (5.33), (5.39), (5.42) and (5.43) to verify (5.34) with $\alpha^* = \min\{1 - \delta, \alpha - \delta, \alpha/2\}$. $\square$

5.4 Approximation of the Scattered Field

In this section we will describe an approximation method for the inverse obstacle scattering problem which was presented in a series of papers by Kirsch and Kress [92, 93, 94]. We confine our analysis to inverse scattering from a three-dimensional sound-soft scatterer. The method can be carried over to the two-dimensional case and also to other boundary conditions. One of its basic ideas is to break up the inverse scattering problem into two parts: the first part deals with the ill-posedness by constructing the scattered wave u^s from a knowledge of its far field pattern u_∞ and the second part deals with the nonlinearity by determining the unknown boundary ∂D of the scatterer as the location of the zeros of the total field $u^i + u^s$.

For the first part, we seek the scattered wave in the form of a surface potential. We choose an auxiliary closed C^2 surface Γ contained in the unknown scatterer D. The knowledge of such an internal surface Γ requires weak a priori information about D. Since the choice of Γ is at our disposal, without loss of generality we may assume that it is chosen such that the Helmholtz equation $\Delta u + k^2 u = 0$ in the interior of Γ with homogeneous Dirichlet boundary condition $u = 0$ on Γ admits only the trivial solution $u = 0$, i.e., k^2 is not a Dirichlet eigenvalue for the negative Laplacian in the interior of Γ. For example, we may choose Γ to be a sphere of radius R such that kR does not coincide with a zero of one of the spherical Bessel functions j_n, $n = 0, 1, 2, \ldots$.

Given the internal surface Γ, we try to represent the scattered field as an acoustic single-layer potential

$$(5.44) \qquad u^s(x) = \int_\Gamma \varphi(y)\Phi(x,y)\,ds(y)$$

with an unknown density $\varphi \in L^2(\Gamma)$. From (2.14) we see that the asymptotic behavior of this single-layer potential is given by

$$u^s(x) = \frac{1}{4\pi}\frac{e^{ik|x|}}{|x|}\int_\Gamma e^{-ik\,\hat{x}\cdot y}\varphi(y)\,ds(y) + O\left(\frac{1}{|x|^2}\right), \quad |x| \to \infty,$$

uniformly for all directions $\hat{x} = x/|x|$. Hence, given the far-field pattern u_∞, we have to solve the integral equation of the first kind

$$(5.45) \qquad F\varphi = u_\infty$$

for the density φ where the integral operator $F : L^2(\Gamma) \to L^2(\Omega)$ is defined by

$$(5.46) \qquad (F\varphi)(\hat{x}) := \frac{1}{4\pi}\int_\Gamma e^{-ik\,\hat{x}\cdot y}\varphi(y)\,ds(y), \quad \hat{x} \in \Omega.$$

Note that we use the same symbol F for a far field operator as in the previous section but now with a different meaning. The integral operator F has an analytic kernel and therefore equation (5.45) is severely ill-posed. We first establish some properties of F.

Theorem 5.13 *The far field integral operator F, defined by (5.46), is injective and has dense range provided k^2 is not a Dirichlet eigenvalue for the negative Laplacian in the interior of Γ.*

Proof. Let $F\varphi = 0$ and define the acoustic single-layer potential by (5.44). Then u^s has far field pattern $u_\infty = 0$, whence $u^s = 0$ in the exterior of Γ follows by Theorem 2.13. Analogous to (3.10), we introduce the normal derivative of the single-layer operator $K' : L^2(\Gamma) \to L^2(\Gamma)$. By the L^2 jump relation (3.20), we find that

$$\varphi - K'\varphi = 0.$$

Employing the argument used in the proof of Theorem 3.20, by the Fredholm alternative we see that the nullspaces of $I - K'$ in $L^2(\Gamma)$ and in $C(\Gamma)$ coincide. Therefore, φ is continuous and by the jump relations for continuous densities u^s represents a solution to the homogeneous Dirichlet problem in the interior of Γ. Hence, by our assumption on the choice of Γ, we have $u^s = 0$ everywhere in $\mathbb{R}^3$. The jump relations of Theorem 3.1 now yield $\varphi = 0$ on Γ, whence F is injective.

The adjoint operator $F^* : L^2(\Omega) \to L^2(\Gamma)$ of F is given by

$$(F^*g)(y) = \frac{1}{4\pi} \int_\Omega e^{ik\,\hat{x}\cdot y} g(\hat{x})\, ds(\hat{x}), \quad y \in \Gamma.$$

Let $F^*g = 0$. Then

$$v(y) := \int_\Omega e^{ik\,\hat{x}\cdot y} g(\hat{x})\, ds(\hat{x}), \quad y \in \mathbb{R}^3,$$

defines a Herglotz wave function which vanishes on Γ, i.e., it solves the homogeneous Dirichlet problem in the interior of Γ. Hence, it vanishes there by our choice of Γ and since v is analytic in $\mathbb{R}^3$ it follows that $v = 0$ everywhere. Theorem 3.15 now yields $g = 0$ on Ω, whence F^* is injective and by Theorem 4.6 the range of F is dense in $L^2(\Omega)$. $\qquad\square$

For later use we state a corresponding theorem for the acoustic single-layer operator $S : L^2(\Gamma) \to L^2(\Lambda)$ defined by

(5.47) $$(S\varphi)(x) := \int_\Gamma \varphi(y)\Phi(x,y)\, ds(y), \quad x \in \Lambda,$$

where Λ denotes a closed C^2 surface containing Γ in its interior. The proof is similar to that of Theorem 5.13 and therefore is left as an exercise for the reader.

Theorem 5.14 *The single-layer operator S, defined by (5.47), is injective and has dense range provided k^2 is not a Dirichlet eigenvalue for the negative Laplacian in the interior of Γ.*

We now know that by our choice of Γ the integral equation of the first kind (5.45) has at most one solution. Its solvability is related to the question of whether or not the scattered wave can be analytically extended as a solution to the Helmholtz equation across the boundary ∂D. Clearly, we can expect (5.45) to have a solution $\varphi \in L^2(\Omega)$ only if u_∞ is the far field of a radiating solution to the Helmholtz equation in the exterior of Γ and by Theorem 3.6 the boundary data must belong to the Sobolev space $H^1(\Gamma)$. Conversely, it can be shown that if u_∞ satisfies these conditions then (5.45) is indeed solvable. Hence, the solvability of (5.45) is related to regularity properties of the scattered field which, in general, cannot be known in advance for an unknown obstacle D.

We wish to illustrate the degree of ill-posedness of the equation (5.45) by looking at the singular values of F in the special case where Γ is the unit sphere. Here, from the Funk–Hecke formula (2.44), we deduce that the singular values of F are given by

$$\mu_n = |j_n(k)|, \quad n = 0, 1, \ldots.$$

Therefore, from the asymptotic formula (2.37) and Stirling's formula, we have the extremely rapid decay

$$\mu_n = O\left(\frac{ek}{2n}\right)^n, \quad n \to \infty,$$

indicating severe ill-posedness.

We may apply the Tikhonov regularization from Section 4.4, that is, we may solve

(5.48) $$\alpha\varphi_\alpha + F^*F\varphi_\alpha = F^*u_\infty$$

with regularization parameter $\alpha > 0$ instead of (5.45). Through the solution φ_α of (5.48) we obtain a corresponding approximation u_α^s for the scattered field by the single-layer potential (5.44) with density φ_α. However, by passing to the limit $\alpha \to 0$ in (5.48), we observe that we can expect convergence of the unique solution φ_α to the regularized equation only if the original equation (5.45) is solvable. Therefore, even if u_∞ is the exact far field pattern of a scatterer D, in general u_α^s will not converge to the exact scattered field u^s since, as mentioned above, the original equation (5.45) may not be solvable.

Given the approximation u_α^s, we can now seek the boundary of the scatterer D as the location of the zeros of $u^i + u_\alpha^s$ in a minimum norm sense, i.e., we can approximate ∂D by minimizing the defect

(5.49) $$\|u^i + u_\alpha^s\|_{L^2(\Lambda)}$$

over some suitable class U of admissible surfaces Λ. In the following, we will choose U to be a compact subset (with respect to the $C^{1,\beta}$ norm, $0 < \beta < 1$,) of the set of all starlike closed C^2 surfaces, described by

(5.50) $$x(\hat{x}) = r(\hat{x})\,\hat{x}, \quad \hat{x} \in \Omega, \quad r \in C^2(\Omega),$$

satisfying the a priori assumption

(5.51) $$0 < r_i(\hat{x}) \leq r(\hat{x}) \leq r_e(\hat{x}), \quad \hat{x} \in \Omega,$$

with given functions r_i and r_e representing surfaces Λ_i and Λ_e such that the internal auxiliary surface Γ is contained in the interior of Λ_i and the boundary ∂D of the unknown scatterer D is contained in the annulus between Λ_i and Λ_e. We recall from Section 5.3 that for a sequence of surfaces we understand convergence $\Lambda_n \to \Lambda$, $n \to \infty$, in the sense that $\|r_n - r\|_{C^{1,\beta}(\Omega)} \to 0$, $n \to \infty$, for the functions r_n and r representing Λ_n and Λ via (5.50).

For a satisfactory reformulation of the inverse scattering problem as an optimization problem, we want some convergence properties when the regularization parameter α tends to zero. Therefore, recalling the definition (5.47) of the single-layer operator S, we combine the minimization of the Tikhonov functional for (5.45) and the defect minimization (5.49) into one cost functional

$$(5.52) \qquad \mu(\varphi, \Lambda; \alpha) := \|F\varphi - u_\infty\|^2_{L^2(\Omega)} + \alpha\|\varphi\|^2_{L^2(\Gamma)} + \gamma\|u^i + S\varphi\|^2_{L^2(\Lambda)}.$$

Here, $\alpha > 0$ denotes the regularization parameter for the Tikhonov regularization of (5.45) represented by the first two terms in (5.52) and $\gamma > 0$ denotes a coupling parameter which has to be chosen appropriately for the numerical implementation in order to make the first and third term in (5.52) of the same magnitude. In the sequel, for theoretical purposes we always may assume $\gamma = 1$.

Definition 5.15 *Given the incident field u^i, a (measured) far field $u_\infty \in L^2(\Omega)$ and a regularization parameter $\alpha > 0$, a surface Λ_0 from the compact set U is called* optimal *if there exists $\varphi_0 \in L^2(\Gamma)$ such that φ_0 and Λ_0 minimize the cost functional (5.52) simultaneously over all $\varphi \in L^2(\Gamma)$ and $\Lambda \in U$, that is, we have*

$$\mu(\varphi_0, \Lambda_0; \alpha) = m(\alpha)$$

where

$$m(\alpha) := \inf_{\varphi \in L^2(\Gamma),\, \Lambda \in U} \mu(\varphi, \Lambda; \alpha).$$

For this reformulation of the inverse scattering problem into a nonlinear optimization problem, we can now state the following results. Note that in the existence Theorem 5.16 we need not assume that u_∞ is an exact far field pattern.

Theorem 5.16 *For each $\alpha > 0$ there exists an optimal surface $\Lambda \in U$.*

Proof. Let (φ_n, Λ_n) be a minimizing sequence in $L^2(\Gamma) \times U$, i.e.,

$$\lim_{n\to\infty} \mu(\varphi_n, \Lambda_n; \alpha) = m(\alpha).$$

Since U is compact, we can assume that $\Lambda_n \to \Lambda \in U$, $n \to \infty$. From

$$\alpha\|\varphi_n\|^2_{L^2(\Gamma)} \leq \mu(\varphi_n, \Lambda_n; \alpha) \to m(\alpha), \quad n \to \infty,$$

and $\alpha > 0$ we conclude that the sequence (φ_n) is bounded, i.e., $\|\varphi_n\|_{L^2(\Gamma)} \leq c$ for all n and some constant c. Hence, we can assume that it converges weakly

$\varphi_n \rightharpoonup \varphi \in L^2(\Gamma)$ as $n \to \infty$. Since $F : L^2(\Gamma) \to L^2(\Omega)$ and $S : L^2(\Gamma) \to L^2(\Lambda)$ represent compact operators, it follows that $F\varphi_n \to F\varphi$ and $S\varphi_n \to S\varphi$ as $n \to \infty$. We indicate the dependence of $S : L^2(\Gamma) \to L^2(\Lambda_n)$ on n by writing S_n. With functions r_n and r representing Λ_n and Λ via (5.50), by Taylor's formula we can estimate

$$|\Phi(r_n(\hat{x})\,\hat{x}, y) - \Phi(r(\hat{x})\,\hat{x}, y)| \leq L\,|r_n(\hat{x}) - r(\hat{x})|$$

for all $\hat{x} \in \Omega$ and all $y \in \Gamma$. Here, L denotes a bound on $\mathrm{grad}_x \Phi$ on $W \times \Gamma$ where W is the closed annular domain between the two surfaces Λ_i and Λ_e. Then, using the Schwarz inequality, we find that

$$\left| \int_\Gamma \{\Phi(r_n(\hat{x})\,\hat{x}, y) - \Phi(r(\hat{x})\,\hat{x}, y)\}\,\varphi_n(y)\,ds(y) \right| \leq cL\,|\Gamma|\,|r_n(\hat{x}) - r(\hat{x})|$$

for all $\hat{x} \in \Omega$. Therefore, from $\|S\varphi_n - S\varphi\|^2_{L^2(\Lambda)} \to 0$, $n \to \infty$, we can deduce that

$$\|u^i + S_n\varphi_n\|^2_{L^2(\Lambda_n)} \to \|u^i + S\varphi\|^2_{L^2(\Lambda)}, \quad n \to \infty.$$

This now implies

$$\alpha\|\varphi_n\|^2_{L^2(\Gamma)} \to m(\alpha) - \|F\varphi - u_\infty\|^2_{L^2(\Omega)} - \|u^i + S\varphi\|^2_{L^2(\Lambda)} \leq \alpha\|\varphi\|^2_{L^2(\Gamma)}$$

for $n \to \infty$. Since we already know weak convergence $\varphi_n \rightharpoonup \varphi$, $n \to \infty$, it follows that

$$\lim_{n \to \infty} \|\varphi_n - \varphi\|^2_{L^2(\Gamma)} = \lim_{n \to \infty} \|\varphi_n\|^2_{L^2(\Gamma)} - \|\varphi\|^2_{L^2(\Gamma)} \leq 0,$$

i.e., we also have norm convergence $\varphi_n \to \varphi$, $n \to \infty$. Finally, by continuity

$$\mu(\varphi, \Lambda; \alpha) = \lim_{n \to \infty} \mu(\varphi_n, \Lambda_n; \alpha) = m(\alpha),$$

and this completes the proof. $\qquad\square$

Theorem 5.17 *Let u_∞ be the exact far field pattern of a domain D such that ∂D belongs to U. Then we have convergence of the cost functional*

$$\text{(5.53)} \qquad\qquad \lim_{\alpha \to 0} m(\alpha) = 0.$$

Proof. By Theorem 5.14, given $\varepsilon > 0$ there exists $\varphi \in L^2(\Gamma)$ such that

$$\|S\varphi + u^i\|_{L^2(\partial D)} < \varepsilon.$$

Since by Theorem 5.9 and the far field representation (2.13) the far field pattern of a radiating solution of the Helmholtz equation depends continuously on the boundary data, we can estimate

$$\|F\varphi - u_\infty\|_{L^2(\Omega)} \leq c\|S\varphi - u^s\|_{L^2(\partial D)}$$

for some constant c. From $u^i + u^s = 0$ on ∂D we then deduce that

$$\mu(\varphi, \partial D; \alpha) \le (1 + c^2)\varepsilon^2 + \alpha \|\varphi\|^2_{L^2(\Gamma)} \to (1 + c^2)\varepsilon^2, \quad \alpha \to 0.$$

Since ε is arbitrary, (5.53) follows.   □

Based on Theorem 5.17, we can state the following convergence result.

Theorem 5.18 *Let* (α_n) *be a null sequence and let* (Λ_n) *be a corresponding sequence of optimal surfaces for the regularization parameter* α_n. *Then there exists a convergent subsequence of* (Λ_n). *Assume that* u_∞ *is the exact far field pattern of a domain* D *such that* ∂D *is contained in* U. *Then every limit point* Λ^* *of* (Λ_n) *represents a surface on which the total field vanishes.*

Proof. The existence of a convergent subsequence of (Λ_n) follows from the compactness of U. Let Λ^* be a limit point. Without loss of generality, we can assume that $\Lambda_n \to \Lambda^*$, $n \to \infty$. Let u^* denote the solution to the direct scattering problem with incident wave u^i for the obstacle with boundary Λ^*, i.e., the boundary condition reads

$$(5.54) \qquad\qquad u^* + u^i = 0 \quad \text{on } \Lambda^*.$$

Since Λ_n is optimal for the parameter α_n, there exists $\varphi_n \in L^2(\Gamma)$ such that

$$\mu(\varphi_n, \Lambda_n; \alpha_n) = m(\alpha_n)$$

for $n = 1, 2, \ldots$. We denote by u_n the single-layer potential with density φ_n and interprete u_n as the solution to the exterior Dirichlet problem with boundary values $S_n\varphi_n$ on the boundary Λ_n. By Theorem 5.17, these boundary data satisfy

$$(5.55) \qquad\qquad \|u_n + u^i\|_{L^2(\Lambda_n)} \to 0, \quad n \to \infty.$$

By Theorem 5.9, from (5.54) and (5.55) we now deduce that the far field patterns $F\varphi_n$ of u_n converge in $L^2(\Omega)$ to the far field pattern u^*_∞ of u^*. By Theorem 5.17, we also have $\|F\varphi_n - u_\infty\|_{L^2(\Omega)} \to 0$, $n \to \infty$. Therefore, we conclude $u_\infty = u^*_\infty$, whence $u^s = u^*$ follows. Because of (5.54), the total field $u^s + u^i$ must vanish on Λ^*.   □

Since we do not have uniqueness either for the inverse scattering problem or for the optimization problem, in the above convergence analysis we cannot expect more than convergent subsequences. In addition, due to the lack of a uniqueness result for one wave number and one incident plane wave, we cannot assume that we always have convergence to the boundary of the unknown scatterer. However, if we have the a priori information that the diameter of the unknown obstacle is less than $2\pi/k$, then by Corollary 5.3 we can sharpen the result of Theorem 5.18 and, by a standard argument, we have convergence of the total sequence to the boundary of the unknown scatterer.

Before we describe some of the details of the numerical implementation of this method, we wish to mention some modifications and extensions.

We can try to achieve more accurate reconstructions by using more incident waves $u_1^i, \ldots, u_n^i$ with different directions $d_1, \ldots, d_n$ and corresponding far field patterns $u_{\infty,1}, \ldots, u_{\infty,n}$. Then we have to minimize the sum

$$(5.56) \qquad \sum_{j=1}^{n} \left\{ \|F\varphi_j - u_{\infty,j}\|_{L^2(\Omega)}^2 + \alpha\|\varphi_j\|_{L^2(\Gamma)}^2 + \gamma\|u_j^i + S\varphi_j\|_{L^2(\Lambda)}^2 \right\}$$

over all $\varphi_1, \ldots, \varphi_n \in L^2(\Gamma)$ and all $\Lambda \in U$. Obviously, the results of the three preceding theorems carry over to the minimization of (5.56). Of course, the use of more than one wave will lead to a tremendous increase in the computational costs. These costs can be reduced by using suitable linear combinations of incident plane waves as suggested by Zinn [197].

In addition to the reconstruction of the scatterer D from far field data, we also can consider the reconstruction from *near field* data, i.e., from measurements of the scattered wave u^s on some closed surface Γ_{meas} containing D in its interior. By the uniqueness for the exterior Dirichlet problem, knowing u^s on the closed surface Γ_{meas} implies knowing the far field pattern u_∞ of u^s. Therefore, the uniqueness results for the reconstruction from far field data immediately carry over to the case of near field data. In particular, since the measurement surface Γ_{meas} trivially provides a priori information on the size of D, a finite number of incident plane waves will always uniquely determine D. In the case of near field measurements, the integral equation (5.45) has to be replaced by

$$(5.57) \qquad\qquad S\varphi = u_{\text{meas}}^s$$

where we recall that the integral operator $S : L^2(\Gamma) \to L^2(\Gamma_{\text{meas}})$ is given by

$$(5.58) \qquad (S\varphi)(x) := \int_\Gamma \varphi(y)\Phi(x,y)\,ds(y), \quad x \in \Gamma_{\text{meas}}.$$

Correspondingly, for given u^i and u_{meas}^s, the optimization problem has to be modified into minimizing the sum

$$(5.59) \qquad \|S\varphi - u_{\text{meas}}^s\|_{L^2(\Gamma_{\text{meas}})}^2 + \alpha\|\varphi\|_{L^2(\Gamma)}^2 + \gamma\|u^i + S\varphi\|_{L^2(\Lambda)}^2$$

simultaneously over all $\varphi \in L^2(\Gamma)$ and $\Lambda \in U$. The results of Theorems 5.16–5.18 again carry over to the near field case.

The integral operator S has an analytic kernel and therefore equation (5.57) is ill-posed. In the special case where Γ is the unit sphere and where Γ_{meas} is a concentric sphere with radius R, from the addition theorem (2.42) we deduce that the singular values of S are given by

$$\mu_n = kR|j_n(k)| \left|h_n^{(1)}(kR)\right|, \quad n = 0, 1, \ldots.$$

Therefore, from the asymptotic formulas (2.37) and (2.38) we have

$$\mu_n = O\left(\frac{R^{-n}}{2n+1}\right), \quad n \to \infty,$$

indicating an ill-posedness which is slightly less severe than the ill-posedness of the corresponding far field case. This reduction in the degree of ill-posedness is also illustrated through stability estimates that were recently obtained by Isakov [73]. However, numerical experiments (see [110, 111]) have shown that unfortunately this does not lead to a highly noticable increase in the accuracy of the reconstructions.

So far we have assumed the far field to be known for all observation directions $\hat{x}$. Due to analyticity, for uniqueness it suffices to know the far field pattern on a subset $\tilde{\Omega} \subset \Omega$ with a nonempty interior. Zinn [197] has shown that after modifying the far field integral operator F given by (5.46) into an operator from $L^2(\Gamma)$ into $L^2(\tilde{\Omega})$ and replacing Ω by $\tilde{\Omega}$ in the Tikhonov part of the cost functional μ given by (5.52) the results of Theorems 5.16–5.18 remain valid. However, as one would expect, the quality of the reconstructions decreases drastically for this so called *limited-aperture problem*. For two-dimensional problems the numerical experiments in [197] indicate that satisfactory reconstructions need an aperture not smaller than 180 degrees and more than one incident wave.

It is obvious that the above method also can be used for other boundary conditions, for example sound-soft and penetrable scatterers. The case of the transmission boundary condition has been studied by Zinn [198, 199].

Finally, we also want to mention that, in principle, we may replace the approximation of the scattered field u^s through a single-layer potential by any other convenient approximation. For example, Angell, Kleinman and Roach [8] have suggested using an expansion with respect to radiating spherical wave functions. Numerical implementations of this approach in two dimensions are given in [6, 7, 80, 81]. Using a single-layer potential approximation on an auxiliary internal surface Γ has the advantage of allowing the incorporation of a priori information on the unknown scatterer by a suitable choice of Γ. Furthermore, by the addition theorem (2.42), from a theoretical point of view we may consider the spherical wave function approach as a special case of the single-layer potential with Γ a sphere. We also want to mention that the above methods may be considered as having some of their roots in the work of Imbriale and Mittra [71] who described the first reconstruction algorithm in inverse obstacle scattering for frequencies in the resonance region.

We conclude this section by describing some details of the numerical implementation. For the numerical solution, we must of course discretize the optimization problem. This is achieved through replacing $L^2(\Gamma)$ and U by finite dimensional subspaces. Denote by (X_n) a sequence of finite dimensional subspaces $X_{n-1} \subset X_n \subset L^2(\Gamma)$ such that $\bigcup_{n=1}^{\infty} X_n$ is dense in $L^2(\Gamma)$. Similarly, let (U_n) be a sequence of finite dimensional subsets $U_{n-1} \subset U_n \subset U$ such that $\bigcup_{n=1}^{\infty} U_n$ is dense in U. We then replace the optimization problem of Definition 5.15 by the finite dimensional problem where we minimize over the finite dimensional set $X_n \times U_n$ instead of $L^2(\Gamma) \times U$.

We call a surface Λ_n from the set U_n optimal with respect to $X_n \times U_n$ if there exists $\varphi_n \in X_n$ such that φ_n and Λ_n minimize the cost functional (5.52) over all $\varphi \in X_n$ and $\Lambda \in U_n$. Clearly, analogous to Theorem 5.16, for each n

there exists a surface Λ_n which is optimal with respect to $X_n \times U_n$. Denoting by

$$m_n(\alpha) := \inf_{\varphi \in X_n, \, \Lambda \in U_n} \mu(\varphi, \Lambda; \alpha)$$

the minimum with respect to the finite dimensional subset, we can establish the following convergence result due to Zinn [198].

Theorem 5.19 *For fixed $\alpha > 0$ we have $m_n(\alpha) \to m(\alpha)$, $n \to \infty$. Let Λ_n be optimal with respect to $X_n \times U_n$. Then there exists a subsequence which converges to a surface Λ which is optimal with respect to $L^2(\Gamma) \times U$.*

Proof. Since we keep α fixed, we suppress the dependence of the cost functional on α. Let $\varphi^* \in L^2(\Gamma)$ and $\Lambda^* \in U$ satisfy

$$\mu(\varphi^*, \Lambda^*) = m(\alpha)$$

and let $\varphi_n \in X_n$ and $\Lambda_n \in U_n$ satisfy

$$\mu(\varphi_n, \Lambda_n) = m_n(\alpha).$$

Denote by φ_n^* the best approximation to φ^* with respect to X_n in the L^2 norm and by Λ_n^* the best approximation to Λ^* with respect to U_n in the norm on U (representing surfaces by (5.50)). Due to our denseness assumptions on X_n and U_n, we have convergence $\varphi_n^* \to \varphi^*$ and $\Lambda_n^* \to \Lambda^*$ as $n \to \infty$. This implies that

$$\lim_{n \to \infty} \mu(\varphi_n^*, \Lambda_n^*) = \mu(\varphi^*, \Lambda^*) = m(\alpha)$$

and consequently, since trivially we have

$$m(\alpha) \le m_n(\alpha) \le \mu(\varphi_n^*, \Lambda_n^*),$$

we obtain convergence $m_n(\alpha) \to m(\alpha)$, $n \to \infty$. This now implies that

$$\lim_{n \to \infty} \mu(\varphi_n, \Lambda_n) = m(\alpha),$$

that is, (φ_n, Λ_n) is a minimizing sequence in $L^2(\Omega) \times U$. The proof of the existence Theorem 5.16 now tells us that we can select a subsequence which converges to an optimal surface Λ. □

The finite dimensional optimization problem is now a nonlinear least squares problem with $\dim X_n + \dim U_n$ unknowns. For its numerical solution, we suggest using a Levenberg–Marquardt algorithm [131] as one of the most efficient nonlinear least squares routines. It does not allow the imposition of constraints but we found in practice that the constraints are unnecessary due to the increase in the cost functional as Λ approaches Γ or tends to infinity.

The numerical evaluation of the cost functional (5.52), including the integral operators F and S, in general requires the numerical evaluation of integrals

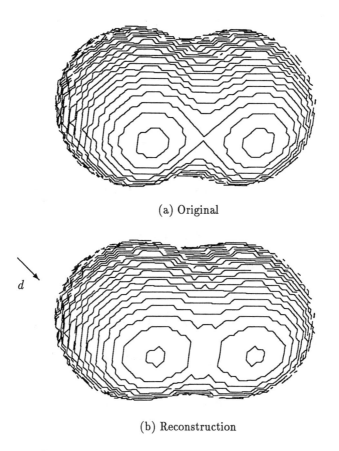

(a) Original

(b) Reconstruction

Fig. 5.1. The peanut and its reconstruction (turned by 90°)

with analytic integrands over analytic surfaces Ω, Γ and Λ. We numerically approximate integrals over the unit sphere by the Gauss trapezoidal product rule (3.90) described in Section 3.6. The integrals over Γ and Λ we transform into integrals over Ω through appropriate substitutions and then again apply (3.90).

Canonical subspaces for the finite dimensional optimization problems are given in terms of spherical harmonics as follows. Denote by Z_n the linear space of all spherical harmonics of order less than or equal to n. Let $q : \Gamma \to \Omega$ be bijective and choose $X_n \subset L^2(\Gamma)$ by

$$X_n := \{\varphi = Y \circ q : Y \in Z_n\}.$$

Choose U_n to be the set of all starlike surfaces described through (5.50) and (5.51) with $r \in Z_n$. Then, by Theorem 2.6, the degree of freedom in the optimization problem is $2(n + 1)^2$.

Fig. 5.1 shows as an example of the method described above the reconstruction of a peanut from the scattering of one incident plane wave with wave number $k = 1$. The peanut is described through its radial distance in terms of the polar angle θ by

$$r(\theta) = \frac{3}{2} \left(\cos^2 \theta + \frac{1}{4} \sin^2 \theta \right)^{1/2}$$

Our synthetic data were obtained through Wienert's Nyström type method from Section 3.6. For the inverse solver, the regularization parameter α and the coupling parameter γ were selected by trial and error. The actual numerical values were $\alpha = 10^{-8}$ and $\gamma = 10^{-6}$. The internal surface Γ was choosen to be an ellipsoid with centre at the origin and major axis 0.6, 0.6 and 1.2. The starting surface Λ for the Levenberg–Marquardt algorithm was a parallel surface with distance 0.2 from Γ, and the starting density was $\varphi = 0$. The arrow in the figure marks the direction of the incident wave. The parameter N for the Gauss trapezoidal product rule (3.90) was $N = 12$ and the dimension of the approximating subspaces was choosen by setting $n = 6$. The reconstruction shown in Fig. 5.1 is based on exact data. However, the examples in [111] illustrate stability of the method for inexact data where random noise is added to the exact far fields.

For further numerical reconstructions, we refer to [96] for two-dimensionsal and to [109, 111] for three-dimensional examples including shapes which are not rotationally symmetric.

5.5 Superposition of the Incident Fields

In the last section of this chapter on inverse acoustic obstacle scattering, we will consider an approximation method which was developed by Colton and Monk [36, 37, 38]. We will present this method in a manner which stresses its close connection to the method of Kirsch and Kress of the previous section. Its analysis is related to the completeness properties for far field patterns of Section 3.3. We again confine our presentation to inverse scattering from a three-dimensional sound-soft scatterer and note that the method can be extended to two dimensions and to other boundary conditions.

As in the method of Kirsch and Kress, the motivation for the method developed by Colton and Monk can also be split into two parts. Whereas in the previous section we first tried to approximate the scattered wave from a knowledge of its far field pattern, here we now look for superpositions of incident fields with different directions which lead to simple far field patterns, for example to far fields belonging to radiating spherical wave functions. To be more precise, we consider as incident wave v^i a superposition of plane waves of the form

(5.60) $$v^i(x) = \int_\Omega e^{ik\,x\cdot d} g(d)\, ds(d), \quad x \in \mathbb{R}^3,$$

with weight function $g \in L^2(\Omega)$, i.e., the incident wave is a Herglotz wave function. By Lemma 3.16, the corresponding far field pattern

$$v_\infty(\hat{x}) = \int_\Omega u_\infty(\hat{x}; d)g(d) \, ds(d), \quad \hat{x} \in \Omega,$$

is obtained by superposing the far field patterns $u_\infty(\cdot \, ; d)$ for the incoming directions d. We note that by the Reciprocity Theorem 3.13 we may also consider (5.60) as a superposition with respect to the observation directions instead of the incident directions. Hence, if we want the scattered wave to become a prescribed radiating solution v^s to the Helmholtz equation with far field pattern v_∞, given the far field patterns $u_\infty(\cdot \, ; d)$ for all incident directions d we need to solve the integral equation of the first kind

(5.61) $Fg = v_\infty$

where the integral operator $F : L^2(\Omega) \to L^2(\Omega)$ now is defined by

(5.62) $(Fg)(\hat{x}) := \int_\Omega u_\infty(\hat{x}; d)g(d) \, ds(d), \quad \hat{x} \in \Omega.$

For example, we may choose a radiating spherical wave function

$$v^s(x) = h_p^{(1)}(k|x|) \, Y_p(\hat{x})$$

of order p with far field pattern

$$v_\infty = \frac{1}{ki^{p+1}} \, Y_p.$$

Once we have constructed the incident field v^i by (5.60) and the solution of (5.61), in the second step we determine the boundary as the location of the zeros of the total field $v^i + v^s$. However, we first proceed with an investigation of the integral operator (5.62).

In this section, we have chosen to call the method developed by Colton and Monk the method of *superposition of incident fields*. On the other hand, if we fix d and superpose with respect to the observation directions, we can view this method as one of determining a linear functional having prescribed values on the set $\mathcal{F}$ of far field patterns. Viewed in this way the method of Colton and Monk is sometimes referred to as a *dual space method* and, for the sake of diversity, we shall use this nomenclature in our later discussion of the inverse scattering problem for an inhomogeneous medium.

The far field operator F again has an analytic kernel and therefore equation (5.61) is ill-posed. We have already investigated the operator F earlier in this book. In particular, by Corollary 3.18, we know that F is injective and has dense range if and only if there does not exist a Dirichlet eigenfunction for D which is a Herglotz wave function. Therefore, for the sequel we will make the restricting assumption that k^2 is not a Dirichlet eigenvalue for the negative Laplacian in the unknown scatterer D. This in particular implies that the interior Dirichlet problem for the Helmholtz equation in D is uniquely solvable.

Now we assume that $\mathbb{R}^3 \setminus D$ is contained in the domain of definition for v^s. In particular, if we choose radiating spherical waves for v^s this means that the origin is contained in D. We associate the following uniquely solvable interior Dirichlet problem

$$(5.63) \qquad \triangle v^i + k^2 v^i = 0 \quad \text{in } D,$$

$$(5.64) \qquad v^i + v^s = 0 \quad \text{on } \partial D$$

to the inverse scattering problem. From Theorem 3.19 we know that the solvability of the integral equation (5.61) is connected to this interior boundary value problem, i.e., (5.61) is solvable for $g \in L^2(\Omega)$ if and only if the solution v^i to (5.63), (5.64) is a Herglotz wave function with kernel g. Therefore, the solvability of (5.61) depends on the question of whether or not the solution to the interior Dirichlet problem (5.63), (5.64) can be analytically extended as a Herglotz wave function across the boundary ∂D and this question again cannot be known in advance for an unknown obstacle D.

We again illustrate the degree of ill-posedness of the equation (5.61) by looking at the singular values in the special case where the scatterer D is the unit ball. Here, from the explicit form (3.30) of the far field pattern and the addition theorem (2.29), we find that the singular values of F are given by

$$\mu_n = \frac{4\pi}{k} \frac{|j_n(k)|}{|h_n^{(1)}(k)|}, \quad n = 0, 1, \ldots,$$

with the asymptotic behavior

$$\mu_n = O\left(\frac{ek}{2n}\right)^{2n}, \quad n \to \infty.$$

This estimate is the square of the corresponding estimate for the singular values of the far field integral operator in the previous section (c.f. p. 129).

We introduce a Herglotz integral operator $H : L^2(\Omega) \to L^2(\Lambda)$ as the restriction of the Herglotz function with kernel g to the surface Λ by

$$(5.65) \qquad (Hg)(x) := \int_\Omega e^{ik\, x \cdot d} g(d)\, ds(d), \quad x \in \Lambda.$$

Theorem 5.20 *The Herglotz integral operator H, defined by (5.65), is injective and has dense range provided k^2 is not a Dirichlet eigenvalue for the negative Laplacian in the interior of Λ.*

Proof. The operator H is the adjoint of the far field integral operator given by (5.46). Therefore, the statement of the theorem is equivalent to Theorem 5.13. $\square$

We can again use Tikhonov regularization with regularization parameter α to obtain an approximate solution g_α of (5.61). This then leads to an approximation v_α^i for the incident wave v^i and we can try to find the boundary of the scatterer D as the set of points where the boundary condition (5.64) is satisfied. We do this by requiring that

(5.66) $v_\alpha^i + v^s = 0$

is satisfied in the minimum norm sense. However, as in the previous section, for a satisfactory reformulation of the inverse scattering problem as an optimization problem we need to combine a regularization for the integral equation (5.61) and the defect minimization for (5.66) into one cost functional. If we use the standard Tikhonov regularization as in Definition 5.15, i.e., if we use as penalty term $\|g\|_{L^2(\Omega)}^2$, then it is easy to prove results analogous to those of Theorems 5.16 and 5.17. However, we would not be able to obtain a convergence result corresponding to Theorem 5.18. Therefore, we follow Blöhbaum [14] and choose a penalty term for (5.61) as follows. We recall the description of the set U of admissable surfaces from the previous section (c.f. p. 129) and pick a closed C^2 surface Γ_e such that Λ_e is contained in the interior of Γ_e. In addition, without loss of generality, we assume that k^2 is not a Dirichlet eigenvalue for the negative Laplacian in the interior of Γ_e. Then we define the combined cost functional by

(5.67) $\mu(g, \Lambda; \alpha) := \|Fg - v_\infty\|_{L^2(\Omega)}^2 + \alpha\|Hg\|_{L^2(\Gamma_e)}^2 + \gamma\|Hg + v^s\|_{L^2(\Lambda)}^2.$

The coupling parameter γ is again necessary for numerical purposes and for the theory we set $\gamma = 1$. Since the operator H in the penalty term does not have a bounded inverse, we have to slightly modify the notion of an optimal surface.

Definition 5.21 *Given the (measured) far field $u_\infty \in L^2(\Omega \times \Omega)$ for all incident and observation directions and a regularization parameter $\alpha > 0$, a surface Λ_0 from the compact set U is called* optimal *if*

$$\inf_{g \in L^2(\Omega)} \mu(g, \Lambda_0; \alpha) = m(\alpha)$$

where

$$m(\alpha) := \inf_{g \in L^2(\Omega), \Lambda \in U} \mu(g, \Lambda; \alpha).$$

Note that the measured far field u_∞ enters in the operator F through (5.61). For this reformulation of the inverse scattering problem into a nonlinear optimization problem, we have results similar to those of the previous section. We note that in the original version of their method, Colton and Monk [36, 37] chose the cost functional (5.67) without the penalty term and minimized over all $g \in L^2(\Omega)$ with $\|g\|_{L^2(\Omega)} \leq \rho$, i.e., the Tikhonov regularization with regularization parameter $\alpha \to 0$ was replaced by the quasi-solution with regularization parameter $\rho \to \infty$.

Theorem 5.22 *For each $\alpha > 0$ there exists an optimal surface $\Lambda \in U$.*

Proof. Let (g_n, Λ_n) be a minimizing sequence from $L^2(\Omega) \times U$, i.e.,

$$\lim_{n \to \infty} \mu(g_n, \Lambda_n; \alpha) = m(\alpha).$$

Since U is compact, we can assume that $\Lambda_n \to \Lambda \in U$, $n \to \infty$. Because of the boundedness

$$\alpha\|Hg_n\|^2_{L^2(\Gamma_e)} \leq \mu(g_n, \Lambda_n; \alpha) \to m(\alpha), \quad n \to \infty,$$

we can assume that the sequence (Hg_n) is weakly convergent in $L^2(\Gamma_e)$. By Theorem 5.4, applied to the interior of Γ'_e, the weak convergence of the boundary data Hg_n on Γ_e then implies that the Herglotz wave functions v_n with kernel g_n converge to a solution v of the Helmholtz equation uniformly on compact subsets of the interior of Γ_e. This, together with $\Lambda_n \to \Lambda \in U$, $n \to \infty$, implies that (indicating the dependence of $H : L^2(\Omega) \to L^2(\Lambda_n)$ on n by writing H_n)

$$\lim_{n\to\infty} \|H_n g_n + v^s\|_{L^2(\Lambda_n)} = \lim_{n\to\infty} \|Hg_n + v^s\|_{L^2(\Lambda)},$$

whence

$$\lim_{n\to\infty} \mu(g_n, \Lambda_n; \alpha) = \lim_{n\to\infty} \mu(g_n, \Lambda; \alpha)$$

follows. This concludes the proof.   □

For the following, we assume that u_∞ is the exact far field pattern and first give a reformulation of the integral equation (5.61) for the Herglotz kernel g in terms of the Herglotz function Hg. For this purpose, we recall the operator $A : L^2(\partial D) \to L^2(\Omega)$ from Theorem 3.21 which maps the boundary values of radiating solutions onto their far field pattern and recall that A is bounded and injective. Clearly, by the definition of A, we have $Av^s = v_\infty$ and, by Lemma 3.16, we conclude that $AHg = -Fg$ for exact far field data u_∞. Hence

$$(5.68) \qquad\qquad v_\infty - Fg = A(Hg + v^s).$$

Theorem 5.23 *For all incident directions d, let $u_\infty(\cdot\,; d)$ be the exact far field pattern of a domain D such that ∂D belongs to U. Then we have convergence of the cost functional*

$$\lim_{\alpha\to 0} m(\alpha) = 0.$$

Proof. By Theorem 5.20, given $\varepsilon > 0$ there exists $g \in L^2(\Omega)$ such that

$$\|Hg + v^s\|_{L^2(\partial D)} < \varepsilon.$$

From (5.68) we have

$$\|Fg - v_\infty\|_{L^2(\Omega)} \leq \|A\|\,\|Hg + v^s\|_{L^2(\partial D)}.$$

Therefore, we have

$$\mu(g, \partial D; \alpha) \leq (1 + \|A\|^2)\varepsilon^2 + \alpha\|Hg\|^2_{L^2(\Gamma_e)} \to (1 + \|A\|^2)\varepsilon^2, \quad \alpha \to 0,$$

and the proof is completed as in Theorem 5.17.   □

Theorem 5.24 *Let (α_n) be a null sequence and let (Λ_n) be a corresponding sequence of optimal surfaces for the regularization parameter α_n. Then there exists a convergent subsequence of (Λ_n). Assume that for all incident directions $u_\infty(\cdot\,; d)$ is the exact far field pattern of a domain D such that ∂D belongs to U. Assume further that the solution v^i to the associated interior Dirichlet problem (5.63), (5.64) can be extended as a solution to the Helmholtz equation across the boundary ∂D into the interior of Γ_e with continuous boundary values on Γ_e. Then every limit point Λ^* of (Λ_n) represents a surface on which the boundary condition (5.64) is satisfied, i.e., $v^i + v^s = 0$ on Λ^*.*

Proof. The existence of a convergent subsequence of (Λ_n) follows from the compactness of U. Let Λ^* be a limit point. Without loss of generality we can assume that $\Lambda_n \to \Lambda^*$, $n \to \infty$.

By Theorem 5.20, there exists a sequence (g_j) in $L^2(\Omega)$ such that

$$\|Hg_j - v^i\|_{L^2(\Gamma_e)} \to 0, \quad j \to \infty.$$

By Theorem 5.4, this implies the uniform convergence of the Herglotz wave functions with kernel g_j to v^i on compact subsets of the interior of Γ_e, whence in view of the boundary condition for v^i on ∂D we obtain

$$\|Hg_j + v^s\|_{L^2(\partial D)} \to 0, \quad j \to \infty.$$

Therefore, by passing to the limit $j \to \infty$ in

$$m(\alpha) \le \|Fg_j - v_\infty\|_{L^2(\Omega)}^2 + \alpha\|Hg_j\|_{L^2(\Gamma_e)}^2 + \|Hg_j + v^s\|_{L^2(\partial D)}^2,$$

with the aid of (5.68) we find that

(5.69) $$m(\alpha) \le \alpha\|v^i\|_{L^2(\Gamma_e)}^2$$

for all $\alpha > 0$.

By Theorem 5.22, for each n there exists $g_n \in L^2(\Omega)$ such that

$$\|Fg_n - v_\infty\|_{L^2(\Omega)}^2 + \alpha_n\|Hg_n\|_{L^2(\Gamma_e)}^2 + \|Hg_n + v^s\|_{L^2(\Lambda_n)}^2 \le m(\alpha_n) + \alpha_n^2.$$

From this inequality and (5.69), we conclude that

$$\|Hg_n\|_{L^2(\Gamma_e)}^2 \le \|v^i\|_{L^2(\Gamma_e)}^2 + \alpha_n$$

for all n and therefore we may assume that the sequence (Hg_n) converges weakly in $L^2(\Gamma_e)$. Then, by Theorem 5.4, the Herglotz wave functions v_n with kernels g_n converge uniformly on compact subsets of the interior of Γ_e to a solution v^* of the Helmholtz equation. By Theorem 5.23, we have convergence of the cost functional $m(\alpha_n) \to 0$, $n \to \infty$. In particular, using (5.68), this yields

$$\|A(v_n + v^s)\|_{L^2(\Omega)}^2 = \|Fg_n - v_\infty\|_{L^2(\Omega)}^2 \le m(\alpha_n) + \alpha_n^2 \to 0, \quad n \to \infty,$$

whence $A(v^* + v^s) = 0$ follows. Since, by Theorem 3.21, the operator A is injective we conclude that $v^* + v^s = 0$ on ∂D, that is, v^i and v^* satisfy the same boundary condition on ∂D. Since k^2 is assumed not to be a Dirichlet eigenvalue for D, v^i and v^* must coincide. Finally, from

$$\|v_n + v^s\|^2_{L^2(\Lambda_n)} \le m(\alpha_n) + \alpha^2_n \to 0, \quad n \to \infty,$$

we see that $v^i + v^s = 0$ on Λ^* and the proof is finished.   □

Since we do not in general have uniqueness for the optimization problem, in the above convergence analysis we cannot expect more than convergent subsequences. From simple examples (see [37]), it can be seen that the function $v^i + v^s$ has surfaces other than ∂D on which the boundary condition $v^i + v^s = 0$ is satisfied. Therefore, despite the fact that an infinite number of incident plane waves by Theorem 5.1 uniquely determines the scatterer, for the scheme of Colton and Monk we cannot in general assume that we always have convergence to the boundary of the unknown scatterer. However, if we have the a priori information that the diameter of the unknown obstacle is less than $2\pi/k$, then by arguments similar to those used in Theorem 5.2 we can sharpen the result of Theorem 5.24 and have convergence of the total sequence to the boundary of the unknown scatterer. A convergence result for the general situation without the restricting assumption that v^i can be extended across ∂D has not yet been proven.

Without going into any detail, we wish to mention that for the method of Colton and Monk we also can consider the reconstruction from near field data as described in the previous section for the method of Kirsch and Kress. The extension of the Colton and Monk method to the limited-aperture problem was studied by Ochs [149] and the extension to penetrable scatterers, that is, to transmission boundary conditions is described in [39].

Since the structure of the cost functional (5.67) is similar to that of (5.52), the numerical implementation described at the end of the previous section can be carried over. However, following [38], we wish to indicate how the implementation can be executed differently. In particular, here the two steps of the method are done separately. The integral equation (5.61) is regularized by using the singular value cut-off described in Theorem 4.10. Then, proceeding analogously to the implementation of the method of Kirsch and Kress from the previous section, the boundary condition (5.64) is satisfied in the least squares sense with respect to a finite dimensional subset of starlike surfaces with the cost functional being evaluated by numerical integration. The resulting finite dimensional least squares problem is solved by a quasi Newton method.

Fig. 5.2 shows the reconstruction of an acorn given through its radial distance in terms of the polar angle θ by

$$r(\theta) = \frac{3}{5}\left(\frac{17}{4} + 2\cos 3\theta\right)^{1/2}.$$

The synthetic data were again obtained through Wienert's method from Section 3.6. We used 72 incident plane waves with wave number $k = 3$. For the finite

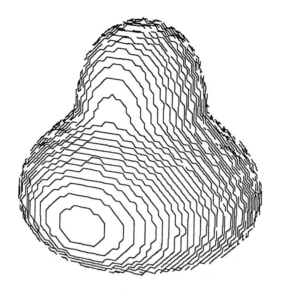

(a) Original

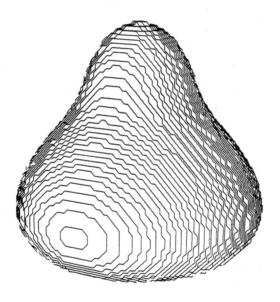

(b) Reconstruction

Fig. 5.2. The acorn and its reconstruction

dimensional subspace used to approximate the unknown boundary, we chose spherical harmonics up to order $n = 4$. Further details on the implementation can be found in [38].

In this example, for the prescribed scattered wave the simplest case of a constant far field pattern $v_\infty = 1$ was used. However, we note that the method of Colton and Monk can also be used with prescribed far fields consisting of higher spherical harmonics and linear combinations thereof. Indeed, it is also possible to combine the method of this section with the method of the previous section by choosing for v^s a single-layer potential on some internal surface within D.

For additional numerical examples we refer to [36, 37, 38]. A comparison of the numerical performance for the methods of Kirsch and Kress and of Colton and Monk in two dimensions is contained in [96]. The method of Colton and Monk has also been used for numerical reconstructions by Ari and Firth [9] and Misici and Zirilli [130].

6. The Maxwell Equations

Up until now, we have considered only the direct and inverse obstacle scattering problem for time-harmonic acoustic waves. In the following two chapters, we want to extend these results to obstacle scattering for time-harmonic electromagnetic waves. As in our analysis on acoustic scattering, we begin with an outline of the solution of the direct problem.

After a brief discussion of the physical background of electromagnetic wave propagation, we derive the Stratton–Chu representation theorems for solutions to the Maxwell equations in a homogeneous medium. We then introduce the Silver–Müller radiation condition, show its connection with the Sommerfeld radiation condition and introduce the electric and magnetic far field patterns. The next section then extends the jump relations and regularity properties of surface potentials from the acoustic to the electromagnetic case. For their appropriate presentation, we find it useful to introduce a weak formulation of the notion of a surface divergence of tangential vector fields.

We then proceed to solve the electromagnetic scattering problem for the perfect conductor boundary condition. Our approach differs from the treatment of the Dirichlet problem in acoustic scattering since we start with a formulation requiring Hölder continuous boundary regularity for both the electric and the magnetic field. We then obtain a solution under the weaker regularity assumption of continuity of the electric field up to the boundary. We have chosen this approach since there is no elementary proof available for uniqueness under the weaker regularity assumption.

For orthonormal expansions of radiating electromagnetic fields and their far field patterns, we need to introduce vector spherical harmonics and vector spherical wave functions as the analogues of the spherical harmonics and spherical wave functions. Here again, we deviate from the route taken for acoustic waves. In particular, in order to avoid lengthy manipulations with special functions, we use the results on the well-posedness of the direct obstacle scattering problem to justify the convergence of the expansions with respect to vector spherical wave functions.

The last section of this chapter presents the reciprocity relation for electromagnetic waves and completeness results for the far field patterns corresponding to the scattering of electromagnetic plane waves with different incident directions and polarizations. For this we need to extend the notion of Herglotz wave functions for acoustic waves to the case of electromagnetic waves.

For the Maxwell equations, we only need to be concerned with the study of three-dimensional problems since the two-dimensional case can be reduced to the two-dimensional Helmholtz equation. In order to numerically solve the boundary value problem for a three-dimensional perfect conductor we suggest using the obvious extension of Wienert's Nyström method described for the three-dimensional Helmholtz equation in Section 3.6.

6.1 Electromagnetic Waves

Consider electromagnetic wave propagation in an isotropic medium in $\mathbb{R}^3$ with space independent electric permittivity ε, magnetic permeability μ and electric conductivity σ. The electromagnetic wave is described by the electric field $\mathcal{E}$ and the magnetic field $\mathcal{H}$ satisfying the *Maxwell equations*

$$\operatorname{curl}\mathcal{E} + \mu\,\frac{\partial \mathcal{H}}{\partial t} = 0, \quad \operatorname{curl}\mathcal{H} - \varepsilon\,\frac{\partial \mathcal{E}}{\partial t} = \sigma\mathcal{E}.$$

For time-harmonic electromagnetic waves of the form

$$\mathcal{E}(x,t) = \operatorname{Re}\left\{\left(\varepsilon + \frac{i\sigma}{\omega}\right)^{-1/2} E(x)\,e^{-i\omega t}\right\}, \quad \mathcal{H}(x,t) = \operatorname{Re}\left\{\mu^{-1/2}H(x)\,e^{-i\omega t}\right\}$$

with frequency $\omega > 0$, we deduce that the complex valued space dependent parts E and H satisfy the *reduced Maxwell equations*

$$(6.1) \qquad\qquad \operatorname{curl}E - ikH = 0, \quad \operatorname{curl}H + ikE = 0$$

where the wave number k is a constant given by

$$k^2 = \left(\varepsilon + \frac{i\sigma}{\omega}\right)\mu\omega^2$$

with the sign of k chosen such that $\operatorname{Im}k \geq 0$. The equations carry the name of the physicist James Clerk Maxwell (1831 – 1879) for his fundamental contributions to electromagnetic theory.

We will consider the scattering of time-harmonic waves by obstacles surrounded by a homogeneous medium with vanishing conductivity $\sigma = 0$, that is, with exterior boundary value problems for the Maxwell equations with a positive wave number. As in the case of acoustic waves, studying the Maxwell equations with constant coefficients is also a prerequisite for studying electromagnetic wave scattering by an inhomogeneous medium.

As for the Helmholtz equation, in electromagnetic obstacle scattering we also must distinguish between the two cases of impenetrable and penetrable objects. For a perfectly conducting obstacle, the tangential component of the electric field of the total wave vanishes on the boundary. Consider the scattering of a given incoming wave E^i, H^i by a perfect conductor D. Then the total wave

$E = E^i + E^s$, $H = H^i + H^s$ where E^s, H^s denotes the scattered wave must satisfy the Maxwell equations in the exterior $\mathbb{R}^3 \setminus \bar{D}$ of D and the *perfect conductor* boundary condition $\nu \times E = 0$ on ∂D where ν is the unit outward normal to the boundary ∂D. The scattering by an obstacle that is not perfectly conducting but that does not allow the electromagnetic wave to penetrate deeply into the obstacle is modelled by an *impedance boundary condition* of the form

$$\nu \times \operatorname{curl} E - i\lambda (\nu \times E) \times \nu = 0 \quad \text{on } \partial D$$

with a positive constant λ. Throughout this book, for two vectors a and b in $\mathbb{R}^3$ or $\mathbb{C}^3$ we will denote the vector product by $a \times b$.

The scattering by a penetrable obstacle D with constant electric permittivity ε_D, magnetic permeability μ_D and electric conductivity σ_D differing from the electric permittivity ε, magnetic permeability μ and electric conductivity $\sigma = 0$ of the surrounding medium $\mathbb{R}^3 \setminus \bar{D}$ leads to a transmission problem. Here, in addition to the superposition of the incoming wave and the scattered wave in $\mathbb{R}^3 \setminus \bar{D}$ satisfying the Maxwell equations with wave number $k^2 = \varepsilon\mu\omega^2$, we also have a transmitted wave in D satisfying the Maxwell equations with wave number $k_D^2 = (\varepsilon_D + i\sigma_D/\omega)\mu_D\omega^2$. The continuity of the tangential components of the electric field $\mathcal{E}$ and the magnetic field $\mathcal{H}$ across the interface leads to transmission conditions on ∂D. Recently, in addition to the transmission conditions, more general *resistive boundary conditions* and *conductive boundary conditions* have been introduced. For their description and treatment we refer to [4].

As in the treatment of acoustic waves, we will consider in detail only one boundary condition namely that of a perfect conductor. For more details on the physical background of electromagnetic waves, we refer to Jones [79], Müller [142] and van Bladel [178].

6.2 Green's Theorem and Formula

We start with a brief outline of some basic properties of solutions to the time-harmonic Maxwell equations (6.1) with positive wave number k. We first note the vector form of Green's integral theorems. Let D be a bounded domain of class C^1 and let ν denote the unit normal vector to the boundary ∂D directed into the exterior of D. Then, for $E \in C^1(\bar{D})$ and $F \in C^2(\bar{D})$, we have *Green's first vector theorem*

(6.2)
$$\int_D \{E \cdot \Delta F + \operatorname{curl} E \cdot \operatorname{curl} F + \operatorname{div} E \operatorname{div} F\} \, dx$$

$$= \int_{\partial D} \{\nu \times E \cdot \operatorname{curl} F + \nu \cdot E \operatorname{div} F\} \, ds$$

and for $E, F \in C^2(\bar{D})$ we have *Green's second vector theorem*

$$\int_D \{E \cdot \Delta F - F \cdot \Delta E\}\, dx$$

(6.3)

$$= \int_{\partial D} \{\nu \times E \cdot \operatorname{curl} F + \nu \cdot E \operatorname{div} F - \nu \times F \cdot \operatorname{curl} E - \nu \cdot F \operatorname{div} E\}\, ds.$$

Both of these integral theorems follow easily from the Gauss divergence integral theorem applied to $E \times \operatorname{curl} F + E \operatorname{div} F$ with the aid of the vector identities $\operatorname{div} uE = \operatorname{grad} u \cdot E + u \operatorname{div} E$ and $\operatorname{div} E \times F = \operatorname{curl} E \cdot F - E \cdot \operatorname{curl} F$ for continuously differentiable scalars u and vector fields E and F and

(6.4) $$\operatorname{curl} \operatorname{curl} E = -\Delta E + \operatorname{grad} \operatorname{div} E$$

for twice continuously differentiable vector fields E. We also note the formula $\operatorname{curl} uE = \operatorname{grad} u \times E + u \operatorname{curl} E$ for later use.

Recalling the fundamental solution to the Helmholtz equation

$$\Phi(x, y) := \frac{1}{4\pi} \frac{e^{ik|x-y|}}{|x-y|}, \quad x \neq y,$$

we now prove a basic representation theorem for vector fields due to Stratton and Chu [170].

Theorem 6.1 *Let D be a bounded domain of class C^2 and let ν denote the unit normal vector to the boundary ∂D directed into the exterior of D. For vector fields $E, H \in C^1(D) \cap C(\bar{D})$ we have the Stratton–Chu formula*

$$E(x) = -\operatorname{curl} \int_{\partial D} \nu(y) \times E(y)\, \Phi(x, y)\, ds(y)$$

$$+ \operatorname{grad} \int_{\partial D} \nu(y) \cdot E(y)\, \Phi(x, y)\, ds(y)$$

$$- ik \int_{\partial D} \nu(y) \times H(y)\, \Phi(x, y)\, ds(y)$$

(6.5)

$$+ \operatorname{curl} \int_D \{\operatorname{curl} E(y) - ikH(y)\}\, \Phi(x, y)\, dy$$

$$- \operatorname{grad} \int_D \operatorname{div} E(y)\, \Phi(x, y)\, dy$$

$$+ ik \int_D \{\operatorname{curl} H(y) + ikE(y)\}\, \Phi(x, y)\, dy, \quad x \in D,$$

where the volume integrals exist as improper integrals. A similar formula holds with the roles of E and H interchanged.

Proof. We first assume that $E, H \in C^1(\bar{D})$. We circumscribe the arbitrary fixed point $x \in D$ with a sphere $\Omega(x; \rho) := \{y \in \mathbb{R}^3 : |x - y| = \rho\}$ contained in D and direct the unit normal ν to $\Omega(x; \rho)$ into the interior of $\Omega(x; \rho)$. From the relation $\text{grad}_x \, \Phi(x, y) = -\text{grad}_y \, \Phi(x, y)$ for vector fields $E, H \in C^1(\bar{D})$, we have

$$\text{curl}_y \{\Phi(x, y) \, E(y)\} = \Phi(x, y) \, \text{curl} \, E(y) - \text{curl}_x \{\Phi(x, y) \, E(y)\},$$

$$\text{div}_y \{\Phi(x, y) \, E(y)\} = \Phi(x, y) \, \text{div} \, E(y) - \text{div}_x \{\Phi(x, y) \, E(y)\},$$

$$\text{curl}_y \{\Phi(x, y) \, H(y)\} = \Phi(x, y) \, \text{curl} \, H(y) - \text{curl}_x \{\Phi(x, y) \, H(y)\}$$

for $x \neq y$. Taking curl_x of the first equation, $-\text{grad}_x$ of the second equation, multiplying the third equation by ik and adding the resulting three equations with the aid of (6.4) now gives

$$\text{curl}_x \, \text{curl}_y \{\Phi(x, y) \, E(y)\} - \text{grad}_x \, \text{div}_y \{\Phi(x, y) \, E(y)\}$$

$$+ik \, \text{curl}_y \{\Phi(x, y) \, H(y)\}$$

$$= \text{curl}_x \{\Phi(x, y) \, [\text{curl} \, E(y) - ikH(y)]\} - \text{grad}_x \{\Phi(x, y) \, \text{div} \, E(y)\}$$

$$+ik\Phi(x, y) \, \{\text{curl} \, H(y) + ikE(y)\}.$$

Integrating this identity over the domain $D_\rho := \{y \in D : |x - y| > \rho\}$ and interchanging differentiation and integration, the Gauss integral theorem yields

$$\text{curl} \int_{\partial D \cup \Omega(x; \rho)} \nu(y) \times E(y) \, \Phi(x, y) \, ds(y)$$

$$-\text{grad} \int_{\partial D \cup \Omega(x; \rho)} \nu(y) \cdot E(y) \, \Phi(x, y) \, ds(y)$$

$$+ik \int_{\partial D \cup \Omega(x; \rho)} \nu(y) \times H(y) \, \Phi(x, y) \, ds(y)$$

(6.6)

$$= \text{curl} \int_{D_\rho} \{\text{curl} \, E(y) - ikH(y)\} \, \Phi(x, y) \, dy$$

$$-\text{grad} \int_{D_\rho} \text{div} \, E(y) \, \Phi(x, y) \, dy$$

$$+ik \int_{D_\rho} \{\text{curl} \, H(y) + ikE(y)\} \, \Phi(x, y) \, dy.$$

Since on $\Omega(x; \rho)$ we have

$$\Phi(x, y) = \frac{e^{ik\rho}}{4\pi\rho}, \quad \text{grad}_x \, \Phi(x, y) = -\left(\frac{1}{\rho} - ik\right) \frac{e^{ik\rho}}{4\pi\rho} \, \nu(y),$$

straightforward calculations show that

$$\text{curl}_x \left\{ \nu(y) \times E(y) \, \Phi(x, y) \right\} - \text{grad}_x \left\{ \nu(y) \cdot E(y) \, \Phi(x, y) \right\}$$

(6.7)
$$= \frac{E(y)}{4\pi \rho^2} + O\left(\frac{1}{\rho}\right), \quad \rho \to 0.$$

Passing to the limit, with the help of the mean value theorem, it now follows from (6.7) that

$$\text{curl} \int_{\Omega(x;\rho)} \nu(y) \times E(y) \, \Phi(x, y) \, ds(y) - \text{grad} \int_{\Omega(x;\rho)} \nu(y) \cdot E(y) \, \Phi(x, y) \, ds(y)$$

$$+ ik \int_{\Omega(x;\rho)} \nu(y) \times H(y) \, \Phi(x, y) \, ds(y) \to E(x), \quad \rho \to 0,$$

and (6.5) is obtained from (6.6).

The case where E and H only belong to $C^1(D) \cap C(\bar{D})$ is treated by first integrating over parallel surfaces to the boundary of D and then passing to the limit ∂D. $\qquad \square$

In the case when E, H solve the Maxwell equations, it is convenient to transform the remaining boundary terms as described in the following theorem.

Theorem 6.2 *Let D be a bounded domain of class C^2 and let ν denote the unit normal vector to the boundary ∂D directed into the exterior of D. Let $E, H \in C^1(D) \cap C(\bar{D})$ be a solution to the Maxwell equations*

$$\text{curl} \, E - ikH = 0, \quad \text{curl} \, H + ikE = 0 \quad \text{in } D.$$

Then we have the Stratton–Chu *formulas*

$$E(x) = - \text{curl} \int_{\partial D} \nu(y) \times E(y) \, \Phi(x, y) \, ds(y)$$

(6.8)
$$+ \frac{1}{ik} \text{curl curl} \int_{\partial D} \nu(y) \times H(y) \, \Phi(x, y) \, ds(y), \quad x \in D,$$

and

$$H(x) = - \text{curl} \int_{\partial D} \nu(y) \times H(y) \, \Phi(x, y) \, ds(y)$$

(6.9)
$$- \frac{1}{ik} \text{curl curl} \int_{\partial D} \nu(y) \times E(y) \, \Phi(x, y) \, ds(y), \quad x \in D.$$

Proof. From

$$\text{div}_x \left\{ \nu(y) \times H(y) \, \Phi(x, y) \right\} = \nu(y) \cdot \text{curl}_y \left\{ H(y) \, \Phi(x, y) \right\} - \Phi(x, y) \, \nu(y) \cdot \text{curl} \, H(y),$$

with the help of the Stokes theorem and the second Maxwell equation we see that

$$\text{div} \int_{\partial D} \nu(y) \times H(y)\,\Phi(x,y)\,ds(y) = ik \int_{\partial D} \nu(y) \cdot E(y)\,\Phi(x,y)\,ds(y).$$

Hence, with the aid of (6.4), we have

(6.10)

$$\frac{1}{ik} \,\text{curl}\,\text{curl} \int_{\partial D} \nu(y) \times H(y)\,\Phi(x,y)\,ds(y)$$

$$= -ik \int_{\partial D} \nu(y) \times H(y)\,\Phi(x,y)\,ds(y) + \text{grad} \int_{\partial D} \nu(y) \cdot E(y)\,\Phi(x,y)\,ds(y).$$

Equation (6.8) now follows by inserting the Maxwell equations into (6.5) and using (6.10). Finally, the representation (6.9) follows easily from (6.8) by using $H = \text{curl}\, E / ik$. □

Theorem 6.2 obviously remains valid for complex values of the wave number k. From the proofs of Theorems 6.1 and 6.2, it can also be seen that the identities

(6.11)

$$\text{curl} \int_{\partial D} \nu(y) \times E(y)\,\Phi(x,y)\,ds(y)$$

$$-\frac{1}{ik}\,\text{curl}\,\text{curl} \int_{\partial D} \nu(y) \times H(y)\,\Phi(x,y)\,ds(y) = 0, \quad x \in \mathbb{R}^3 \setminus \bar{D},$$

and

(6.12)

$$\text{curl} \int_{\partial D} \nu(y) \times H(y)\,\Phi(x,y)\,ds(y)$$

$$+\frac{1}{ik}\,\text{curl}\,\text{curl} \int_{\partial D} \nu(y) \times E(y)\,\Phi(x,y)\,ds(y) = 0, \quad x \in \mathbb{R}^3 \setminus \bar{D},$$

are valid since for $x \in \mathbb{R}^3 \setminus \bar{D}$ the integrands are twice continuously differentiable in D.

Analogous to Theorem 2.2, we now can state the following theorem.

Theorem 6.3 *Any continuously differentiable solution to the Maxwell equations has analytic cartesian components.*

In particular, the cartesian components of solutions to the Maxwell equations are automatically two times continuously differentiable. Therefore, we can employ the vector identity (6.4) to prove the following result.

Theorem 6.4 *Let E, H be a solution to the Maxwell equations. Then E and H are divergence free and satisfy the* vector Helmholtz equation

$$\Delta E + k^2 E = 0 \quad and \quad \Delta H + k^2 H = 0.$$

Conversely, let E (or H) be a solution to the vector Helmholtz equation satisfying $\mathrm{div}\, E = 0$ (or $\mathrm{div}\, H = 0$). Then E and $H := \mathrm{curl}\, E / ik$ (or H and $E := -\mathrm{curl}\, H / ik$) satisfy the Maxwell equations.

We now formulate the Silver–Müller radiation conditions (see Müller [136] and Silver [168]) as the counterpart of the Sommerfeld radiation condition for electromagnetic waves.

Definition 6.5 *A solution E, H to the Maxwell equations whose domain of definition contains the exterior of some sphere is called* radiating *if it satisfies one of the* Silver–Müller radiation conditions

$$(6.13) \qquad \lim_{r \to \infty} (H \times x - rE) = 0$$

or

$$(6.14) \qquad \lim_{r \to \infty} (E \times x + rH) = 0$$

where $r = |x|$ and where the limit is assumed to hold uniformly in all directions $x/|x|$.

Theorem 6.6 *Assume the bounded set D is the open complement of an unbounded domain of class C^2 and let ν denote the unit normal vector to the boundary ∂D directed into the exterior of D. Let $E, H \in C^1(\mathbb{R}^3 \setminus \bar{D}) \cap C(\mathbb{R}^3 \setminus D)$ be a radiating solution to the Maxwell equations*

$$\mathrm{curl}\, E - ikH = 0, \quad \mathrm{curl}\, H + ikE = 0 \quad in\ \mathbb{R}^3 \setminus \bar{D}.$$

Then we have the Stratton–Chu formulas

$$(6.15) \qquad \begin{aligned} E(x) &= \mathrm{curl} \int_{\partial D} \nu(y) \times E(y)\, \Phi(x, y)\, ds(y) \\ &\quad - \frac{1}{ik} \mathrm{curl}\,\mathrm{curl} \int_{\partial D} \nu(y) \times H(y)\, \Phi(x, y)\, ds(y), \quad x \in \mathbb{R}^3 \setminus \bar{D}, \end{aligned}$$

and

$$(6.16) \qquad \begin{aligned} H(x) &= \mathrm{curl} \int_{\partial D} \nu(y) \times H(y)\, \Phi(x, y)\, ds(y) \\ &\quad + \frac{1}{ik} \mathrm{curl}\,\mathrm{curl} \int_{\partial D} \nu(y) \times E(y)\, \Phi(x, y)\, ds(y), \quad x \in \mathbb{R}^3 \setminus \bar{D}. \end{aligned}$$

Proof. We first assume that condition (6.13) is satisfied and show that

$$(6.17) \qquad\qquad \int_{\Omega_r} |E|^2 ds = O(1), \quad r \to \infty,$$

where Ω_r denotes the sphere of radius r and center at the origin. To accomplish this, we observe that from (6.13) it follows that

$$\int_{\Omega_r} \{|H \times \nu|^2 + |E|^2 - 2\,\mathrm{Re}(\nu \times E \cdot \bar{H})\}\, ds = \int_{\Omega_r} |H \times \nu - E|^2 ds \to 0, \quad r \to \infty,$$

where ν is the unit outward normal to Ω_r. We take r large enough so that D is contained in the interior of Ω_r and apply Gauss' divergence theorem in the domain $D_r := \{y \in \mathbb{R}^3 \setminus \bar{D} : |y| < r\}$ to obtain

$$\int_{\Omega_r} \nu \times E \cdot \bar{H}\, ds = \int_{\partial D} \nu \times E \cdot \bar{H}\, ds + ik \int_{D_r} \{|H|^2 - |E|^2\}\, dy.$$

We now insert the real part of the last equation into the previous equation and find that

$$(6.18) \qquad \lim_{r \to \infty} \int_{\Omega_r} \{|H \times \nu|^2 + |E|^2\}\, ds = 2\,\mathrm{Re} \int_{\partial D} \nu \times E \cdot \bar{H}\, ds.$$

Both terms on the left hand side of (6.18) are nonnegative. Hence, they must be individually bounded as $r \to \infty$ since their sum tends to a finite limit. Therefore, (6.17) is proven.

From (6.17) and the radiation conditions

$$\mathrm{grad}_y\, \Phi(x,y) \times \nu(y) = O\left(\frac{1}{r^2}\right), \quad r \to \infty,$$

$$\frac{\partial \Phi(x,y)}{\partial \nu(y)} - ik\Phi(x,y) = O\left(\frac{1}{r^2}\right), \quad r \to \infty,$$

which for fixed $x \in \mathbb{R}^3$ are valid uniformly for $y \in \Omega_r$, by the Schwarz inequality we see that

$$I_1 := \int_{\Omega_r} E(y) \times \{\mathrm{grad}_y\, \Phi(x,y) \times \nu(y)\}\, ds(y) \to 0, \quad r \to \infty,$$

$$I_2 := \int_{\Omega_r} E(y) \left\{ \frac{\partial \Phi(x,y)}{\partial \nu(y)} - ik\Phi(x,y) \right\} ds(y) \to 0, \quad r \to \infty.$$

The radiation condition (6.13) and $\Phi(x,y) = O(1/r)$ for $y \in \Omega_r$ yield

$$I_3 := ik \int_{\Omega_r} \Phi(x,y) \{\nu(y) \times H(y) + E(y)\}\, ds(y) \to 0, \quad r \to \infty.$$

Analogous to (6.10), we derive

$$\text{curl} \int_{\Omega_r} \nu(y) \times E(y) \, \Phi(x,y) \, ds(y)$$

$$-\frac{1}{ik} \text{curl} \, \text{curl} \int_{\Omega_r} \nu(y) \times H(y) \, \Phi(x,y) \, ds(y)$$

$$= \text{curl} \int_{\Omega_r} \nu(y) \times E(y) \, \Phi(x,y) \, ds(y)$$

$$- \text{grad} \int_{\Omega_r} \nu(y) \cdot E(y) \, \Phi(x,y) \, ds(y)$$

$$+ik \int_{\Omega_r} \nu(y) \times H(y) \, \Phi(x,y) \, ds(y)$$

$$= I_1 + I_2 + I_3 \to 0, \quad r \to \infty.$$

The proof is now completed by applying Theorem 6.2 in the bounded domain D_r and passing to the limit $r \to \infty$.

Finally, let (6.14) be satisfied. Then $\tilde{E} := -H$ and $\tilde{H} := E$ solve the Maxwell equations and satisfy

$$\lim_{r \to \infty} (\tilde{H} \times x - r\tilde{E}) = 0.$$

Hence, establishing the representation (6.15) and (6.16) under the assumption of the radiation condition (6.14) is reduced to the case of assuming the radiation condition (6.13). □

From Theorem 6.6 we deduce that radiating solutions E, H to the Maxwell equations automatically satisfy the finiteness condition

$$(6.19) \qquad E(x) = O\left(\frac{1}{|x|}\right), \quad H(x) = O\left(\frac{1}{|x|}\right), \quad |x| \to \infty,$$

uniformly for all directions and that the validity of the Silver–Müller radiation conditions (6.13) and (6.14) is invariant under translations of the origin. Our proof has followed Wilcox [194] who first established the Stratton–Chu formulas (6.15) and (6.16) without assuming the conditions (6.19) of finiteness. From the proof of Theorem 6.6 it is obvious that (6.13) and (6.14) can be replaced by the weaker formulation

$$\int_{\Omega_r} |H \times \nu - E|^2 ds \to 0, \quad \int_{\Omega_r} |E \times \nu + H|^2 ds \to 0, \quad r \to \infty.$$

Let a be a constant vector. Then

$$(6.20) \qquad E_m(x) := \operatorname{curl}_x a\Phi(x,y), \quad H_m(x) := \frac{1}{ik} \operatorname{curl} E_m(x)$$

represent the electromagnetic field generated by a magnetic dipole located at the point y and solve the Maxwell equations for $x \neq y$. Similarly,

$$(6.21) \qquad H_e(x) := \operatorname{curl}_x a\Phi(x,y), \quad E_e(x) := -\frac{1}{ik} \operatorname{curl} H_e(x)$$

represent the electromagnetic field generated by an electric dipole. Theorems 6.2 and 6.6 obviously give representations of solutions to the Maxwell equations in terms of electric and magnetic dipoles distributed over the boundary. In this sense, the fields (6.20) and (6.21) may be considered as fundamental solutions to the Maxwell equations. By straightforward calculations, it can be seen that both pairs E_m, H_m and E_e, H_e satisfy

$$H(x) \times x - rE(x) = O\left(\frac{|a|}{|x|}\right), \quad E(x) \times x + rH(x) = O\left(\frac{|a|}{|x|}\right), \quad r = |x| \to \infty,$$

uniformly for all directions $x/|x|$ and all $y \in \partial D$. Hence, from the representations (6.15) and (6.16) we can deduce that the radiation condition (6.13) implies the radiation condition (6.14) and vice versa.

Straightforward calculations show that the cartesian components of the fundamental solutions (6.20) and (6.21) satisfy the Sommerfeld radiation condition (2.7) uniformly for all $y \in \partial D$. Therefore, again from (6.15) and (6.16), we see that the cartesian components of solutions to the Maxwell equations satisfying the Silver–Müller radiation condition also satisfy the Sommerfeld radiation condition. Similarly, elementary asymptotics show that

$$\operatorname{curl} a\Phi(x,y) \times x + x \operatorname{div} a\Phi(x,y) - ik|x|a\Phi(x,y) = O\left(\frac{|a|}{|x|}\right), \quad |x| \to \infty,$$

uniformly for all directions $x/|x|$ and all $y \in \partial D$. The same inequality also holds with $\Phi(x,y)$ replaced by $\partial\Phi(x,y)/\partial\nu(y)$. Hence, from Theorems 2.4 and 6.4, we conclude that solutions of the Maxwell equations for which the cartesian components satisfy the Sommerfeld radiation condition also satisfy the Silver–Müller radiation condition. Therefore we have proven the following result.

Theorem 6.7 *For solutions to the Maxwell equations, the Silver–Müller radiation condition is equivalent to the Sommerfeld radiation condition for the cartesian components.*

Solutions to the Maxwell equations which are defined in all of $\mathbb{R}^3$ are called *entire solutions*. An entire solution to the Maxwell equations satisfying the Silver–Müller radiation condition must vanish identically. This is a consequence of Theorems 6.4 and 6.7 and the fact that entire solutions to the Helmholtz equation satisfying the Sommerfeld radiation condition must vanish identically.

The following theorem deals with the *far field pattern* or *scattering amplitude* of radiating electromagnetic waves.

Theorem 6.8 *Every radiating solution E, H to the Maxwell equations has the asymptotic form*

(6.22)

$$E(x) = \frac{e^{ik|x|}}{|x|} \left\{ E_\infty(\hat{x}) + O\left(\frac{1}{|x|}\right) \right\}, \quad |x| \to \infty,$$

$$H(x) = \frac{e^{ik|x|}}{|x|} \left\{ H_\infty(\hat{x}) + O\left(\frac{1}{|x|}\right) \right\}, \quad |x| \to \infty,$$

uniformly in all directions $\hat{x} = x/|x|$ where the vector fields E_∞ and H_∞ defined on the unit sphere Ω are known as the electric far field pattern *and* magnetic far field pattern, *respectively. They satisfy*

(6.23)
$$H_\infty = \nu \times E_\infty \quad and \quad \nu \cdot E_\infty = \nu \cdot H_\infty = 0$$

with the unit outward normal ν on Ω. Under the assumptions of Theorem 6.6, we have

$$E_\infty(\hat{x}) = \frac{ik}{4\pi} \hat{x} \times \int_{\partial D} \{\nu(y) \times E(y) + [\nu(y) \times H(y)] \times \hat{x}\} \, e^{-ik\hat{x}\cdot y} \, ds(y),$$

(6.24)

$$H_\infty(\hat{x}) = \frac{ik}{4\pi} \hat{x} \times \int_{\partial D} \{\nu(y) \times H(y) - [\nu(y) \times E(y)] \times \hat{x}\} \, e^{-ik\hat{x}\cdot y} \, ds(y).$$

Proof. As in the proof of Theorem 2.5, for a constant vector a we derive

(6.25) $\quad \operatorname{curl}_x a \dfrac{e^{ik|x-y|}}{|x-y|} = ik \dfrac{e^{ik|x|}}{|x|} \left\{ e^{-ik\hat{x}\cdot y} \hat{x} \times a + O\left(\dfrac{|a|}{|x|}\right) \right\},$

(6.26) $\quad \operatorname{curl}_x \operatorname{curl}_x a \dfrac{e^{ik|x-y|}}{|x-y|} = k^2 \dfrac{e^{ik|x|}}{|x|} \left\{ e^{-ik\hat{x}\cdot y} \hat{x} \times (a \times \hat{x}) + O\left(\dfrac{|a|}{|x|}\right) \right\}$

as $|x| \to \infty$ uniformly for all $y \in \partial D$. Inserting this into (6.15) and (6.16) we obtain (6.24). Now (6.22) and (6.23) are obvious from (6.24). $\qquad \square$

Rellich's lemma establishes a one-to-one correspondence between radiating electromagnetic waves and their far field patterns.

Theorem 6.9 *Assume the bounded domain D is the open complement of an unbounded domain and let $E, H \in C^1(\mathbb{R}^3 \setminus \bar{D})$ be a radiating solution to the Maxwell equations for which the electric or magnetic far field pattern vanishes identically. Then $E = H = 0$ in $\mathbb{R}^3 \setminus \bar{D}$.*

Proof. This is a consequence of the corresponding Theorem 2.13 for the Helmholtz equation and Theorems 6.4 and 6.7. $\qquad \square$

Rellich's lemma also ensures uniqueness for solutions to exterior boundary value problems through the following theorem.

Theorem 6.10 *Assume the bounded set D is the open complement of an unbounded domain of class C^2 with unit normal ν to the boundary ∂D directed into the exterior of D and let $E, H \in C^1(\mathbb{R}^3 \setminus \bar{D}) \cap C(\mathbb{R}^3 \setminus D)$ be a radiating solution to the Maxwell equations with wave number $k > 0$ satisfying*

$$\mathrm{Re} \int_{\partial D} \nu \times E \cdot \bar{H} \, ds \le 0.$$

Then $E = H = 0$ in $\mathbb{R}^3 \setminus \bar{D}$.

Proof. From the identity (6.18) and the assumption of the theorem, we conclude that (2.46) is satisfied for the cartesian components of E. Hence, $E = 0$ in $\mathbb{R}^3 \setminus \bar{D}$ by Rellich's Lemma 2.11. $\qquad \square$

6.3 Vector Potentials

For the remainder of this chapter, if not stated otherwise, we always will assume that D is the open complement of an unbounded domain of class C^2. In this section, we extend our review of the basic jump relations and regularity properties of surface potentials from the scalar to the vector case. Given an integrable vector field a on the boundary ∂D, the integral

$$A(x) := \int_{\partial D} a(y)\Phi(x, y) \, ds(y), \quad x \in \mathbb{R}^3 \setminus \partial D,$$

is called the *vector potential* with density a. Analogous to Theorem 3.1, we have the following *jump relations* for the behavior at the boundary.

Theorem 6.11 *Let ∂D be of class C^2 and let a be a continuous tangential field. Then the vector potential A with density a is continuous throughout $\mathbb{R}^3$. On the boundary, we have*

(6.27) $$A(x) = \int_{\partial D} a(y)\Phi(x, y) \, ds(y),$$

(6.28) $$\nu(x) \times \mathrm{curl}\, A_\pm(x) = \int_{\partial D} \nu(x) \times \mathrm{curl}_x \{a(y)\Phi(x, y)\} \, ds(y) \pm \frac{1}{2}\, a(x)$$

for $x \in \partial D$ where

$$\nu(x) \times \mathrm{curl}\, A_\pm(x) := \lim_{h \to +0} \nu(x) \times \mathrm{curl}\, A(x \pm h\nu(x))$$

is to be understood in the sense of uniform convergence on ∂D and where the integrals exist as improper integrals. Furthermore,

(6.29) $$\lim_{h \to +0} \nu(x) \times [\mathrm{curl}\,\mathrm{curl}\, A(x + h\nu(x)) - \mathrm{curl}\,\mathrm{curl}\, A(x - h\nu(x))] = 0$$

uniformly for all $x \in \partial D$.

Proof. The continuity of the vector potential is an immediate consequence of Theorem 3.1. The proof of the jump relation for the curl of the vector potential follows in the same manner as for the double-layer potential after observing that the kernel

$$\nu(x) \times \text{curl}_x \left\{a(y)\Phi(x,y)\right\} = \text{grad}_x\,\Phi(x,y)[\nu(x) - \nu(y)] \cdot a(y) - a(y)\,\frac{\partial\Phi(x,y)}{\partial\nu(x)}$$

has the same type of singularity for $x = y$ as the kernel of the double-layer potential. It is essential that a is a tangential vector, that is, $\nu \cdot a = 0$ on ∂D. For the details, we refer to [32]. The proof for the continuity (6.29) of the double curl can be found in [33].   □

For convenience we denote by $T(\partial D)$ and $T^{0,\alpha}(\partial D)$, $0 < \alpha \leq 1$, the spaces of all continuous and uniformly Hölder continuous tangential fields a equipped with the supremum norm and the Hölder norm, respectively. In the Hölder space setting, we can deduce from Theorem 3.3 the following result.

Theorem 6.12 *Let ∂D be of class C^2 and let $0 < \alpha < 1$. Then the vector potential A with a (not necessarily tangential) density $a \in C(\partial D)$ is uniformly Hölder continuous throughout $\mathbb{R}^3$ and*

$$\|A\|_{\alpha,\mathbb{R}^3} \leq C_\alpha\,\|a\|_{\infty,\partial D}$$

for some constant C_α depending on ∂D and α. For densities $a \in T^{0,\alpha}(\partial D)$, the first derivatives of the vector potential can be uniformly Hölder continuously extended from D to $\bar{D}$ and from $\mathbb{R}^3 \setminus \bar{D}$ to $\mathbb{R}^3 \setminus D$ with boundary values

$$\text{div}\,A_\pm(x) = \int_{\partial D} \text{grad}_x\,\Phi(x,y) \cdot a(y)\,ds(y) \mp \frac{1}{2}\,\nu(x) \cdot a(x), \quad x \in \partial D,$$

$$\text{curl}\,A_\pm(x) = \int_{\partial D} \text{grad}_x\,\Phi(x,y) \times a(y)\,ds(y) \mp \frac{1}{2}\,\nu(x) \times a(x), \quad x \in \partial D,$$

where

$$\text{div}\,A_\pm(x) := \lim_{h \to +0} \text{div}\,A(x \pm h\nu(x)), \quad \text{curl}\,A_\pm(x) := \lim_{h \to +0} \text{curl}\,A(x \pm h\nu(x)).$$

Furthermore, we have

$$\|\text{div}\,A\|_{\alpha,\bar{D}} \leq C_\alpha\,\|a\|_{\alpha,\partial D}, \quad \|\text{div}\,A\|_{\alpha,\mathbb{R}^3 \setminus D} \leq C_\alpha\,\|a\|_{\alpha,\partial D}$$

and

$$\|\text{curl}\,A\|_{\alpha,\bar{D}} \leq C_\alpha\,\|a\|_{\alpha,\partial D}, \quad \|\text{curl}\,A\|_{\alpha,\mathbb{R}^3 \setminus D} \leq C_\alpha\,\|a\|_{\alpha,\partial D}.$$

for some constant C_α depending on ∂D and α.

For the tangential component of the curl on the boundary ∂D, we have more regularity which can be expressed in terms of mapping properties for the magnetic dipole operator M given by

$$(6.30) \quad (Ma)(x) := 2 \int_{\partial D} \nu(x) \times \mathrm{curl}_x \{a(y)\Phi(x,y)\}\,ds(y), \quad x \in \partial D.$$

The operator M describes the tangential component of the electric field of a magnetic dipole distribution.

Theorem 6.13 *The operator M is bounded from $T(\partial D)$ into $T^{0,\alpha}(\partial D)$.*

Proof. For a proof, we refer to Theorem 2.32 of [32].   □

In order to develop further regularity properties of the vector potential, we need to introduce the concept of the surface divergence of tangential vector fields. For a continuously differentiable function φ on ∂D, the *surface gradient* $\mathrm{Grad}\,\varphi$ is defined as the vector pointing into the direction of the maximal increase of φ with the modulus given by the value of this increase. In terms of a parametric representation

$$x(u) = (x_1(u_1, u_2), x_2(u_1, u_2), x_3(u_1, u_2))$$

of a surface patch of ∂D, the surface gradient can be expressed by

$$(6.31) \qquad\qquad \mathrm{Grad}\,\varphi = \sum_{i,j=1}^{2} g^{ij}\, \frac{\partial \varphi}{\partial u_i}\, \frac{\partial x}{\partial u_j}$$

where g^{ij} is the inverse of the first fundamental matrix

$$g_{ij} := \frac{\partial x}{\partial u_i} \cdot \frac{\partial x}{\partial u_j}, \quad i,j = 1,2,$$

of differential geometry. We note that for a continuously differentiable function φ defined in a neighborhood of ∂D we have the relation

$$(6.32) \qquad\qquad \mathrm{grad}\,\varphi = \mathrm{Grad}\,\varphi + \frac{\partial \varphi}{\partial \nu}\, \nu$$

between the spatial gradient grad and the surface gradient Grad.

Let S be a connected surface contained in ∂D with C^2 boundary ∂S and let ν_0 denote the unit normal vector to ∂S that is perpendicular to the surface normal ν to ∂D and directed into the exterior of S. Then for any continuously differentiable tangential field a with the representation

$$a = a_1 \frac{\partial x}{\partial u_1} + a_2 \frac{\partial x}{\partial u_2},$$

by Gauss' integral theorem applied in the parameter domain (c.f. [126], p. 74) it can be readily shown that

$$(6.33) \qquad \int_S \text{Div}\, a\, ds = \int_{\partial S} \nu_0 \cdot a\, ds$$

with the *surface divergence* Div a given by

$$(6.34) \qquad \text{Div}\, a = \frac{1}{\sqrt{g}} \left\{ \frac{\partial}{\partial u_1}\left(\sqrt{g}\, a_1\right) + \frac{\partial}{\partial u_2}\left(\sqrt{g}\, a_2\right) \right\}.$$

Here, g denotes the determinant of the matrix g_{ij}. In particular, from (6.33) we have the special case

$$(6.35) \qquad \int_{\partial D} \text{Div}\, a\, ds = 0.$$

We call (6.33) the *Gauss surface divergence theorem* and may view it as a coordinate independent definition of the surface divergence.

For our purposes, this definition is not yet adequate and has to be generalized. This generalization, in principle, can be done in two ways. One possibility (see [32, 142]) is to use (6.33) as motivation for a definition in the limit integral sense by letting the surface S shrink to a point. Here, as a second possibility, we use the concept of weak derivatives. From (6.31) and (6.34) we see that for continuously differentiable fucntions φ and tangential fields a we have the product rule Div $\varphi a = \text{Grad}\, \varphi \cdot a + \varphi\, \text{Div}\, a$ and consequently by (6.35) we have

$$(6.36) \qquad \int_{\partial D} \varphi\, \text{Div}\, a\, ds = - \int_{\partial D} \text{Grad}\, \varphi \cdot a\, ds.$$

This now leads to the following definition.

Definition 6.14 *We say that an integrable tangential field a has a* weak surface divergence *if there exists an integrable scalar denoted by* Div a *such that (6.36) is satisfied for all $\varphi \in C^1(\partial D)$.*

It is left as an exercise to show that the weak surface divergence, if it exists, is unique. In the sequel, we will in general suppress the adjective weak and just speak of the surface divergence.

Let $E \in C^1(\mathbb{R}^3 \setminus \bar{D}) \cap C(\mathbb{R}^3 \setminus D)$ and assume that

$$\nu(x) \cdot \text{curl}\, E(x) := \lim_{h \to +0} \nu(x) \cdot \text{curl}\, E(x + h\nu(x)), \quad x \in \partial D,$$

exists in the sense of uniform convergence on ∂D. Then by applying Stokes' integral theorem on parallel surfaces and passing to the limit it can be seen that

$$(6.37) \qquad \int_{\partial D} \nu \cdot \text{curl}\, E\, ds = 0.$$

By setting $\psi(x + h\nu(x)) := \varphi(x)$, $x \in \partial D$, $-h_0 \le h \le h_0$, with h_0 sufficiently small, any continuously differentiable function φ on ∂D can be considered as the restriction of a function ψ which is continuously differentiable in a neighborhood of ∂D. Then from the product rule $\operatorname{curl} \psi E = \operatorname{grad} \psi \times E + \psi \operatorname{curl} E$, (6.32) and Stokes' theorem (6.37) applied to ψE, we find that

$$\int_{\partial D} \varphi \, \nu \cdot \operatorname{curl} E \, ds = \int_{\partial D} \operatorname{Grad} \varphi \cdot \nu \times E \, ds$$

for all $\varphi \in C^1(\partial D)$, whence the important identity

$$(6.38) \qquad\qquad \operatorname{Div}(\nu \times E) = -\nu \cdot \operatorname{curl} E$$

follows.

We introduce normed spaces of tangential fields possessing a surface divergence by

$$T_d(\partial D) := \{ a \in T(\partial D) : \operatorname{Div} a \in C(\partial D) \}$$

and

$$T_d^{0,\alpha}(\partial D) := \{ a \in T^{0,\alpha}(\partial D) : \operatorname{Div} a \in C^{0,\alpha}(\partial D) \}$$

equipped with the norms

$$\|a\|_{T_d} := \|a\|_\infty + \|\operatorname{Div} a\|_\infty, \quad \|a\|_{T_d^{0,\alpha}} := \|a\|_{0,\alpha} + \|\operatorname{Div} a\|_{0,\alpha}.$$

Theorem 6.15 *Let $0 < \alpha < \beta \le 1$. Then the imbedding operators*

$$I^\beta : T_d^{0,\beta}(\partial D) \to T_d(\partial D), \quad I^{\beta,\alpha} : T_d^{0,\beta}(\partial D) \to T_d^{0,\alpha}(\partial D)$$

are compact.

Proof. Let (a_n) be a bounded sequence in $T_d^{0,\beta}(\partial D)$. Then by Theorem 3.2 there exists a subsequence $(a_{n(j)})$, a tangential field $a \in T(\partial D)$ and a scalar $\psi \in C(\partial D)$ such that $\|a_{n(j)} - a\|_\infty \to 0$ and $\|\operatorname{Div} a_{n(j)} - \psi\|_\infty \to 0$ as $j \to \infty$. Passing to the limit in

$$\int_{\partial D} \varphi \operatorname{Div} a_{n(j)} \, ds = -\int_{\partial D} \operatorname{Grad} \varphi \cdot a_{n(j)} \, ds$$

for $\varphi \in C^1(\partial D)$ shows that $a \in T_d(\partial D)$ with $\operatorname{Div} a = \psi$. This finishes the proof for the compactness of I^β since we now have $\|a_{n(j)} - a\|_{T_d} \to 0$ as $j \to \infty$. The proof for $I^{\beta,\alpha}$ is analogous. $\qquad\square$

We now extend Theorem 6.13 by proving the following result.

Theorem 6.16 *The operator M is bounded from $T_d(\partial D)$ into $T_d^{0,\alpha}(\partial D)$.*

Proof. This follows from the boundedness of S and K' from $C(\partial D)$ into $C^{0,\alpha}(\partial D)$ and of M from $T(\partial D)$ into $T^{0,\alpha}(\partial D)$ with the aid of

$$(6.39) \qquad\qquad \operatorname{Div} Ma = -k^2 \, \nu \cdot Sa - K' \operatorname{Div} a$$

for all $a \in T_d(\partial D)$. To establish (6.39), we first note that using the symmetry relation $\mathrm{grad}_x \, \Phi(x,y) = -\mathrm{grad}_y \, \Phi(x,y)$, (6.32) and (6.36) for the vector potential A with density a in $T_d(\partial D)$ we can derive

$$\mathrm{div}\, A(x) = \int_{\partial D} \mathrm{Div}\, a(y) \Phi(x,y)\, ds(y), \quad x \in \mathbb{R}^3 \setminus \partial D.$$

Then, using the identity (6.4), we find that

$$\mathrm{curl\, curl}\, A(x) = k^2 \int_{\partial D} a(y) \Phi(x,y)\, ds(y)$$

(6.40)

$$+ \mathrm{grad} \int_{\partial D} \mathrm{Div}\, a(y) \Phi(x,y)\, ds(y), \quad x \in \mathbb{R}^3 \setminus \partial D.$$

Applying the jump relations of Theorems 3.1 and 6.11, we find that

(6.41) $$2\nu \times \mathrm{curl}\, A_\pm = Ma \pm a \quad \text{on } \partial D$$

and

(6.42) $$2\nu \cdot \mathrm{curl\, curl}\, A_\pm = k^2 \nu \cdot Sa + K' \mathrm{Div}\, a \mp \mathrm{Div}\, a \quad \text{on } \partial D.$$

Hence, by using the vector identity (6.38) we now obtain (6.39) by (6.41) and (6.42).   □

In our analysis of boundary value problems, we will also need the electric dipole operator N given by

(6.43) $$(Nb)(x) := 2\,\nu(x) \times \mathrm{curl\, curl} \int_{\partial D} \nu(y) \times b(y)\, \Phi(x,y)\, ds(y), \quad x \in \partial D.$$

The operator N describes the tangential component of the electric field of an electric dipole distribution. After introducing the normed space

$$T_r^{0,\alpha}(\partial D) := \left\{ b \in T^{0,\alpha}(\partial D) : \nu \times b \in T_d^{0,\alpha}(\partial D) \right\}$$

with the norm

$$\|b\|_{T_r^{0,\alpha}} := \|\nu \times b\|_{T_d^{0,\alpha}}$$

we can state the following mapping property.

Theorem 6.17 *The operator N is bounded from $T_r^{0,\alpha}(\partial D)$ into $T_d^{0,\alpha}(\partial D)$.*

Proof. From the decomposition (6.40) and Theorems 3.3 and 6.12, we observe that N is bounded from $T_r^{0,\alpha}(\partial D)$ into $T^{0,\alpha}(\partial D)$. Furthermore, from (6.38) and (6.40), we also deduce that

$$\mathrm{Div}\, Nb = k^2 \, \mathrm{Div}(\nu \times S(\nu \times b)).$$

Hence, in view of Theorem 6.12 and (6.38), there exists a constant C such that

$$\| \operatorname{Div} Nb\|_{0,\alpha} \leq C \|b\|_{0,\alpha}$$

and this finally implies that N is also bounded from $T_r^{0,\alpha}(\partial D)$ into $T_d^{0,\alpha}(\partial D)$. $\square$

By interchanging the order of integration, we see that the adjoint operator M' of the weakly singular operator M with respect to the bilinear form

$$\langle a, b \rangle := \int_{\partial D} a \cdot b \, ds$$

is given by

(6.44) $$M'a := \nu \times M(\nu \times a).$$

After defining the operator R by

$$Ra := a \times \nu,$$

we may rewrite (6.44) as $M' = RMR$.

To show that N is self adjoint, let $a, b \in T_r^{0,\alpha}(\partial D)$ and denote by A and B the vector potentials with densities $\nu \times a$ and $\nu \times b$, respectively. Then by the jump relations of Theorem 6.11, Green's vector theorem (6.3) applied to $E = \operatorname{curl} A$ and $F = \operatorname{curl} B$, and the radiation condition we find that

$$\int_{\partial D} Na \cdot b \, ds = 2 \int_{\partial D} \nu \times \operatorname{curl} \operatorname{curl} A \cdot (\operatorname{curl} B_+ - \operatorname{curl} B_-) \, ds$$

$$= 2 \int_{\partial D} \nu \times \operatorname{curl} \operatorname{curl} B \cdot (\operatorname{curl} A_+ - \operatorname{curl} A_-) \, ds = \int_{\partial D} Nb \cdot a \, ds,$$

that is, N indeed is self adjoint. Furthermore, applying Green's vector theorem (6.3) to $E = \operatorname{curl} \operatorname{curl} A$ and $F = \operatorname{curl} B$, we derive

$$\int_{\partial D} Na \cdot Nb \times \nu \, ds = 4 \int_{\partial D} \nu \times \operatorname{curl} \operatorname{curl} A \cdot \operatorname{curl} \operatorname{curl} B \, ds$$

$$= 4k^2 \int_{\partial D} \nu \times \operatorname{curl} B_- \cdot \operatorname{curl} A_- \, ds = k^2 \int_{\partial D} (I - M)(\nu \times b) \cdot (I + M')a \, ds,$$

whence

$$\int_{\partial D} a \cdot N(Nb \times \nu) \, ds = k^2 \int_{\partial D} a \cdot (I - M^2)(\nu \times b) \, ds$$

follows for all $a, b \in T_r^{0,\alpha}(\partial D)$. Thus, setting $c = \nu \times b$, we have proven the relation

$$N(N(c \times \nu) \times \nu) = k^2(I - M^2)c$$

for all $c \in T_d^{0,\alpha}(\partial D)$, that is,

(6.45) $$NRNR = k^2(I - M^2).$$

As in the scalar case, corresponding mapping properties for the two vector operators M and N in a Sobolev space setting can again be deduced from the classical results in the Hölder space setting by using Lax's Theorem 3.5. For details we refer to Hähner [65] and Kirsch [89].

Using Lax's theorem, the jump relations of Theorem 6.11 can be extended from continuous densities to L^2 densities. As a simple consequence of the L^2 jump relation (3.23), Hähner [65] has shown that for the vector potential with tangential L^2 density a the jump relation (6.28) has to be replaced by

$$(6.46) \qquad \lim_{h \to +0} \int_{\partial D} |2\,\nu \times \mathrm{curl}\, A(\cdot \pm h\nu) - Ma \mp a|^2 ds = 0$$

and he has further verified that (6.29) can be replaced by

$$(6.47) \quad \lim_{h \to +0} \int_{\partial D} |\nu \times [\mathrm{curl}\,\mathrm{curl}\, A(\cdot + h\nu) - \mathrm{curl}\,\mathrm{curl}\, A(\cdot - h\nu)]|^2 ds = 0,$$

since the singularity of the double curl is similar to the singularity of the normal derivative of the double-layer potential in (3.4) and (3.22).

6.4 Scattering from a Perfect Conductor

The scattering of time-harmonic electromagnetic waves by a perfectly conducting body leads to the following problem.

Direct Electromagnetic Obstacle Scattering Problem. *Given an entire solution E^i, H^i to the Maxwell equations representing an incident electromagnetic field, find a solution*

$$E = E^i + E^s, \quad H = H^i + H^s$$

to the Maxwell equations in $\mathbb{R}^3 \setminus \bar{D}$ such that the scattered field E^s, H^s satisfies the Silver–Müller radiation condition and the total electric field E satisfies the boundary condition

$$\nu \times E = 0 \quad \text{on } \partial D$$

where ν is the unit outward normal to ∂D.

Clearly, after renaming the unknown fields, this direct scattering problem is a special case of the following problem.

Exterior Maxwell Problem. *Given a tangential field $c \in T_d^{0,\alpha}(\partial D)$, find a radiating solution $E, H \in C^1(\mathbb{R}^3 \setminus \bar{D}) \cap C(\mathbb{R}^3 \setminus D)$ to the Maxwell equations*

$$\mathrm{curl}\, E - ikH = 0, \quad \mathrm{curl}\, H + ikE = 0 \quad \text{in } \mathbb{R}^3 \setminus \bar{D}$$

which satisfies the boundary condition

$$\nu \times E = c \quad \text{on } \partial D.$$

From the vector formula (6.38), we observe that the continuity of the magnetic field H up to the boundary requires the differentiability of the tangential component $\nu \times E$ of the electric field, that is, the given tangential field c must have a continuous surface divergence. The Hölder continuity of the boundary data is necessary for our integral equation approach to solving the exterior Maxwell problem.

Theorem 6.18 *The exterior Maxwell problem has at most one solution.*

Proof. This follows from Theorem 6.10.   □

Theorem 6.19 *The exterior Maxwell problem has a unique solution. The solution depends continuously on the boundary data in the sense that the operator mapping the given boundary data onto the solution is continuous from $T_d^{0,\alpha}(\partial D)$ into $C^{0,\alpha}(\mathbb{R}^3 \setminus D) \times C^{0,\alpha}(\mathbb{R}^3 \setminus D)$.*

Proof. We seek the solution in the form of the electromagnetic field of a combined magnetic and electric dipole distribution

$$E(x) = \operatorname{curl} \int_{\partial D} a(y)\Phi(x,y)\,ds(y)$$

(6.48)
$$+i\eta \operatorname{curl}\operatorname{curl} \int_{\partial D} \nu(y) \times (S_0^2 a)(y)\Phi(x,y)\,ds(y),$$

$$H(x) = \frac{1}{ik}\operatorname{curl} E(x), \quad x \in \mathbb{R}^3 \setminus \partial D,$$

with a density $a \in T_d^{0,\alpha}(\partial D)$ and a real coupling parameter $\eta \neq 0$. By S_0 we mean the single-layer operator (3.8) in the potential theoretic limit case $k = 0$. From Theorems 6.4 and 6.7 and the jump relations, we see that E, H defined by (6.48) in $\mathbb{R}^3 \setminus \bar{D}$ solves the exterior Maxwell problem provided the density solves the integral equation

(6.49)
$$a + Ma + i\eta NPS_0^2 a = 2c.$$

Here, the operator P stands for the projection of a vector field defined on ∂D onto the tangent plane, that is,

$$Pb := (\nu \times b) \times \nu.$$

By Theorems 3.2 and 3.4, the operator S_0 is compact from $C^{0,\alpha}(\partial D)$ into $C^{0,\alpha}(\partial D)$ and, with the aid of Theorem 3.3 and the identity (6.38), the operator $PS_0 : C^{0,\alpha}(\partial D) \to T_r^{0,\alpha}(\partial D)$ can be seen to be bounded. Therefore, combining Theorems 6.15–6.17, the operator $M + i\eta NPS_0^2 : T_d^{0,\alpha}(\partial D) \to T_d^{0,\alpha}(\partial D)$ turns out to be compact. Hence, the existence of a solution to (6.49) can be established by the Riesz–Fredholm theory.

Let $a \in T_d^{0,\alpha}(\partial D)$ be a solution to the homogeneous form of (6.49). Then the electromagnetic field E, H given by (6.48) satisfies the homogeneous boundary condition $\nu \times E_+ = 0$ on ∂D whence $E = H = 0$ in $\mathbb{R}^3 \setminus \bar{D}$ follows by Theorem 6.18. The jump relations together with the decomposition (6.40) now yield

$$-\nu \times E_- = a, \quad -\nu \times \operatorname{curl} E_- = i\eta k^2 \nu \times S_0^2 a \quad \text{on } \partial D.$$

Hence, from Gauss' divergence theorem we have

$$i\eta k^2 \int_{\partial D} |S_0 a|^2 ds = i\eta k^2 \int_{\partial D} \bar{a} \cdot S_0^2 a \, ds$$

$$= \int_{\partial D} \nu \times \bar{E}_- \cdot \operatorname{curl} E_- \, ds = \int_D \left\{ |\operatorname{curl} E|^2 - k^2 |E|^2 \right\} dx,$$

whence $S_0 a = 0$ follows. This implies $a = 0$ as in the proof of Theorem 3.10. Thus, we have established injectivity of the operator $I + M + i\eta N P S_0^2$ and, by the Riesz–Fredholm theory, the inverse operator $(I + M + i\eta N P S_0^2)^{-1}$ exists and is bounded from $T_d^{0,\alpha}(\partial D)$ into $T_d^{0,\alpha}(\partial D)$. This, together with the regularity results of Theorems 3.3 and 6.12 and the decomposition (6.40), shows that E and H both belong to $C^{0,\alpha}(\mathbb{R}^3 \setminus D)$ and depend continuously on c in the norm of $T_d^{0,\alpha}(\partial D)$.   □

From

$$ikH(x) = \operatorname{curl} \operatorname{curl} \int_{\partial D} a(y) \Phi(x, y) \, ds(y)$$

$$+ i\eta k^2 \operatorname{curl} \int_{\partial D} \nu(y) \times (S_0^2 a)(y) \Phi(x, y) \, ds(y), \quad x \in \mathbb{R}^3 \setminus \bar{D},$$

and the jump relations we find that

$$2ik \, \nu \times H = -NRa + i\eta k^2 (RS_0^2 a + MRS_0^2 a)$$

with the bounded operator

$$NR - i\eta k^2 (I + M) RS_0^2 : T_d^{0,\alpha}(\partial D) \to T_d^{0,\alpha}(\partial D).$$

Therefore, we can write

$$\nu \times H = A(\nu \times E)$$

where

$$A := \frac{i}{k} \left\{ NR - i\eta k^2 (I + M) RS_0^2 \right\} (I + M + i\eta N P S_0^2)^{-1} : T_d^{0,\alpha}(\partial D) \to T_d^{0,\alpha}(\partial D)$$

is bounded. The operator A transfers the tangential component of the electric field on the boundary onto the tangential component of the magnetic field and therefore we call it the *electric to magnetic boundary component map*. It is bijective and has a bounded inverse since it satisfies

$$A^2 = -I.$$

This equation is a consequence of the fact that for any radiating solution E, H of the Maxwell equations the fields $\tilde{E} := -H$ and $\tilde{H} := E$ solve the Maxwell equations and satisfy the Silver–Müller radiation condition. Hence, by the uniqueness Theorem 6.18, the solution to the exterior Maxwell problem with given electric boundary data Ac has magnetic boundary data $-c$ so that $A^2 c = -c$. Thus, we can state the following result.

Theorem 6.20 *The electric to magnetic boundary component map is a bijective bounded operator from* $T_d^{0,\alpha}(\partial D)$ *onto* $T_d^{0,\alpha}(\partial D)$ *with bounded inverse.*

Note that, analogous to the acoustic case, the integral equation (6.49) is not uniquely solvable if $\eta = 0$ and k is a *Maxwell eigenvalue*, i.e., a value of k such that there exists a nontrivial solution E, H to the Maxwell equations in D satisfying the homogeneous boundary condition $\nu \times E = 0$ on ∂D. An existence proof for the exterior Maxwell problem based on seeking the solution as the electromagnetic field of a combined magnetic and electric dipole distribution was first accomplished by Knauff and Kress [99, 102] in order to overcome the non-uniqueness difficulties of the classical approach by Müller [137, 140] and Weyl [191]. The combined field approach was also independently suggested by Jones [78] and Mautz and Harrington [127]. The idea to incorporate a smoothing operator in (6.48) analogous to (3.27) was first used by Kress [104].

Since the electric and the magnetic fields occur in the Maxwell equations in a symmetric fashion, it seems appropriate to assume the same regularity properties for both fields up to the boundary as we have done in our definition of the exterior Maxwell problem. On the other hand, only the electric field is involved in the formulation of the boundary condition. Therefore, it is quite natural to ask whether the boundary condition can be weakened by seeking a solution $E, H \in C^1(\mathbb{R}^3 \setminus \bar{D})$ to the Maxwell equation such that

(6.50)
$$\lim_{h \to +0} \nu(x) \times E(x + h\nu(x)) = c(x)$$

uniformly for all $x \in \partial D$. But then as far as proving uniqueness is concerned we are in a similar position as for the Dirichlet problem: there is a gap between the regularity properties of the solution and the requirements for the application of the Gauss divergence theorem which is involved in the proof of Theorem 6.10. Unfortunately, for the Maxwell problem there is no elegant circumvention of this difficulty available as in Lemma 3.8. Nevertheless, due to an idea going back to Calderón [19], which has been rediscovered and extended more recently by Hähner [66], it is still possible to establish uniqueness for the exterior Maxwell problem with the boundary condition (6.50). The main idea is to represent the solution with homogeneous boundary condition by surface potentials of the form

(6.48) with magnetic and electric dipole distributions on surfaces parallel to ∂D as suggested by the formulation (6.50) and then pass to the limit $h \to +0$. Of course, this procedure requires the existence analysis in the Hölder space setting which we developed in Theorem 6.19. Since the details of this analysis are beyond the aims of this book, the reader is referred to [19] and [66].

Once uniqueness under the weaker conditions is established, existence can be obtained by solving the integral equation (6.49) in the space $T(\partial D)$ of continuous tangential fields. By Theorems 3.2 and 3.4, the operator S_0^2 is bounded from $C(\partial D)$ into $C^{1,\alpha}(\partial D)$. Therefore, again as a consequence of Theorems 6.15–6.17, the operator $M + i\eta N P S_0^2 : T(\partial D) \to T(\partial D)$ is seen to be compact. Hence, the existence of a continuous solution to (6.49) for any given continuous right hand side c can be established by the Fredholm alternative. We now proceed to do this.

The adjoint operator of $M + i\eta N P S_0^2$ with respect to the L^2 bilinear form is given by $M' + i\eta P S_0^2 N$ and, by Theorems 3.2, 3.4 and 6.15–6.17, the adjoint is seen to be compact from $T_r^{0,\alpha}(\partial D) \to T_r^{0,\alpha}(\partial D)$. Working with the two dual systems $\langle T_d^{0,\alpha}(\partial D), T_r^{0,\alpha}(\partial D) \rangle$ and $\langle T(\partial D), T_r^{0,\alpha}(\partial D) \rangle$ as in the proof of Theorem 3.20, the Fredholm alternative tells us that the operator $I + M + i\eta N P S_0^2$ has a trivial nullspace in $T(\partial D)$. Hence, by the Riesz–Fredholm theory it has a bounded inverse $(I + M + i\eta N P S_0^2)^{-1} : T(\partial D) \to T(\partial D)$. Then, given $c \in T(\partial D)$ and the unique solution $a \in T(\partial D)$ of (6.49), the electromagnetic field defined by (6.48) is seen to solve the exterior Maxwell problem in the sense of (6.50). For this we note that the jump relations of Theorem 6.11 are valid for continuous densities and that $P S_0^2 a$ belongs to $T_r^{0,\alpha}(\partial D)$. From the representation (6.48) of the solution, the continuous dependence of the density a on the boundary data c shows that the exterior Maxwell problem is also well-posed in this setting, i.e., small deviations in c in the maximum norm ensure small deviations in E and H and all their derivatives in the maximum norm on closed subsets of $\mathbb{R}^3 \setminus \bar{D}$. Thus, leaving aside the uniqueness part of the proof, we have established the following theorem.

Theorem 6.21 *The exterior Maxwell problem with continuous boundary data has a unique solution and the solution depends continuously on the boundary data with respect to uniform convergence of the solution and all its derivatives on closed subsets of* $\mathbb{R}^3 \setminus \bar{D}$.

For the scattering problem, the boundary values are the restriction of an analytic field to the boundary and therefore they are as smooth as the boundary. Hence, for domains D of class C^2 there exists a solution in the sense of Theorem 6.19. Therefore, we can apply the Stratton–Chu formulas (6.15) and (6.16) for the scattered field E^s, H^s and the Stratton–Chu formulas (6.11) and (6.12) for the incident field E^i, H^i. Then, adding both formulas and using the boundary condition $\nu \times (E^i + E^s) = 0$ on ∂D, we have the following theorem known as *Huygen's principle*. The representation for the far field pattern is obtained with the aid of (6.25) and (6.26).

Theorem 6.22 *For the scattering of an entire electromagnetic field* E^i, H^i *by a perfect conductor D we have*

$$E(x) = E^i(x) - \frac{1}{ik} \operatorname{curl} \operatorname{curl} \int_{\partial D} \nu(y) \times H(y) \, \Phi(x,y) \, ds(y),$$

(6.51)

$$H(x) = H^i(x) + \operatorname{curl} \int_{\partial D} \nu(y) \times H(y) \, \Phi(x,y) \, ds(y)$$

for $x \in \mathrm{IR}^3 \setminus \bar{D}$ where E, H is the total field. The far field pattern is given by

$$E_\infty(\hat{x}) = \frac{ik}{4\pi} \, \hat{x} \times \int_{\partial D} [\nu(y) \times H(y)] \times \hat{x} \, e^{-ik\,\hat{x}\cdot y} \, ds(y),$$

(6.52)

$$H_\infty(\hat{x}) = \frac{ik}{4\pi} \, \hat{x} \times \int_{\partial D} \nu(y) \times H(y) \, e^{-ik\,\hat{x}\cdot y} \, ds(y)$$

for $\hat{x} \in \Omega$.

6.5 Vector Wave Functions

For any orthonormal system Y_n^m, $m = -n, \ldots, n$, of spherical harmonics of order $n > 0$, the tangential fields on the unit sphere

(6.53) $$U_n^m := \frac{1}{\sqrt{n(n+1)}} \operatorname{Grad} Y_n^m, \quad V_n^m := \nu \times U_n^m$$

are called *vector spherical harmonics* of order n. Since in spherical coordinates (θ, φ) we have

$$\operatorname{Div} \operatorname{Grad} f = \frac{1}{\sin\theta} \frac{\partial}{\partial\theta} \sin\theta \frac{\partial f}{\partial\theta} + \frac{1}{\sin^2\theta} \frac{\partial^2 f}{\partial\varphi^2},$$

we can rewrite (2.16) in the form

(6.54) $$\operatorname{Div} \operatorname{Grad} Y_n + n(n+1) Y_n = 0$$

for spherical harmonics Y_n of order n. Then, from (6.36) we deduce that

$$\int_\Omega \operatorname{Grad} Y_n^m \cdot \operatorname{Grad} \overline{Y_{n'}^{m'}} \, ds = n(n+1) \int_\Omega Y_n^m \overline{Y_{n'}^{m'}} \, ds.$$

Hence, in view of Stokes' theorem

$$\int_\Omega \nu \times \operatorname{Grad} U \cdot \operatorname{Grad} V \, ds = 0$$

for functions $U, V \in C^1(\Omega)$, the vector spherical harmonics are seen to form an orthonormal system in the space

$$T^2(\Omega) := \{a : \Omega \to \mathbb{C}^3 : a \in L^2(\Omega), \, a \cdot \nu = 0\}$$

of tangential L^2 fields on the unit sphere. Analogous to Theorem 2.7, we wish to establish that this system is complete.

We first show that for a function $f \in C^2(\Omega)$ the Fourier expansion

$$f = \sum_{n=0}^{\infty} \sum_{m=-n}^{n} a_n^m Y_n^m$$

with Fourier coefficients

$$a_n^m = \int_{\Omega} f \overline{Y_n^m} \, ds$$

converges uniformly. Setting $\hat{y} = \hat{x}$ in the addition theorem (2.29) yields

$$\text{(6.55)} \qquad \sum_{m=-n}^{n} |Y_n^m(\hat{x})|^2 = \frac{2n+1}{4\pi}$$

whence, by applying Div Grad and using (6.54),

$$\text{(6.56)} \qquad \sum_{m=-n}^{n} |\operatorname{Grad} Y_n^m(\hat{x})|^2 = \frac{1}{4\pi} \, n(n+1)(2n+1)$$

readily follows. From (6.36) and (6.54) we see that

$$n(n+1) \int_{\Omega} f \overline{Y_n^m} \, ds = - \int_{\Omega} \operatorname{Div} \operatorname{Grad} f \, \overline{Y_n^m} \, ds$$

and therefore Parseval's equality, applied to Div Grad f, shows that

$$\sum_{n=0}^{\infty} n^2(n+1)^2 \sum_{m=-n}^{n} |a_n^m|^2 = \int_{\Omega} |\operatorname{Div} \operatorname{Grad} f|^2 ds.$$

The Schwarz inequality

$$\left[\sum_{n=1}^{N} \sum_{m=-n}^{n} |a_n^m Y_n^m(\hat{x})| \right]^2 \leq \sum_{n=1}^{N} n^2(n+1)^2 \sum_{m=-n}^{n} |a_n^m|^2 \sum_{n=1}^{N} \frac{1}{n^2(n+1)^2} \sum_{m=-n}^{n} |Y_n^m(\hat{x})|^2$$

with the aid of (6.55) now yields a uniformly convergent majorant for the Fourier series of f.

Theorem 6.23 *The vector spherical harmonics U_n^m and V_n^m for $m = -n, \ldots, n$, $n = 1, 2, \ldots,$ form a complete orthonormal system in $T^2(\Omega)$.*

Proof. Assume that the tangential field a belongs to $C^3(\Omega)$ and denote by

$$\alpha_n^m := \int_{\Omega} \operatorname{Div} a \, \overline{Y_n^m} \, ds$$

the Fourier coefficients of $\operatorname{Div} a$. Since $\operatorname{Div} a \in C^2(\Omega)$, we have uniform convergence of the series representation

$$(6.57) \qquad \operatorname{Div} a = \sum_{n=1}^{\infty} \sum_{m=-n}^{n} \alpha_n^m Y_n^m.$$

Note that $\alpha_0^0 = 0$ since $\int_\Omega \operatorname{Div} a \, ds = 0$. Now define

$$(6.58) \qquad u := -\sum_{n=1}^{\infty} \frac{1}{n(n+1)} \sum_{m=-n}^{n} \alpha_n^m Y_n^m.$$

Then proceeding as above in the series for f using Parseval's equality for $\operatorname{Div}\operatorname{Grad}\operatorname{Div} a$ and (6.56) we can show that the term by term derivatives of the series for u are uniformly convergent. Hence, (6.58) defines a function $u \in C^1(\Omega)$. From the uniform convergence of the series (6.57) for $\operatorname{Div} a$ and (6.54), we observe that

$$(6.59) \qquad \operatorname{Div}\operatorname{Grad} u = \operatorname{Div} a$$

in the sense of Definition 6.14.

Analogously, with the Fourier coefficients

$$\beta_n^m := \int_\Omega \operatorname{Div}(\nu \times a) \overline{Y_n^m} \, ds$$

of the uniformly convergent expansion

$$(6.60) \qquad \operatorname{Div}(\nu \times a) = \sum_{n=1}^{\infty} \sum_{m=-n}^{n} \beta_n^m Y_n^m$$

we define a second function $v \in C^1(\Omega)$ by

$$(6.61) \qquad v := \sum_{n=1}^{\infty} \frac{1}{n(n+1)} \sum_{m=-n}^{n} \beta_n^m Y_n^m.$$

It satisfies

$$(6.62) \qquad \operatorname{Div}\operatorname{Grad} v = -\operatorname{Div}(\nu \times a).$$

Then the tangential field

$$b := \operatorname{Grad} u + \nu \times \operatorname{Grad} v$$

is continuous and in view of (6.38), (6.59) and (6.62) it satisfies

$$(6.63) \qquad \operatorname{Div} a = \operatorname{Div} b, \quad \operatorname{Div}(\nu \times a) = \operatorname{Div}(\nu \times b).$$

By Gauss's surface divergence theorem (6.33), for any connected surface $S \subset \Omega$ with smooth boundary ∂S and any tangential field $c \in C^1(\Omega)$ we have

$$\int_S \operatorname{Div}(\nu \times c) \, ds = -\int_{\partial S} t \cdot c \, ds$$

where t stands for the unit tangent to ∂S in counterclockwise orientation. This, together with the uniform convergence of (6.60), can be used to show that the second equation of (6.63) implies that the circulation of $a - b$ around any closed contour on the unit sphere Ω is equal to zero. Hence, $a - b$ can be put in the form $a - b = \text{Grad}\, f$ for some $f \in C^1(\Omega)$. Then the first equation of (6.63) requires f to satisfy $\text{Div Grad}\, f = 0$ and from (6.36) we see that

$$\int_\Omega |\text{Grad}\, f|^2 ds = 0$$

which implies $a = b$.

Thus, we have established that three times continuously differentiable tangential fields can be expanded into a uniformly convergent series with respect to the vector spherical harmonics. The proof is now completed by a denseness argument. □

We now formulate the analogue of Theorem 2.9 for *spherical vector wave functions*.

Theorem 6.24 *Let Y_n be a spherical harmonic of order $n \geq 1$. Then the pair*

$$M_n(x) = \text{curl}\,\{x j_n(k|x|)\, Y_n(\hat{x})\}\,, \quad \frac{1}{ik}\,\text{curl}\, M_n(x)$$

is an entire solution to the Maxwell equations and

$$N_n(x) = \text{curl}\,\{x h_n^{(1)}(k|x|)\, Y_n(\hat{x})\}\,, \quad \frac{1}{ik}\,\text{curl}\, N_n(x)$$

is a radiating solution to the Maxwell equations in $\mathbb{R}^3 \setminus \{0\}$.

Proof. We use (6.4) and Theorems 2.9 and 6.4 to verify by straightforward calculations that the Maxwell equations are satisfied in $\mathbb{R}^3$ and $\mathbb{R}^3 \setminus \{0\}$, respectively. We also compute

(6.64)
$$M_n(x) = j_n(k|x|)\, \text{Grad}\, Y_n(\hat{x}) \times \hat{x},$$

$$N_n(x) = h_n^{(1)}(k|x|)\, \text{Grad}\, Y_n(\hat{x}) \times \hat{x},$$

and

(6.65)
$$\hat{x} \times \text{curl}\, M_n(x) = \frac{1}{|x|}\,\{j_n(k|x|) + k|x| j_n'(k|x|)\}\, \hat{x} \times \text{Grad}\, Y_n(\hat{x}),$$

$$\hat{x} \times \text{curl}\, N_n(x) = \frac{1}{|x|}\,\{h_n^{(1)}(k|x|) + k|x| h_n^{(1)\prime}(k|x|)\}\, \hat{x} \times \text{Grad}\, Y_n(\hat{x}).$$

Hence, the Silver–Müller radiation condition for N_n, $\text{curl}\, N_n/ik$ follows with the aid of the asymptotic behavior (2.41) of the spherical Hankel functions. □

Theorem 6.25 *Let E, H be a radiating solution to the Maxwell equations for $|x| > R > 0$. Then E has an expansion with respect to spherical vector wave functions of the form*

(6.66)

$$E(x) = \sum_{n=1}^{\infty} \sum_{m=-n}^{n} a_n^m \operatorname{curl} \left\{ x h_n^{(1)}(k|x|) Y_n^m(\hat{x}) \right\}$$

$$+ \sum_{n=1}^{\infty} \sum_{m=-n}^{n} b_n^m \operatorname{curl} \operatorname{curl} \left\{ x h_n^{(1)}(k|x|) Y_n^m(\hat{x}) \right\}$$

that (together with its derivatives) converges uniformly on compact subsets of $|x| > R$. Conversely, if the tangential component of the series (6.66) converges in the mean square sense on the sphere $|x| = R$ then the series itself converges (together with its derivatives) uniformly on compact subsets of $|x| > R$ and $E, H = \operatorname{curl} E/ik$ represent a radiating solution to the Maxwell equations.

Proof. By Theorem 6.3, the tangential component of the electric field E on a sphere $|x| = \tilde{R}$ with $\tilde{R} > R$ is analytic. Hence, as shown before the proof of Theorem 6.23, it can be expanded in a uniformly convergent Fourier series with repect to spherical vector harmonics. The spherical Hankel functions $h_n^{(1)}(t)$ and $h_n^{(1)}(t) + t h_n^{(1)\prime}(t)$ do not have real zeros since the Wronskian (2.36) does not vanish. Therefore, in view of (6.64) and (6.65), we may write the Fourier expansion in the form

$$\hat{x} \times E(x) = \sum_{n=1}^{\infty} \sum_{m=-n}^{n} a_n^m \, \hat{x} \times \operatorname{curl} \left\{ x h_n^{(1)}(k|x|) Y_n^m(\hat{x}) \right\}$$

$$+ \sum_{n=1}^{\infty} \sum_{m=-n}^{n} b_n^m \, \hat{x} \times \operatorname{curl} \operatorname{curl} \left\{ x h_n^{(1)}(k|x|) Y_n^m(\hat{x}) \right\}$$

for $|x| = \tilde{R}$. But now, by the continuous dependence of the solution to the exterior Maxwell problem with continuous tangential component in the maximum norm (see Theorem 6.21), the uniform convergence of the latter series implies the uniform convergence of the series (6.66) (together with its derivatives) on compact subsets of $|x| > \tilde{R}$ and the first statement of the theorem is proven.

Conversely, proceeding as in the proof of Theorem 2.14, with the help of (6.56) it can be shown that L^2 convergence of the tangential component of the series (6.66) on the sphere $|x| = R$ implies uniform convergence of the tangential component on any sphere $|x| = \tilde{R} > R$. Therefore, the second part of the theorem follows from the first part applied to the solution to the exterior Maxwell problem with continuous tangential components given on a sphere $|x| = \tilde{R}$. □

Theorem 6.26 *The electric far field pattern of the radiating solution to the Maxwell equations with the expansion (6.66) is given by*

$$(6.67) \qquad E_\infty = \frac{1}{k} \sum_{n=1}^\infty \frac{1}{i^{n+1}} \sum_{m=-n}^n \{ ikb_n^m \operatorname{Grad} Y_n^m - a_n^m \, \nu \times \operatorname{Grad} Y_n^m \}.$$

The coefficients in this expansion satisfy the growth condition

$$(6.68) \qquad \sum_{n=1}^\infty \left(\frac{2n}{ker} \right)^{2n} \sum_{m=-n}^n |a_n^m|^2 + |b_n^m|^2 < \infty$$

for all $r > R$.

Proof. Since by Theorem 6.8 the far field pattern E_∞ is analytic, we have an expansion

$$E_\infty = \sum_{n=1}^\infty \sum_{m=-n}^n \{ c_n^m \operatorname{Grad} Y_n^m + d_n^m \, \nu \times \operatorname{Grad} Y_n^m \}$$

with coefficients

$$n(n+1) \, c_n^m = \int_\Omega E_\infty(\hat{x}) \cdot \operatorname{Grad} \overline{Y_n^m(\hat{x})} \, ds(\hat{x}),$$

$$n(n+1) \, d_n^m = \int_\Omega E_\infty(\hat{x}) \cdot \hat{x} \times \operatorname{Grad} \overline{Y_n^m(\hat{x})} \, ds(\hat{x}).$$

On the other hand, in view of (6.64) and (6.65), the coefficients a_n^m and b_n^m in the expansion (6.66) satisfy

$$n(n+1) \, a_n^m \, h_n^{(1)}(k|x|) = \int_\Omega E(r\hat{x}) \cdot \operatorname{Grad} \overline{Y_n^m(\hat{x})} \times \hat{x} \, ds(\hat{x}),$$

$$n(n+1) \, b_n^m \left\{ k h_n^{(1)\prime}(k|x|) + \frac{1}{|x|} h_n^{(1)}(k|x|) \right\} = \int_\Omega E(r\hat{x}) \cdot \operatorname{Grad} \overline{Y_n^m(\hat{x})} \, ds(\hat{x}).$$

Therefore, with the aid of (2.41) we find that

$$n(n+1) \, d_n^m = \int_\Omega \lim_{r \to \infty} r \, e^{-ikr} E(r\hat{x}) \cdot \hat{x} \times \operatorname{Grad} \overline{Y_n^m(\hat{x})} \, ds(\hat{x})$$

$$= \lim_{r \to \infty} r \, e^{-ikr} \int_\Omega E(r\hat{x}) \cdot \hat{x} \times \operatorname{Grad} \overline{Y_n^m(\hat{x})} \, ds(\hat{x}) = -\frac{n(n+1) \, a_n^m}{k \, i^{n+1}}$$

and

$$n(n+1) \, c_n^m = \int_\Omega \lim_{r \to \infty} r \, e^{-ikr} E(r\hat{x}) \operatorname{Grad} \overline{Y_n^m(\hat{x})} \, ds(\hat{x})$$

$$= \lim_{r \to \infty} r \, e^{-ikr} \int_\Omega E(r\hat{x}) \operatorname{Grad} \overline{Y_n^m(\hat{x})} \, ds(\hat{x}) = \frac{n(n+1) \, b_n^m}{i^n}.$$

In particular, this implies that the expansion (6.67) is valid in the L^2 sense. Parseval's equality for the expansion (6.66) reads

$$r^2 \sum_{n=1}^{\infty} \sum_{m=-n}^{n} n(n+1) |a_n^m|^2 \left| h_n^{(1)}(kr) \right|^2$$

$$+ \ r^2 \sum_{n=1}^{\infty} \sum_{m=-n}^{n} n(n+1) |b_n^m|^2 \left| k h_n^{(1)\prime}(kr) + \frac{1}{r} h_n^{(1)}(kr) \right|^2$$

$$= \int_{|x|=r} |\nu \times E|^2 ds(x).$$

From this, using the asymptotic behavior (2.39) of the Hankel functions for large order n, the condition (6.68) follows.   □

Analogous to Theorem 2.16, it can be shown that the growth condition (6.68) for the Fourier coefficients of a tangential field $E_\infty \in L^2(\Omega)$ is sufficient for E_∞ to be the far field pattern of a radiating solution to the Maxwell equations.

Concluding this section, we wish to apply Theorem 6.25 to derive the vector analogue of the addition theorem (2.42). From (6.64) and (6.65), we see that the coefficients in the expansion (6.66) can be expressed by

$$(6.69) \quad n(n+1) R^2 h_n^{(1)}(kR) a_n^m = \int_{|x|=R} E(x) \cdot \operatorname{Grad} \overline{Y_n^m(\hat{x})} \times \hat{x} \, ds(x),$$

$$(6.70) \quad n(n+1) R\{ h_n^{(1)}(kR) + kR h_n^{(1)\prime}(kR) \} b_n^m = \int_{|x|=R} E(x) \cdot \operatorname{Grad} \overline{Y_n^m(\hat{x})} \, ds(x).$$

Given a set Y_n^m, $m = -n, \ldots, n$, $n = 0, 1, \ldots$, of orthonormal spherical harmonics, we recall the functions

$$u_n^m(x) := j_n(k|x|) Y_n^m(\hat{x}), \quad v_n^m(x) := h_n^{(1)}(k|x|) Y_n^m(\hat{x})$$

and set

$$M_n^m(x) := \operatorname{curl}\{ x u_n^m(x) \}, \quad N_n^m(x) := \operatorname{curl}\{ x v_n^m(x) \}.$$

For convenience, we also write

$$\tilde{M}_n^m(x) := \operatorname{curl}\{ x j_n(k|x|) \overline{Y_n^m(\hat{x})} \}, \quad \tilde{N}_n^m(x) := \operatorname{curl}\{ x h_n^{(1)}(k|x|) \overline{Y_n^m(\hat{x})} \}.$$

From the Stratton–Chu formula (6.8) applied to $\tilde{M}_n^m$ and $\operatorname{curl} \tilde{M}_n^m / ik$, we have

$$\frac{1}{k^2} \operatorname{curl} \operatorname{curl} \int_{|x|=R} \nu(x) \times \operatorname{curl} \tilde{M}_n^m(x) \, \Phi(y, x) \, ds(x)$$

$$+ \operatorname{curl} \int_{|x|=R} \nu(x) \times \tilde{M}_n^m(x) \, \Phi(y, x) \, ds(x) = -\tilde{M}_n^m(y), \quad |y| < R,$$

and from the Stratton–Chu formula (6.15) applied to $\tilde{N}_n^m$ and $\operatorname{curl}\tilde{N}_n^m/ik$, we have

$$\frac{1}{k^2}\operatorname{curl}\operatorname{curl}\int_{|x|=R}\nu(x)\times\operatorname{curl}\tilde{N}_n^m(x)\,\Phi(y,x)\,ds(x)$$

$$+\operatorname{curl}\int_{|x|=R}\nu(x)\times\tilde{N}_n^m(x)\,\Phi(y,x)\,ds(x)=0,\quad |y|<R.$$

Using (6.64), (6.65) and the Wronskian (2.36), from the last two equations we see that

$$\frac{i}{k^3R^2}\operatorname{curl}\operatorname{curl}\int_{|x|=R}\nu(x)\times\operatorname{Grad}\overline{Y_n^m(\hat{x})}\,\Phi(y,x)\,ds(x)=h_n^{(1)}(kR)\tilde{M}_n^m(y)$$

for $|y|<R$ whence

(6.71)
$$\int_{|x|=R}\operatorname{Grad}\overline{Y_n^m(\hat{x})}\times\hat{x}\cdot\operatorname{curl}_x\operatorname{curl}_x\{p\Phi(y,x)\}\,ds(x)$$

$$=ik^3R^2h_n^{(1)}(kR)\,p\cdot\overline{M_n^m(y)},\quad |y|<R,$$

follows for all $p\in\mathbb{R}^3$ with the aid of the vector identity

$$p\cdot\operatorname{curl}_y\operatorname{curl}_y\{c(x)\Phi(y,x)\}=c(x)\cdot\operatorname{curl}_x\operatorname{curl}_x\{p\Phi(y,x)\}.$$

Analogously, from the Stratton–Chu formulas (6.9) and (6.16) for the magnetic field, we can derive that

(6.72)
$$\int_{|x|=R}\operatorname{Grad}\overline{Y_n^m(\hat{x})}\cdot\operatorname{curl}_x\operatorname{curl}_x\{p\Phi(y,x)\}\,ds(x)$$

$$=ikR\{h_n^{(1)}(kR)+kRh_n^{(1)\prime}(kR)\}\,p\cdot\operatorname{curl}\overline{M_n^m(y)},\quad |y|<R.$$

Therefore, from (6.69)–(6.72) we can derive the expansion

$$\frac{1}{k^2}\operatorname{curl}_x\operatorname{curl}_x\{p\cdot\Phi(y,x)\}$$

(6.73)
$$=ik\sum_{n=1}^{\infty}\frac{1}{n(n+1)}\sum_{m=-n}^{n}N_n^m(x)\,\overline{M_n^m(y)}\cdot p$$

$$+\frac{i}{k}\sum_{n=1}^{\infty}\frac{1}{n(n+1)}\sum_{m=-n}^{n}\operatorname{curl}N_n^m(x)\,\operatorname{curl}\overline{M_n^m(y)}\cdot p$$

which, for fixed y, converges uniformly (together with its derivatives) with respect to x on compact subsets of $|x|>|y|$. Interchanging the roles of x and y and using a corresponding expansion of solutions to the Maxwell equations in the interior of a sphere, it can be seen that for fixed x the series (6.73) also converges uniformly (together with its derivatives) with respect to y on compact subsets of $|x| > |y|$.

Using the vector identity

$$\operatorname{div}_x p \cdot \Phi(y,x) = -p \cdot \operatorname{grad}_y \Phi(y,x),$$

from the addition theorem (2.42) we find that

(6.74)
$$\frac{1}{k^2} \operatorname{grad}_x \operatorname{div}_x \{ p \cdot \Phi(y,x) \}$$

$$= -\frac{i}{k} \sum_{n=0}^{\infty} \sum_{m=-n}^{n} \operatorname{grad} v_n^m(x) \, \operatorname{grad} \overline{u_n^m(y)} \cdot p.$$

In view of the continuous dependence results of Theorem 3.9, it can be seen the series (6.74) has the same convergence properties as (6.73). Finally, using the vector identity (6.4), we can use (6.73) and (6.74) to establish the following *vector addition theorem* for the fundamental solution.

Theorem 6.27 *We have*

(6.75)
$$\Phi(x,y)p = ik \sum_{n=1}^{\infty} \frac{1}{n(n+1)} \sum_{m=-n}^{n} N_n^m(x) \, \overline{M_n^m(y)} \cdot p$$

$$+ \frac{i}{k} \sum_{n=1}^{\infty} \frac{1}{n(n+1)} \sum_{m=-n}^{n} \operatorname{curl} N_n^m(x) \, \operatorname{curl} \overline{M_n^m(y)} \cdot p$$

$$+ \frac{i}{k} \sum_{n=0}^{\infty} \sum_{m=-n}^{n} \operatorname{grad} v_n^m(x) \, \operatorname{grad} \overline{u_n^m(y)} \cdot p$$

where the series and its term by term derivatives are uniformly convergent for fixed y with respect to x and, conversely, for fixed x with respect to y on compact subsets of $|x| > |y|$.

6.6 The Reciprocity Relation

We consider the scattering of electromagnetic plane waves

(6.76)
$$E^i(x; d, p) = \frac{i}{k} \operatorname{curl} \operatorname{curl} p \, e^{ik\,x\cdot d} = ik \, (d \times p) \times d \, e^{ik\,x\cdot d},$$

$$H^i(x; d, p) = \operatorname{curl} p \, e^{ik\,x\cdot d} = ik \, d \times p \, e^{ik\,x\cdot d},$$

where the unit vector d describes the direction of propagation and where the constant vector p gives the polarization. We will indicate the dependence of the scattered field, of the total field and of the far field pattern on the incident direction d and the polarization p by writing $E^s(x; d, p)$, $H^s(x; d, p)$, $E(x; d, p)$, $H(x; d, p)$ and $E_\infty(\hat{x}; d, p)$, $H_\infty(\hat{x}; d, p)$, respectively. Note that the dependence on the polarization is linear. Analogous to Theorem 3.13, we establish the following reciprocity result for the electromagnetic case.

Theorem 6.28 *The electric far field pattern for the scattering of plane electromagnetic waves by a perfect conductor satisfies the* reciprocity relation

$$(6.77) \qquad q \cdot E_\infty(\hat{x}; d, p) = p \cdot E_\infty(-d; -\hat{x}, q)$$

for all $\hat{x}, d \in \Omega$ and all $p, q \in \mathbb{R}^3$.

Proof. From Gauss' divergence theorem, the Maxwell equations for the incident and the scattered fields and the radiation condition for the scattered field we have

$$\int_{\partial D} \{ \nu \times E^i(\cdot; d, p) \cdot H^i(\cdot; -\hat{x}, q) + \nu \times H^i(\cdot; d, p) \cdot E^i(\cdot; -\hat{x}, q) \} \, ds = 0$$

and

$$\int_{\partial D} \{ \nu \times E^s(\cdot; d, p) \cdot H^s(\cdot; -\hat{x}, q) + \nu \times H^s(\cdot; d, p) \cdot E^s(\cdot; -\hat{x}, q) \} \, ds = 0.$$

With the aid of

$$\operatorname{curl}_y q \, e^{-ik \, \hat{x} \cdot y} = ik \, q \times \hat{x} \, e^{-ik \, \hat{x} \cdot y},$$

$$\operatorname{curl}_y \operatorname{curl}_y q \, e^{-ik \, \hat{x} \cdot y} = k^2 \, \hat{x} \times \{ q \times \hat{x} \} \, e^{-ik \, \hat{x} \cdot y},$$

from (6.24) we derive

$$4\pi \, q \cdot E_\infty(\hat{x}; d, p)$$

$$= \int_{\partial D} \{ \nu \times E^s(\cdot; d, p) \cdot H^i(\cdot; -\hat{x}, q) + \nu \times H^s(\cdot; d, p) \cdot E^i(\cdot; -\hat{x}, q) \} \, ds$$

and from this by interchanging the roles of d and $\hat{x}$ and of p and q, respectively,

$$4\pi \, p \cdot E_\infty(-d; -\hat{x}, q)$$

$$= \int_{\partial D} \{ \nu \times E^s(\cdot; -\hat{x}, q) \cdot H^i(\cdot; d, p) + \nu \times H^s(\cdot; -\hat{x}, q) \cdot E^i(\cdot; d, p) \} \, ds.$$

We now subtract the last integral from the sum of the three preceding integrals to obtain

$$4\pi \{ q \cdot E_\infty(\hat{x}; d, p) - p \cdot E_\infty(-d; -\hat{x}, q) \}$$

$$(6.78) \\ = \int_{\partial D} \{ \nu(y) \times E(\cdot; d, p) \cdot H(\cdot; -\hat{x}, q) + \nu(y) \times H(\cdot; d, p) \cdot E(\cdot; -\hat{x}, q) \} \, ds,$$

whence the reciprocity relation (6.77) follows from the boundary condition $\nu \times E(\cdot\,; d, p) = \nu \times E(\cdot\,; -\hat{x}, q) = 0$ on ∂D.   □

Since in the derivation of (6.78) we only made use of the Maxwell equations for the incident wave in $\mathbb{R}^3$ and for the scattered wave in $\mathbb{R}^3 \setminus \bar{D}$ and the radiation condition, it is obvious that the reciprocity condition is also valid for the impedance boundary condition and the transmission boundary condition.

Before we can state results on the completeness of electric far field patterns corresponding to Theorem 3.17, we have to introduce the concept of electromagnetic Herglotz pairs. Consider the vector Herglotz wave function

$$E(x) = \int_\Omega e^{ik\,x\cdot d} a(d)\, ds(d), \quad x \in \mathbb{R}^3,$$

with a vector Herglotz kernel $a \in L^2(\Omega)$, that is, the cartesian components of E are Herglotz wave functions. From

$$\text{div}\, E(x) = ik \int_\Omega e^{ik\,x\cdot d}\, d \cdot a(d)\, ds(d), \quad x \in \mathbb{R}^3,$$

and Theorem 3.15 we see that the property of the kernel a to be tangential is equivalent to $\text{div}\, E = 0$ in $\mathbb{R}^3$.

Definition 6.29 *An electromagnetic Herglotz pair is a pair of vector fields of the form*

$$(6.79) \quad E(x) = \int_\Omega e^{ik\,x\cdot d} a(d)\, ds(d), \quad H(x) = \frac{1}{ik}\, \text{curl}\, E(x), \quad x \in \mathbb{R}^3,$$

where the square integrable tangential field a on the unit sphere Ω is called the Herglotz kernel of E, H.

Herglotz pairs obviously represent entire solutions to the Maxwell equations. For any electromagnetic Herglotz pair E, H with kernel a, the pair $H, -E$ is also an electromagnetic Herglotz pair with kernel $d \times a$. Using Theorem 3.22, we can characterize Herglotz pairs through a growth condition in the following theorem.

Theorem 6.30 *An entire solution E, H to the Maxwell equations possesses the growth property*

$$(6.80) \qquad \sup_{R>0} \frac{1}{R} \int_{|x| \leq R} \{|E(x)|^2 + |H(x)|^2\}\, dx < \infty$$

if and only if it is an electromagnetic Herglotz pair.

Proof. For a pair of entire solutions E, H to the vector Helmholtz equation, by Theorem 3.22, the growth condition (6.80) is equivalent to the property that E

can be represented in the form (6.79) with a square integrable field a and the property div $E = 0$ in $\mathbb{R}^3$ is equivalent to a being tangential. $\square$

In the following analysis, we will utilize the analogue of Lemma 3.16 for the superposition of solutions to the electromagnetic scattering problem.

Lemma 6.31 *Given a L^2 field g on Ω, the solution to the perfect conductor scattering problem for the incident wave*

$$\tilde{E}^i(x) = \int_\Omega E^i(x; d, g(d)) \, ds(d), \quad \tilde{H}^i(x) = \int_\Omega H^i(x; d, g(d)) \, ds(d)$$

is given by

$$\tilde{E}^s(x) = \int_\Omega E^s(x; d, g(d)) \, ds(d), \quad \tilde{H}^s(x) = \int_\Omega H^s(x; d, g(d)) \, ds(d)$$

for $x \in \mathbb{R}^3 \setminus \bar{D}$ and has the far field pattern

$$\tilde{E}_\infty(\hat{x}) = \int_\Omega E_\infty(\hat{x}; d, g(d)) \, ds(d), \quad \tilde{H}_\infty(\hat{x}) = \int_\Omega H_\infty(\hat{x}; d, g(d)) \, ds(d)$$

for $\hat{x} \in \Omega$.

Proof. Multiply (6.48) and (6.49) by g, integrate with respect to d over Ω and interchange orders of integration. $\square$

We note that for a tangential field $g \in T^2(\Omega)$ we can write

$$\tilde{E}^i(x) = ik \int_\Omega g(d) \, e^{ik\,x\cdot d} \, ds(d), \quad \tilde{H}^i(x) = \text{curl} \int_\Omega g(d) \, e^{ik\,x\cdot d} \, ds(d), \quad x \in \mathbb{R}^3,$$

that is, $\tilde{E}^i, \tilde{H}^i$ represents an electromagnetic Herglotz pair with kernel ikg.

Variants of the following completeness results were first obtained by Colton and Kress [33] and by Blöhbaum [14].

Theorem 6.32 *Let (d_n) be a sequence of unit vectors that is dense on Ω and define the set $\mathcal{F}$ of electric far field patterns by*

$$\mathcal{F} := \{ E_\infty(\cdot \,; d_n, e_j) : n = 1, 2, \ldots, j = 1, 2, 3 \}$$

with the cartesian unit vectors e_j. Then $\mathcal{F}$ is complete in $T^2(\Omega)$ if and only if there does not exist a nontrivial electromagnetic Herglotz pair E, H satisfying $\nu \times E = 0$ on ∂D.

Proof. By the continuity of E_∞ as a function of d and the Reciprocity Theorem 6.28, the completeness condition

$$\int_\Omega h(\hat{x}) \cdot E_\infty(\hat{x}; d_n, e_j) \, ds(\hat{x}) = 0, \quad n = 1, 2, \ldots, j = 1, 2, 3,$$

for a tangential field $h \in T^2(\Omega)$ is equivalent to

$$\int_\Omega p \cdot E_\infty(\hat{x}; d, g(d)) \, ds(d) = 0, \quad \hat{x} \in \Omega, \ p \in \mathbb{R}^3,$$

that is,

(6.81) $$\int_\Omega E_\infty(\hat{x}; d, g(d)) \, ds(d) = 0, \quad \hat{x} \in \Omega,$$

for $g \in T^2(\Omega)$ with $g(d) = h(-d)$.

By Theorem 3.15 and Lemma 6.31, the existence of a nontrivial tangential field g satisfying (6.81) is equivalent to the existence of a nontrivial Herglotz pair $\tilde{E}^i, \tilde{H}^i$ (with kernel ikg) for which the electric far field pattern of the corresponding scattered wave $\tilde{E}^s$ fulfills $\tilde{E}_\infty = 0$. By Theorem 6.9, the vanishing electric far field $\tilde{E}_\infty = 0$ on Ω is equivalent to $\tilde{E}^s = 0$ in $\mathbb{R}^3 \setminus \bar{D}$. This in turn, by the boundary condition $\nu \times \tilde{E}^i + \nu \times \tilde{E}^s = 0$ on ∂D and the uniqueness of the solution to the exterior Maxwell problem, is equivalent to $\nu \times \tilde{E}^i = 0$ on ∂D and the proof is finished. $\square$

We note that the set $\mathcal{F}$ of electric far field patterns is linearly dependent. Since for $p = d$ the incident fields (6.76) vanish, the corresponding electric far field pattern also vanishes. This implies

(6.82) $$\sum_{j=1}^{3} d_j E_\infty(\,\cdot\,; d, e_j) = 0$$

for $d = (d_1, d_2, d_3)$. Of course, the completeness result of Theorem 6.32 also holds for the magnetic far field patterns.

A nontrivial electromagnetic Herglotz pair for which the tangential component of the electric field vanishes on the boundary ∂D is a Maxwell eigensolution, i.e., a nontrivial solution E, H to the Maxwell equations in D with homogeneous boundary condition $\nu \times E = 0$ on ∂D. Therefore, as in the acoustic case, we have the surprising result that the eigensolutions are connected to the exterior scattering problem.

From Theorem 6.30, with the aid of the differentiation formula (2.34), the integral (3.45) and the representations (6.64) and (6.65) for M_n and curl M_n, we conclude that M_n, curl M_n/ik provide examples of electromagnetic Herglotz pairs. From (6.64) we observe that the spherical vector wave functions M_n describe Maxwell eigensolutions for a ball of radius R centered at the origin with the eigenvalues k given by the zeros of the spherical Bessel functions. The pairs curl M_n, $-ikM_n$ are also electromagnetic Herglotz pairs. From (6.65) we see that the spherical vector wave functions curl M_n also yield Maxwell eigensolutions for the ball with the eigenvalues k given through $j_n(kR) + kRj_n'(kR) = 0$. By expansion of an arbitrary eigensolution with respect to vector spherical harmonics, and arguing as in the proof of Rellich's Lemma 2.11, it can be seen that all Maxwell eigensolutions in a ball must be spherical vector wave functions. Therefore, the eigensolutions for balls are always electromagnetic Herglotz pairs and by Theorem 6.32 the electric far field patterns for plane waves are not complete for a ball D when k is a Maxwell eigenvalue.

We can express the result of Theorem 6.32 also in terms of a far field operator in the following corollary.

Corollary 6.33 *The linear operator $F : T^2(\Omega) \to T^2(\Omega)$ defined by*

$$(Fg)(\hat{x}) := \int_\Omega E_\infty(\hat{x}; d, g(d)) \, ds(d), \quad \hat{x} \in \Omega,$$

is injective and has dense range if and only if there does not exist a Maxwell eigensolution for D which is an electromagnetic Herglotz pair.

Proof. From the reciprocity relation (6.77), we easily derive that the L^2 adjoint $F^* : T^2(\Omega) \to T^2(\Omega)$ of F satisfies

$$(F^*h)(d) = \overline{(Fg)(-d)}, \quad d \in \Omega,$$

where $g(\hat{x}) = \overline{h(-\hat{x})}$ and the proof is completed as in Corollary 3.18. $\square$

The question of when we can find a superposition of incident electromagnetic plane waves of the form (6.76) such that the resulting far field pattern becomes a prescribed far field $\tilde{E}_\infty$ is examined in the following theorem.

Theorem 6.34 *Let $\tilde{E}^s, \tilde{H}^s$ be a radiating solution to the Maxwell equations with electric far field pattern $\tilde{E}_\infty$. Then the linear integral equation of the first kind*

$$(6.83) \qquad \int_\Omega E_\infty(\hat{x}; d, g(d)) \, ds(d) = \tilde{E}_\infty(\hat{x}), \quad \hat{x} \in \Omega,$$

possesses a solution $g \in T^2(\Omega)$ if and only if $\tilde{E}^s, \tilde{H}^s$ is defined in $\mathbb{R}^3 \setminus \bar{D}$ and continuous in $\mathbb{R}^3 \setminus D$ and the interior boundary value problem for the Maxwell equations

$$(6.84) \qquad \operatorname{curl} \tilde{E}^i - ik\tilde{H}^i = 0, \quad \operatorname{curl} \tilde{H}^i + ik\tilde{E}^i = 0 \quad \text{in } D,$$

$$(6.85) \qquad \nu \times (\tilde{E}^i + \tilde{E}^s) = 0 \quad \text{on } \partial D$$

is solvable and a solution $\tilde{E}^i, \tilde{H}^i$ is an electromagnetic Herglotz pair.

Proof. By Theorem 3.15 and Lemma 6.31, the solvability of the integral equation (6.83) for g is equivalent to the existence of a Herglotz pair $\tilde{E}^i, \tilde{H}^i$ (with kernel ikg) for which the electric far field pattern for the scattering by the perfect conductor D coincides with the given $\tilde{E}_\infty$, that is, the scattered electromagnetic field coincides with the given $\tilde{E}^s, \tilde{H}^s$. $\square$

By reciprocity, the solvability of (6.83) is equivalent to the solvability of

$$(6.86) \qquad \int_\Omega E_\infty(\hat{x}; d, e_j) h(\hat{x}) \, ds(\hat{x}) = e_j \cdot \tilde{E}_\infty(-d), \quad d \in \Omega, \; j = 1, 2, 3,$$

where $h(\hat{x}) = g(-\hat{x})$. In the special cases where the prescribed scattered wave is given by the electromagnetic field

$$\tilde{E}^s(x) = \operatorname{curl} a\Phi(x,0), \quad \tilde{H}^s(x) = \frac{1}{ik} \operatorname{curl} \tilde{E}^s(x)$$

of a magnetic dipole at the origin with electric far field pattern

$$\tilde{E}_\infty(\hat{x}) = \frac{ik}{4\pi} \hat{x} \times a,$$

or by the electromagnetic field

$$\tilde{E}^s(x) = \operatorname{curl}\operatorname{curl} a\Phi(x,0), \quad \tilde{H}^s(x) = \frac{1}{ik} \operatorname{curl} \tilde{E}^s(x)$$

of an electric dipole with electric far field pattern

$$\tilde{E}_\infty(\hat{x}) = \frac{k^2}{4\pi} \hat{x} \times (a \times \hat{x}),$$

the connection between the solution to the integral equation (6.86) and the interior Maxwell problem (6.84), (6.85) was first obtained by Blöhbaum [14]. Corresponding completeness results for the impedance boundary condition were derived by Angell, Colton and Kress [3].

We now follow Colton and Kirsch [29] and indicate how a suitable linear combination of the electric and the magnetic far field patterns with orthogonal polarization vectors are complete without exceptional interior eigenvalues.

Theorem 6.35 *Let (d_n) be a sequence of unit vectors that is dense on Ω and let λ and μ be fixed nonzero real numbers. Then the set*

$$\{\lambda E_\infty(\cdot\,; d_n, e_j) + \mu H_\infty(\cdot\,; d_n, e_j \times d_n) : n = 1, 2, \ldots, j = 1, 2, 3\}$$

is complete in $T^2(\Omega)$.

Proof. We use the continuity of E_∞ and H_∞ as a function of d, the orthogonality (6.23), that is, $H_\infty(\hat{x}; \cdot\,, \cdot) = \hat{x} \times E_\infty(\hat{x}; \cdot\,, \cdot)$, and the reciprocity relation (6.77) to find that for a tangential field $h \in T^2(\Omega)$ the completeness condition

$$\int_\Omega h(\hat{x}) \cdot \{\lambda E_\infty(\hat{x}; d_n, e_j) + \mu H_\infty(\hat{x}; d_n, e_j \times d_n)\}\, ds(\hat{x}) = 0$$

for $n = 1, 2, \ldots$ and $j = 1, 2, 3$ is equivalent to

$$\int_\Omega \{\lambda\, p \cdot E_\infty(-d; -\hat{x}, h(\hat{x})) + \mu\, p \times d \cdot E_\infty(-d; -\hat{x}, h(\hat{x}) \times \hat{x})\}\, ds(\hat{x}) = 0$$

for all $p \in \mathbb{R}^3$, that is,

$$\int_\Omega \{\lambda E_\infty(-d; -\hat{x}, h(\hat{x})) - \mu H_\infty(-d; -\hat{x}, h(\hat{x}) \times \hat{x})\}\, ds(\hat{x}) = 0, \quad d \in \Omega.$$

Relabeling and using the linearity of the far field pattern with respect to polarization yields

$$(6.87) \quad \int_\Omega \{\lambda E_\infty(\hat{x}; d, g(d)) + \mu H_\infty(\hat{x}; d, g(d) \times d)\}\, ds(d) = 0, \quad \hat{x} \in \Omega,$$

with $g(d) = h(-d)$.

We define incident waves E_1^i, H_1^i as Herglotz pair with kernel g and E_2^i, H_2^i as Herglotz pair with kernel $g \times d$ and note that from

$$\frac{1}{ik}\, \text{curl} \int_\Omega g(d) \times d\, e^{ik\,x\cdot d}\, ds(d) = \int_\Omega g(d)\, e^{ik\,x\cdot d}\, ds(d)$$

we have

$$(6.88) \qquad\qquad H_2^i = E_1^i \quad \text{and} \quad E_2^i = -H_1^i.$$

By Lemma 6.31, the condition (6.87) implies that $\lambda E_1^s + \mu H_2^s = 0$ in $\mathbb{R}^3 \setminus \bar{D}$ whence by applying the curl we have

$$(6.89) \qquad\qquad \lambda H_1^s - \mu E_2^s = 0 \quad \text{in } \mathbb{R}^3 \setminus \bar{D}.$$

Using Gauss' divergence theorem, the perfect conductor boundary condition $\nu \times (E_1^i + E_1^s) = 0$ and $\nu \times (E_2^i + E_2^s) = 0$ on ∂D and the equations (6.88) and (6.89) we now deduce that

$$\lambda \int_{\partial D} \nu \times E_1^s \cdot \overline{H_1^s}\, ds = \mu \int_D \text{div}\left\{ E_1^i \times \overline{E_2^i} \right\} dx = i\mu k \int_D \left\{ |E_1^i|^2 - |H_1^i|^2 \right\} dx,$$

and this implies $E_1^s = 0$ in $\mathbb{R}^3 \setminus \bar{D}$ by Theorem 6.10. Then (6.89) also yields $E_2^s = 0$ in $\mathbb{R}^3 \setminus \bar{D}$. As a consequence of the boundary conditions, we obtain $\nu \times E_1^i = 0$ on ∂D and $\nu \times H_1^i = \nu \times E_2^i = 0$ on ∂D and from this we have $E_1^i = 0$ in D by the representation Theorem 6.2. It now follows that $E_1^i = 0$ in $\mathbb{R}^3$ by analyticity (Theorem 6.3). From this we finally arrive at $g = 0$, that is, $h = 0$ by Theorem 3.15. □

We conclude this section with the vector analogue of the completeness Theorem 3.20.

Theorem 6.36 *Let (d_n) be a sequence of unit vectors that is dense on Ω. Then the tangential components of the total magnetic fields*

$$\{\nu \times H(\cdot; d_n, e_j) : n = 1, 2, \ldots, j = 1, 2, 3\}$$

for the incident plane waves of the form (6.76) with directions d_n are complete in $T^2(\partial D)$.

Proof. From the decomposition (6.40), it can be deduced that the operator N maps tangential fields from the Sobolev space $H^1(\partial D)$ boundedly into $T^2(\partial D)$.

Therefore, the composition $NPS_0^2 : T^2(\partial D) \to T^2(\partial D)$ is compact since by Theorem 3.6 the operator S_0 is bounded from $L^2(\partial D)$ into $H^1(\partial D)$. Since M has a weakly singular kernel, we thus have compactness of $M + iNPS_0^2$ from $T^2(\partial D)$ into $T^2(\partial D)$ and proceeding as in the proof of Theorem 6.21 we can show that the operator $I + M + iNPS_0^2$ has a trivial nullspace in $T^2(\partial D)$. Hence, by the Riesz–Fredholm theory, $I + M + iNPS_0^2 : T^2(\partial D) \to T^2(\partial D)$ is bijective and has a bounded inverse.

From the representation formulas (6.51), the boundary condition $\nu \times E = 0$ on ∂D and the jump relations of Theorem 6.11, in view of the definitions (6.43) and (6.44) we deduce that $b = \{\nu \times H\} \times \nu$ solves the integral equation

$$b + M'b + iPS_0^2 Nb = 2\{\nu \times H^i\} \times \nu - 2kPS_0^2\{\nu \times E^i\}.$$

Now let $g \in T^2(\partial D)$ satisfy

$$\int_{\partial D} g \cdot b(\cdot\,; d_n, e_j)\, ds = 0, \quad n = 1, 2, \dots, j = 1, 2, 3.$$

This, by the continuity of the electric to magnetic boundary component map (Theorem 6.20), implies that

$$\int_{\partial D} g \cdot b(\cdot\,; d, p)\, ds = 0, \quad d \in \Omega, p \in \mathbb{R}^3.$$

We now set

$$a := (I + M + iNPS_0^2)^{-1} g$$

and obtain

$$\int_{\partial D} a \cdot (I + M' + iPS_0^2 N) b\, ds = 0$$

and consequently

$$\int_{\partial D} a \cdot \{\{\nu \times H^i(\cdot\,; d, p)\} \times \nu - kS_0^2\{\nu \times E^i(\cdot\,; d, p)\}\}\, ds = 0,$$

that is,

$$\int_{\partial D} \{a \cdot H^i(\cdot\,; d, p) + k\{\nu \times S_0^2 a\} \cdot E^i(\cdot\,; d, p)\}\, ds = 0$$

for all $d \in \Omega$ and $p \in \mathbb{R}^3$. This in turn, by elementary vector algebra, implies that

$$(6.90) \quad \int_{\partial D} \{a(y) \times d + k\, d \times [\{\nu(y) \times (S_0^2 a)(y)\} \times d]\}\, e^{ik\, y \cdot d}\, ds(y) = 0$$

for all $d \in \Omega$.

Now consider the electric field of the combined magnetic and electric dipole distribution

$$E(x) = \mathrm{curl} \int_{\partial D} a(y)\Phi(x, y)\, ds(y)$$

$$+ i\, \mathrm{curl\, curl} \int_{\partial D} \nu(y) \times (S_0^2 a)(y)\Phi(x, y)\, ds(y), \quad x \in \mathbb{R}^3 \setminus \partial D.$$

By (6.25) and (6.26), its far field pattern is given by

$$E_\infty(\hat{x}) = \frac{ik}{4\pi} \int_{\partial D} \left\{ \hat{x} \times a(y) + k\,\hat{x} \times \left[\{ \nu(y) \times (S_0^2 a)(y) \} \times \hat{x} \right] \right\} e^{-ik\,\hat{x}\cdot y}\, ds(y)$$

for $\hat{x} \in \Omega$. Hence, the condition (6.90) implies that $E_\infty = 0$ on Ω and Theorem 6.9 yields $E = 0$ in $\mathbb{R}^3 \setminus \bar{D}$. From this, with the help of the decomposition (6.40) and the L^2 jump relations (3.23) and (6.46), we derive that

$$a + Ma + iNPS_0^2 a = 0$$

whence $g = 0$ follows and the proof is finished. $\qquad\qquad\qquad\qquad\qquad\square$

With the tools involved in the proof of the preceding theorem we can establish the following result which we shall later need in our analysis of the inverse problem.

Theorem 6.37 *The operator $B : T_d^{0,\alpha}(\partial D) \to L^2(\Omega)$ which maps the electric tangential components of radiating solutions $E, H \in C^1(\mathbb{R}^3 \setminus D)$ to the Maxwell equations onto the electric far field pattern E_∞ can be extended to an injective bounded linear operator $B : T^2(\partial D) \to L^2(\Omega)$.*

Proof. From the form (6.48) of the solution to the exterior Maxwell problem and (6.25) and (6.26), we derive

$$E_\infty(\hat{x}) = \frac{ik}{2\pi} \int_{\partial D} \left\{ \hat{x} \times a(y) + k\,\hat{x} \times \left[\{ \nu(y) \times (S_0^2 a)(y) \} \times \hat{x} \right] \right\} e^{-ik\,\hat{x}\cdot y}\, ds(y)$$

for $\hat{x} \in \Omega$ where $a = (I + M + iNPS_0^2)^{-1}(\nu \times E)$. The theorem now follows by using the analysis of the previous proof. $\qquad\qquad\qquad\qquad\qquad\square$

7. Inverse Electromagnetic Obstacle Scattering

This last chapter on obstacle scattering is concerned with the extension of the results from Chapter 5 on inverse acoustic scattering to inverse electromagnetic scattering. In order to avoid repeating ourselves, we keep this chapter short by referring back to the corresponding parts of Chapter 5 when appropriate. In particular, for notations and for the motivation of our analysis we urge the reader to get reacquainted with the corresponding analysis in Chapter 5 on acoustics. We again follow the general guideline of our book and consider only one of the many possible inverse electromagnetic obstacle problems: given the electric far field pattern for one or several incident plane electromagnetic waves and knowing that the scattering obstacle is perfectly conducting, find the shape of the scatterer.

We begin the chapter with a uniqueness result. Due to the lack of an appropriate selection theorem, we do not follow Schiffer's proof as in acoustics. Instead of this, we prove a uniqueness result following Isakov's approach and, in addition, we use a method based on differentiation with respect to the wave number. We also include the electromagnetic version of Karp's theorem.

We then proceed to establish a continuous dependence result on the boundary based on the integral equation approach. The final sections of this chapter give an outline of the extensions of the methods due to Kirsch and Kress and to Colton and Monk from acoustics to electromagnetics.

7.1 Uniqueness

For the investigation of uniqueness in inverse electromagnetic obstacle scattering, as in the case of the Neumann boundary condition in acoustics, Schiffer's method cannot be applied since the appropriate selection theorem in electromagnetics requires the boundary to be sufficiently smooth (see [119]). However, the methods used in Theorem 5.6 for inverse acoustic scattering from sound-hard obstacles can be extended to the case of inverse electromagnetic scattering from perfect conductors.

Theorem 7.1 *Assume that D_1 and D_2 are two perfect conductors such that for a fixed wave number the electric far field patterns for both scatterers coincide for all incident directions and all polarizations. Then $D_1 = D_2$.*

Proof. The proof is completely analogous to that of Theorem 5.6 which was based on Theorems 5.4 and 5.5. Instead of the well-posedness for the exterior Neumann problem (Theorem 3.10) we have to use the well-posedness for the exterior Maxwell problem (Theorem 6.21), and instead of point sources $\Phi(\cdot, x_n)$ as incident fields we use electric dipoles $\operatorname{curl} \operatorname{curl} p\Phi(\cdot, x_n)$.   □

For diversity, we now prove a uniqueness theorem for fixed direction and polarization.

Theorem 7.2 *Assume that D_1 and D_2 are two perfect conductors such that for one fixed incident direction and polarization the electric far field patterns of both scatterers coincide for all wave numbers contained in some interval $0 < k_1 < k < k_2 < \infty$. Then $D_1 = D_2$.*

Proof. We will use the fact that the scattered wave depends analytically on the wave number k. Deviating from our usual notation, we indicate the dependence on the wave number by writing $E^i(x; k)$, $E^s(x; k)$, and $E(x; k)$. Since the fundamental solution to the Helmholtz equation depends analytically on k, the integral operator $I + M + iNPS_0^2$ in the integral equation (6.49) is also analytic in k. (For the reader who is not familiar with analytic operators, we refer to Section 8.5.) From the fact that for each $k > 0$ the inverse operator of $I + M + iNPS_0^2$ exists, by using a Neumann series argument it can be deduced that the inverse $(I + M + iNPS_0^2)^{-1}$ is also analytic in k. Therefore, the analytic dependence of the right hand side $c = 2E^i(\cdot; k) \times \nu$ of (6.49) for the scattering problem implies that the solution a also depends analytically on k and consequently from the representation (6.48) it can be seen that the scattered field $E^s(\cdot; k)$ also depends analytically on k. In addition, from (6.48) it also follows that the derivatives of E^s with respect to the space variables and with respect to the wave number can be interchanged. Therefore, from the vector Helmholtz equation $\Delta E + k^2 E = 0$ for the total field $E = E^i + E^s$ we derive the inhomogeneous vector Helmholtz equation

$$\Delta F + k^2 F = -2kE$$

for the derivative

$$F := \frac{\partial E}{\partial k}.$$

Let k_0 be an accumulation point of the wave numbers for the incident waves and assume that $D_1 \neq D_2$. By Theorem 6.9, the electric far field pattern uniquely determines the scattered field. Hence, for any incident wave E^i the scattered wave E^s for both obstacles coincide in the unbounded component G of the complement of $D_1 \cup D_2$. Without loss of generality, we assume that

$(\mathbb{R}^3 \setminus G) \setminus \bar{D}_2$ is a nonempty open set and denote by D^* a connected component of $(\mathbb{R}^3 \setminus G) \setminus \bar{D}_2$. Then E is defined in D^* since it describes the total wave for D_2, that is, E satisfies the vector Helmholtz equation in D^* and fulfills homogeneous boundary conditions $\nu \times E = 0$ and $\operatorname{div} E = 0$ on ∂D^* for each k with $k_1 < k < k_2$. By differentiation with respect to k, it follows that $F(\cdot\,; k_0)$ satisfies the same homogeneous boundary conditions. Therefore, from Green's vector theorem (6.3) applied to $E(\cdot\,; k_0)$ and $F(\cdot\,; k_0)$ we find that

$$2k_0 \int_{D^*} |E|^2 dx = \int_{D^*} \{\bar{F} \bigtriangleup E - E \bigtriangleup \bar{F}\} dx = 0,$$

whence $E = 0$ first in D^* and then by analyticity everywhere outside $D_1 \cup D_2$. This implies that E^i satisfies the radiation condition whence $E^i = 0$ in $\mathbb{R}^3$ follows (c.f. p. 156). This is a contradiction. $\square$

We include in this section on uniqueness the electromagnetic counterpart of Karp's theorem for acoustics. If the perfect conductor D is a ball centered at the origin, it is obvious from symmetry considerations that the electric far field pattern for incoming plane waves of the from (6.76) satisfies

(7.1) $E_\infty(Q\hat{x}; Qd, Qp) = QE_\infty(\hat{x}; d, p)$

for all $\hat{x}, d \in \Omega$, all $p \in \mathbb{R}^3$ and all rotations Q, i.e., for all real orthogonal matrices Q with $\det Q = 1$. As shown by Colton and Kress [34], the converse of this statement is also true. We include a simplified version of the original proof.

The vectors $\hat{x}$, $p \times \hat{x}$ and $\hat{x} \times (p \times \hat{x})$ form a basis in $\mathbb{R}^3$ provided $p \times \hat{x} \neq 0$. Hence, since the electric far field pattern is orthogonal to $\hat{x}$, we can write

$$E_\infty(\hat{x}; d, p) = e_1(\hat{x}; d, p)\, p \times \hat{x} + e_2(\hat{x}; d, p)\, \hat{x} \times (p \times \hat{x})$$

and the condition (7.1) is equivalent to

$$e_j(Q\hat{x}; Qd, Qp) = e_j(\hat{x}; d, p), \quad j = 1, 2.$$

This implies that

$$\int_\Omega e_j(\hat{x}; d, p)\, ds(d) = \int_\Omega e_j(Q\hat{x}; d, Qp)\, ds(d)$$

and therefore

(7.2) $\displaystyle\int_\Omega E_\infty(\hat{x}; d, p)\, ds(d) = c_1(\theta)\, p \times \hat{x} + c_2(\theta)\, \hat{x} \times (p \times \hat{x})$

for all $\hat{x} \in \Omega$ and all $p \in \mathbb{R}^3$ with $p \times \hat{x} \neq 0$ where c_1 and c_2 are functions depending only on the angle θ between $\hat{x}$ and p. Given $p \in \mathbb{R}^3$ such that $0 < \theta < \pi/2$, we also consider the vector

$$q := 2\,\hat{x} \cdot p\,\hat{x} - p$$

which clearly makes the same angle with $\hat{x}$ as p. From the linearity of the electric far field pattern with respect to polarization, we have

$$(7.3) \qquad E_\infty(\hat{x}; d, \lambda p + \mu q) = \lambda E_\infty(\hat{x}; d, p) + \mu E_\infty(\hat{x}; d, q)$$

for all $\lambda, \mu \in \mathbb{R}$. Since $q \times \hat{x} = -p \times \hat{x}$, from (7.2) and (7.3) we can conclude that

$$c_j(\theta_{\lambda\mu}) = c_j(\theta), \quad j = 1, 2,$$

for all $\lambda, \mu \in \mathbb{R}$ with $\lambda \neq \mu$ where $\theta_{\lambda\mu}$ is the angle between $\hat{x}$ and $\lambda p + \mu q$. This now implies that both functions c_1 and c_2 are constants since, by choosing λ and μ appropriately, we can make $\theta_{\lambda\mu}$ to be any angle between 0 and π. With these constants, by continuity, (7.2) is valid for all $\hat{x} \in \Omega$ and all $p \in \mathbb{R}^3$.

Choosing a fixed but arbitrary vector $p \in \mathbb{R}^3$ and using the Funk–Hecke formula (2.44), we consider the superposition of incident plane waves given by

$$(7.4) \qquad \tilde{E}^i(x) = \frac{i}{k} \operatorname{curl} \operatorname{curl} p \int_\Omega e^{ik\,x\cdot d} ds(d) = \frac{4\pi i}{k^2} \operatorname{curl} \operatorname{curl} p \, \frac{\sin k|x|}{|x|}.$$

Then, by Lemma 6.31 and (7.2), the corresponding scattered wave $\tilde{E}^s$ has the electric far field pattern

$$\tilde{E}_\infty(\hat{x}) = c_1 \, p \times \hat{x} + c_2 \, \hat{x} \times (p \times \hat{x}).$$

From this, with the aid of (6.25) and (6.26), we conclude that

$$(7.5) \qquad \tilde{E}^s(x) = \frac{ic_1}{k} \operatorname{curl} p \, \frac{e^{ik|x|}}{|x|} + \frac{c_2}{k^2} \operatorname{curl} \operatorname{curl} p \, \frac{e^{ik|x|}}{|x|}$$

follows. Using (7.4) and (7.5) and setting $r = |x|$, the boundary condition $\nu \times (\tilde{E}^i + \tilde{E}^s) = 0$ on ∂D can be brought into the form

$$(7.6) \qquad \nu(x) \times \{g_1(r)\, p + g_2(r)\, p \times x + g_3(r)\, (p \cdot x)\, x\} = 0, \quad x \in \partial D,$$

for some functions g_1, g_2, g_3. In particular,

$$g_1(r) = \frac{4\pi i}{k^2 r} \left\{ \frac{d}{dr} \frac{\sin kr}{r} + k^2 \sin kr + \frac{c_2}{4\pi i} \frac{d}{dr} \frac{e^{ikr}}{r} + \frac{c_2 k^2}{4\pi i} e^{ikr} \right\}.$$

For a fixed, but arbitrary $x \in \partial D$ with $x \neq 0$ we choose p to be orthogonal to x and take the scalar product of (7.6) with $p \times x$ to obtain

$$g_1(r)\, x \cdot \nu(x) = 0, \quad x \in \partial D.$$

Assume that $g_1(r) \neq 0$. Then $x \cdot \nu(x) = 0$ and inserting $p = x \times \nu(x)$ into (7.6) we arrive at the contradiction $g_1(r)\, x = 0$. Hence, since $x \in \partial D$ can be chosen arbitrarily, we have that

$$g_1(r) = 0$$

for all $x \in \partial D$ with $x \neq 0$. Since g_1 does not vanish identically and is analytic, it can have only discrete zeros. Therefore, $r = |x|$ must be constant for all $x \in \partial D$, i.e., D is a ball with center at the origin.

7.2 Continuous Dependence on the Boundary

In this section, as in the case of acoustic obstacle scattering, we wish to study some of the properties of the far field operator

$$F : \partial D \mapsto E_\infty$$

which for a fixed incident plane wave E^i maps the boundary ∂D of the perfect conductor D onto the electric far field pattern E_∞ of the scattered wave.

We first briefly wish to indicate why the weak solution methods used in the proof of Theorems 5.7 and 5.8 have no immediate counterpart for the electromagnetic case. Let B be the electric to magnetic boundary component map for radiating solutions to the Maxwell equations in the exterior of the sphere Ω_R of radius R centered at the origin. It transforms the tangential components of the electric field into the tangential components of the magnetic field

$$B : \nu \times E \mapsto (\nu \times \operatorname{curl} E) \times \nu \quad \text{on } \Omega_R.$$

Note that the definition of the operator B is slightly different from the one used in Theorem 6.20. For the discussion of B, the appropriate Sobolev space corresponding to $T_d^{0,\alpha}(\Omega_R)$ is given by the space of tangential fields $a \in H^{-1/2}(\Omega_R)$ that have a surface divergence $\operatorname{Div} a \in H^{-1/2}(\Omega_R)$, that is,

$$H^{-1/2}(\operatorname{Div}; \Omega_R) := \left\{ a \in H^{-1/2}(\Omega_R) : \operatorname{Div} a \in H^{-1/2}(\Omega_R) \right\}.$$

Recall the definition (6.53) of the vector spherical harmonics U_n^m and V_n^m. Then for the tangential field

$$a = \sum_{n=1}^{\infty} \sum_{m=-n}^{n} \{ a_n^m U_n^m + b_n^m V_n^m \}$$

with Fourier coefficients a_n^m and b_n^m the norm on $H^{-1/2}(\Omega_R)$ can be written as

$$\|a\|_{H^{-1/2}(\Omega_R)}^2 = \sum_{n=1}^{\infty} \frac{1}{n} \sum_{m=-n}^{n} \{ |a_n^m|^2 + |b_n^m|^2 \}.$$

Since $\operatorname{Div} U_n^m = -\sqrt{n(n+1)}\, Y_n^m$ and $\operatorname{Div} V_n^m = 0$, the norm on the Sobolev space $H^{-1/2}(\operatorname{Div}; \Omega_R)$ is equivalent to

$$\|a\|_{H^{-1/2}(\operatorname{Div};\Omega_R)}^2 = \sum_{n=1}^{\infty} \left\{ n \sum_{m=-n}^{n} |a_n^m|^2 + \frac{1}{n} \sum_{m=-n}^{n} |b_n^m|^2 \right\}.$$

From the expansion (6.66) for radiating solutions to the Maxwell equations, we see that B maps the tangential field a with Fourier coefficients a_n^m and b_n^m onto

$$(7.7) \qquad Ba = \sum_{n=1}^{\infty} \left\{ \delta_n \sum_{m=-n}^{n} a_n^m U_n^m - \frac{k^2}{\delta_n} \sum_{m=-n}^{n} b_n^m V_n^m \right\}$$

where

$$\delta_n := \frac{k h_n^{(1)\prime}(kR)}{h_n^{(1)}(kR)} + \frac{1}{R}, \qquad n = 1, 2, \ldots.$$

Comparing this with (5.14), we note that

$$\delta_n = \gamma_n + \frac{1}{R},$$

that is, we can use the results from the proof of Theorem 5.7 on the coefficients γ_n. There does not exist a positive t such that $h_n^{(1)}(t) = 0$ or $h_n^{(1)}(t) + t h_n^{(1)\prime}(t) = 0$ since the Wronskian (2.36) does not vanish. Therefore, the operator B is bijective. Furthermore, from

$$c_1 n \leq |\delta_n| \leq c_2 n$$

which is valid for all n and some constants $0 < c_1 < c_2$, we see that B is bounded from $H^{-1/2}(\mathrm{Div}; \Omega_R)$ into the dual space $H^{-1/2}(\mathrm{Div}; \Omega_R)^*$ with norm

$$\|a\|_{H^{-1/2}(\mathrm{Div};\Omega_R)^*} = \sum_{n=1}^{\infty} \left\{ \frac{1}{n} \sum_{m=-n}^{n} |a_n^m|^2 + n \sum_{m=-n}^{n} |b_n^m|^2 \right\}.$$

However, different from the acoustic case, due to the factor k^2 in the second term of the expansion (7.7) the operator B in the limiting case $k = 0$ no longer remains bijective. This reflects the fact that for $k = 0$ the Maxwell equations decouple. Therefore, there is no obvious way of splitting B into a strictly coercive and a compact operator as was done for the Dirichlet to Neumann map A in the proof of Theorem 5.7.

Hence, for the continuous dependence on the boundary in electromagnetic obstacle scattering, we rely on the integral equation approach. For this, we describe a modification of the boundary integral equations used for proving existence of a solution to the exterior Maxwell problem in Theorem 6.19 which was introduced by Werner [185] and simplified by Hähner [65, 67]. In addition to surface potentials, it also contains volume potentials which makes it less satisfactory from a numerical point of view. However, it will make the investigation of the continuous dependence on the boundary easier since it avoids dealing with the more complicated second term in the approach (6.48) containing the double curl. We recall the notations introduced in Sections 6.3 and 6.4. After choosing an open ball B such that $\bar{B} \subset D$, we try to find the solution to the exterior Maxwell problem in the form

$$E(x) = \mathrm{curl} \int_{\partial D} a(y) \Phi(x, y) \, ds(y)$$

(7.8)
$$- \int_{\partial D} \varphi(y) \nu(y) \Phi(x, y) \, ds(y) - \int_{B} b(y) \Phi(x, y) \, dy,$$

$$H(x) = \frac{1}{ik} \, \mathrm{curl} \, E(x), \qquad x \in \mathbb{R}^3 \setminus \partial D.$$

We assume that the densities $a \in T_d^{0,\alpha}(\partial D)$, $\varphi \in C^{0,\alpha}(\partial D)$ and $b \in C^{0,\alpha}(B)$ satisfy the three integral equations

$$a + M_{11}a + M_{12}\varphi + M_{13}b = 2c$$

(7.9)
$$\varphi + M_{22}\varphi + M_{23}b = 0$$

$$b + M_{31}a + M_{32}\varphi + M_{33}b = 0$$

where the operators are given by $M_{11} := M$, $M_{22} := K$, and

$$M_{12}\varphi := -\nu \times S(\nu\varphi),$$

$$(M_{13}b)(x) := -2\nu(x) \times \int_B b(y)\Phi(x,y)\,dy, \quad x \in \partial D,$$

$$(M_{23}b)(x) := -2\int_B b(y) \cdot \operatorname{grad}_x \Phi(x,y)\,dy, \quad x \in \partial D,$$

$$(M_{31}a)(x) := i\eta(x)\operatorname{curl}\int_{\partial D} a(y)\Phi(x,y)\,ds(y), \quad x \in B,$$

$$(M_{32}\varphi)(x) := -i\eta(x)\int_{\partial D} \varphi(y)\nu(y)\Phi(x,y)\,ds(y), \quad x \in B,$$

$$(M_{33}b)(x) := -i\eta(x)\int_B b(y)\Phi(x,y)\,dy, \quad x \in B,$$

and where $\eta \in C^{0,\alpha}(\mathbb{R}^3)$ is a function with $\eta > 0$ in B and $\operatorname{supp}\eta = \bar{B}$.

First assume that we have a solution to these integral equations. Then clearly div E is a radiating solution to the Helmholtz equation in $\mathbb{R}^3 \setminus \bar{D}$ and, by the jump relations, the second integral equation implies div $E = 0$ on ∂D. Hence, div $E = 0$ in $\mathbb{R}^3 \setminus \bar{D}$ because of the uniqueness for the exterior Dirichlet problem. Now, with the aid of Theorems 6.4 and 6.7, we conclude that E, H is a radiating solution to the Maxwell equations in $\mathbb{R}^3 \setminus \bar{D}$. By the jump relations, the first integral equation ensures the boundary condition $\nu \times E = c$ on ∂D is satisfied.

We now establish that the system (7.9) of integral equations is uniquely solvable. For this, we first observe that all the integral operators M_{ij} are compact. The compactness of $M_{11} = M$ and $M_{22} = K$ is stated in Theorems 6.16 and 3.4 and the compactness of M_{33} follows from the fact that the volume potential operator maps $C(\bar{B})$ boundedly into $C^{1,\alpha}(\bar{B})$ (see Theorem 8.1) and the imbedding Theorem 3.2. The compactness of $M_{12} : C^{0,\alpha}(\partial D) \to T_d^{0,\alpha}(\partial D)$ follows from Theorem 3.4 and the representation

$$(\operatorname{Div} M_{12}\varphi)(x) = 2\nu(x) \cdot \int_{\partial D} \varphi(y)\{\nu(x) - \nu(y)\} \times \operatorname{grad}_x \Phi(x,y)\,ds(y), \quad x \in \partial D,$$

which can be derived with the help of the vector formula (6.38). The term $\nu(x) - \nu(y)$ makes the kernel weakly singular in a way such that Corollary 2.9

from [32] can be applied. For the other terms, compactness is obvious since the kernels are sufficiently smooth. Hence, by the Riesz–Fredholm theory it suffices to show that the homogeneous system only allows the nontrivial solution.

Assume that a, φ, b solve the homogeneous form of (7.9) and define E, H by (7.8). Then, by the above analysis, we already know that E, H solve the homogeneous exterior Maxwell problem whence $E = H = 0$ in $\mathbb{R}^3 \setminus \bar{D}$ follows. The jump relations then imply that

(7.10) $$\nu \times \operatorname{curl} E_- = 0, \quad \nu \cdot E_- = 0 \quad \text{on } \partial D.$$

From the third integral equation and the conditions on η, we observe that we may view b as a field in $C^{0,\alpha}(\mathbb{R}^3)$ with support in $\bar{B}$. Therefore, by the jump relations for volume potentials (see Theorem 8.1), we have $E \in C^2(D)$ and, in view of the third integral equation,

(7.11) $$\triangle E + k^2 E = b = -i\eta E \quad \text{in } D.$$

From (7.10) and (7.11) with the aid of Green's vector theorem (6.2), we now derive

$$\int_D \{|\operatorname{curl} E|^2 + |\operatorname{div} E|^2 - (k^2 + i\eta)|E|^2\} dx = 0,$$

whence, taking the imaginary part,

$$\int_B \eta |E|^2 dx = 0$$

follows. This implies $E = 0$ in B and from (7.11) we obtain $b = \triangle E + k^2 E = 0$ in D. Since solutions to the Helmholtz equation are analytic, from $E = 0$ in B we obtain $E = 0$ in D. The jump relations now finally yield $a = \nu \times E_+ - \nu \times E_- = 0$ and $\varphi = \operatorname{div} E_+ - \operatorname{div} E_- = 0$. Thus, we have established unique solvability for the system (7.9).

We are now ready to outline the proof of the electromagnetic analogue to the continuous dependence result of Theorem 5.9. We again consider surfaces Λ which are starlike with respect to the origin and are described by $x = r(\hat{x})\hat{x}$ where r denotes a positive function from $C^{1,\alpha}(\Omega)$ with $0 < \alpha < 1$. We recall from Section 5.3 what we mean by convergence of surfaces $\Lambda_n \to \Lambda$, $n \to \infty$, and L^2 convergence of functions f_n from $L^2(\Lambda_n)$ to a function f in $L^2(\Lambda)$. (For consistency with the rest of our book, we again choose C^2 surfaces Λ_n and Λ instead of $C^{1,\alpha}$ surfaces.)

Theorem 7.3 *Let (Λ_n) be a sequence of starlike C^2 surfaces which converges with respect to the $C^{1,\alpha}$ norm to a C^2 surface Λ as $n \to \infty$ and let E_n, H_n and E, H be radiating solutions to the Maxwell equations in the exterior of Λ_n and Λ, respectively. Assume that the continuous tangential components of E_n on Λ_n are L^2 convergent to the tangential components of E on Λ. Then the sequence (E_n), together with all its derivatives, converges to E uniformly on compact subsets of the open exterior of Λ.*

Proof. As in the proof of Theorem 5.9, we transform the boundary integral equations in (7.9) onto a fixed reference surface by substituting $x = x_r = r(\hat{x})\,\hat{x}$ to obtain integral equations over the unit sphere for the surface densities

$$\tilde{a}(\hat{x}) := \hat{x} \times a(r(\hat{x})\,\hat{x}), \quad \tilde{\varphi}(\hat{x}) := \varphi(r(\hat{x})\,\hat{x}).$$

Since the weak singularities of the operators M_{11}, M_{22} and M_{12} are similar in structure to those of K and S which enter into the combined double- and single-layer approach to the exterior Dirichlet problem, proceeding as in Lemma 5.12 it is possible to establish an estimate of the form (5.27) for the boundary integral terms in the transformed equations corresponding to (7.9). For the mixed terms like M_{31} and M_{13}, estimates of the type (5.27) follow trivially from Taylor's formula and the smoothness of the kernels. Finally, the volume integral term corresponding to M_{33} does not depend on the boundary at all. Based on these estimates, the proof is now completely analogous to that of Theorem 5.9.   □

7.3 Approximation of the Scattered Field

In this section we will describe the electromagnetic version of the method proposed by Kirsch and Kress for inverse acoustic obstacle scattering. We confine our analysis to inverse scattering from a perfectly conducting obstacle. Extensions to other boundary conditions are also possible.

We again first construct the scattered wave E^s from a knowledge of its electric far field pattern E_∞. To this end, we choose an auxiliary closed C^2 surface Γ with unit outward normal ν contained in the unknown scatterer D such that k is not a Maxwell eigenvalue for the interior of Γ. For example, we can choose Γ a sphere of radius R such that $j_n(kR) \neq 0$ and $j_n(kR) + kRj_n'(kR) \neq 0$ for $n = 0, 1, \ldots$. Given the internal surface Γ, we try to represent the scattered field as the electromagnetic field

$$(7.12) \qquad E^s(x) = \operatorname{curl} \int_\Gamma a(y)\Phi(x,y)\,ds(y), \quad H^s(x) = \frac{1}{ik}\operatorname{curl} E^s(x)$$

of a magnetic dipole distribution a from the space $T^2(\Gamma)$ of tangential L^2 fields on Γ. From (6.25) we see that the electric far field pattern of E^s is given by

$$E_\infty(\hat{x}) = \frac{ik}{4\pi}\,\hat{x} \times \int_\Gamma e^{-ik\,\hat{x}\cdot y} a(y)\,ds(y), \quad \hat{x} \in \Omega.$$

Hence, given the (measured) electric far field pattern E_∞, we have to solve the ill-posed integral equation of the first kind

$$(7.13) \qquad\qquad\qquad Fa = E_\infty$$

for the density a where the integral operator $F : T^2(\Gamma) \to T^2(\Omega)$ is defined by

$$(7.14) \qquad (Fa)(\hat{x}) := \frac{ik}{4\pi}\,\hat{x} \times \int_\Gamma e^{-ik\,\hat{x}\cdot y} a(y)\,ds(y), \quad \hat{x} \in \Omega.$$

As for the corresponding operator (5.46) in acoustics, the operator (7.14) has an analytic kernel and therefore the integral equation (7.13) is severely ill-posed. We now establish some properties of F.

Theorem 7.4 *The far field operator F defined by (7.14) is injective and has dense range provided k is not a Maxwell eigenvalue for the interior of Γ.*

Proof. Let $Fa = 0$ and define an electromagnetic field by (7.12). Then E^s has vanishing electric far field pattern $E_\infty = 0$, whence $E^s = 0$ in the exterior of Γ follows by Theorem 6.9. After introducing, analogous to (6.30), the magnetic dipole operator $M : L^2(\Gamma) \to L^2(\Gamma)$, by the L^2 jump relation (6.46) we find that

$$a + Ma = 0.$$

Employing the argument used in the proof of Theorem 6.21, by the Fredholm alternative we see that the nullspaces of $I + M$ in $L^2(\Gamma)$ and in $C(\Gamma)$ coincide. Therefore, a is continuous and, by the jump relations of Theorem 6.11 for continuous densities, $H^s, -E^s$ represents a solution to the Maxwell equations in the interior of Γ satisfying the homogeneous boundary condition $\nu \times H^s = 0$ on Γ. Hence, by our assumption on the choice of Γ we have $H^s = E^s = 0$ everywhere in $\mathbb{R}^3$. The jump relations now yield $a = 0$ on Γ, whence F is injective.

The adjoint operator $F^* : T^2(\Omega) \to T^2(\Gamma)$ of F is given by

$$(F^*g)(y) = \left(\nu(y) \times \frac{ik}{4\pi} \int_\Omega e^{ik\hat{x}\cdot y}\, \hat{x} \times g(\hat{x})\, ds(\hat{x}) \right) \times \nu(y), \quad y \in \Gamma.$$

Let $F^*g = 0$. Then

$$E(y) := \int_\Omega e^{ik\hat{x}\cdot y}\, \hat{x} \times g(\hat{x})\, ds(\hat{x}), \quad H(y) := \frac{1}{ik}\, \operatorname{curl} E(y), \quad y \in \mathbb{R}^3,$$

defines an electromagnetic Herglotz pair with vanishing tangential component $\nu \times E = 0$ on Γ. Hence, $E = H = 0$ in the interior of Γ by our assumption on the choice of Γ. Since E and H are analytic in $\mathbb{R}^3$, it follows that $E = H = 0$ everywhere. Theorem 3.15 now yields $g = 0$ on Γ, whence F^* is also injective and by Theorem 4.6 the range of F is dense in $L^2(\Omega)$. $\quad\square$

We now define a magnetic dipole operator $\tilde{M} : T^2(\Gamma) \to T^2(\Lambda)$ by

$$(7.15) \qquad (\tilde{M}a)(x) := \nu(x) \times \operatorname{curl} \int_\Gamma a(y)\Phi(x,y)\, ds(y), \quad x \in \Lambda,$$

where Λ denotes a closed C^2 surface with unit outward normal ν containing Γ in its interior. The proof of the following theorem is similar to that of Theorem 7.4.

Theorem 7.5 *The operator $\tilde{M}$ defined by (7.15) is injective and has dense range provided k is not a Maxwell eigenvalue for the interior of Γ.*

Now we know that by our choice of Γ the integral equation of the first kind (7.13) has at most one solution. Analogous to the acoustic case, its solvability is related to the question of whether or not the scattered wave can be analytically extended as a solution to the Maxwell equations across the boundary ∂D.

For the same reasons as in the acoustic case, we combine a Tikhonov regularization for the integral equation (7.13) and a defect minimization for the boundary search into one cost functional. We proceed analogously to Definition 5.15 and choose a compact (with respect to the $C^{1,\beta}$ norm, $0 < \beta < 1$,) subset U of the set of all starlike closed C^2 surfaces described by

$$x(\hat{x}) := r(\hat{x})\,\hat{x}, \quad \hat{x} \in \Omega, \quad r \in C^2(\Omega),$$

satisfying the a priori assumption

$$0 < r_i(\hat{x}) \le r(\hat{x}) \le r_e(\hat{x}), \quad \hat{x} \in \Omega,$$

with given functions r_i and r_e representing surfaces Λ_i and Λ_e such that the internal auxiliary surface Γ is contained in the interior of Λ_i and that the boundary ∂D of the unknown scatterer D is contained in the annulus between Λ_i and Λ_e. We now introduce the cost functional

$$(7.16) \quad \mu(a, \Lambda; \alpha) := \|Fa - E_\infty\|^2_{T^2(\Omega)} + \alpha\|a\|^2_{T^2(\Gamma)} + \gamma\|\nu \times E^i + \tilde{M}a\|^2_{T^2(\Lambda)}.$$

Again, $\alpha > 0$ denotes the regularization parameter for the Tikhonov regularization of (7.13) and $\gamma > 0$ denotes a suitable coupling parameter which for theoretical purposes we always assume equal to one.

Definition 7.6 *Given the incident field E^i, a (measured) electric far field pattern $E_\infty \in T^2(\Omega)$, and a regularization parameter $\alpha > 0$, a surface Λ_0 from the compact set U is called* optimal *if there exists $a_0 \in T^2(\Gamma)$ such that a_0 and Λ_0 minimize the cost functional (7.16) simultaneously over all $a \in T^2(\Gamma)$ and $\Lambda \in U$, that is, we have*

$$\mu(a_0, \Lambda_0; \alpha) = m(\alpha)$$

where

$$m(\alpha) := \inf_{a \in T^2(\Gamma),\, \Lambda \in U} \mu(a, \Lambda; \alpha).$$

For this reformulation of the electromagnetic inverse obstacle problem as a nonlinear optimization problem, we can state the following counterparts of Theorems 5.16–5.18. We omit the proofs since, except for minor adjustments, they literally coincide with those for the acoustic case. The use of Theorems 5.9 and 5.14, of course, has to be replaced by the corresponding electromagnetic versions given in Theorems 7.3 and 7.5.

Theorem 7.7 *For each $\alpha > 0$ there exists an optimal surface $\Lambda \in U$.*

Theorem 7.8 *Let E_∞ be the exact electric far field pattern of a domain D such that ∂D belongs to U. Then we have convergence of the cost functional $\lim_{\alpha \to 0} m(\alpha) = 0$.*

Theorem 7.9 *Let (α_n) be a null sequence and let (Λ_n) be a corresponding sequence of optimal surfaces for the regularization parameter α_n. Then there exists a convergent subsequence of (Λ_n). Assume that E_∞ is the exact electric far field pattern of a domain D such that ∂D is contained in U. Then every limit point Λ^* of (Λ_n) represents a surface on which the total field vanishes.*

Variants of these results were first established by Blöhbaum [14]. We will not repeat all the possible modifications mentioned in Section 5.4 for acoustic waves such as using more than one incoming wave, the limited aperture problem or using near field data. It is also straightforward to replace the magnetic dipole distribution on the internal surface for the approximation of the scattered field by an electric dipole distribution.

The above method for the electromagnetic inverse obstacle problem has not yet been numerically implemented and tested. However, we expect that this can be done analogously to the implementation in the acoustic case with reconstructions of similar accuracy.

7.4 Superposition of the Incident Fields

In this last section on inverse obstacle scattering we will present the electromagnetic version of the method of Colton and Monk. We again confine ourselves to the scattering from a perfect conductor and note that there are straightforward extensions to other boundary conditions.

As in the acoustic case, we try to find a superposition of incident plane electromagnetic fields with different directions and polarizations which lead to simple scattered fields and far field patterns. Starting from incident plane waves of the form (6.76), we consider as incident wave a superposition of the form

$$(7.17) \quad \tilde{E}^i(x) = \int_\Omega ik\, e^{ik\,x \cdot d} g(d)\, ds(d), \quad \tilde{H}^i(x) = \frac{1}{ik}\, \mathrm{curl}\, \tilde{E}^i(x), \quad x \in \mathbb{R}^3,$$

with a square integrable tangential field $g \in T^2(\Omega)$, i.e., the incident wave is an electromagnetic Herglotz pair. By Lemma 6.31, the corresponding electric far field pattern

$$\tilde{E}_\infty(\hat{x}) = \int_\Omega E_\infty(\hat{x}; d, g(d))\, ds(d), \quad \hat{x} \in \Omega,$$

is obtained by superposing the far field patterns $E_\infty(\,\cdot\,; d, g(d))$ for the incoming directions d with polarization $g(d)$. We note that by the Reciprocity Theorem 6.28 we may consider (7.17) also as a superposition with respect to the observation directions instead of the incident directions and in this case the method we are considering is sometimes referred to as dual space method.

If we want the scattered wave to become a prescribed radiating solution $\tilde{E}^s, \tilde{H}^s$ with explicitly known electric far field pattern $\tilde{E}_\infty$, given the (measured) far field patterns for all incident directions and polarizations, we need to solve the linear integral equation of the first kind

$$(7.18) \qquad\qquad Fg = \tilde{E}_\infty$$

where the integral operator $F : T^2(\Omega) \to T^2(\Omega)$ is given by

$$(7.19) \qquad (Fg)(\hat{x}) := \int_\Omega E_\infty(\hat{x}; d, g(d))\, ds(d), \quad \hat{x} \in \Omega.$$

For a reformulation of this definition, let

$$E_{\infty,j}(\hat{x}; d) := E_\infty(\hat{x}; d, e_j), \quad j = 1, 2, 3,$$

denote the electric far field patterns for the cartesian unit vectors e_j as polarization and recall the linear dependence (6.82), that is,

$$(7.20) \qquad\qquad \sum_{j=1}^{3} d_j E_{\infty,j}(\cdot\,; d) = 0$$

for all $d = (d_1, d_2, d_3)$. With the aid of the $E_{\infty,j}$ we can now write

$$(7.21) \qquad (Fg)(\hat{x}) = \sum_{j=1}^{3} \int_\Omega E_{\infty,j}(\cdot\,; d)\, e_j \cdot g(d)\, ds(d), \quad \hat{x} \in \Omega.$$

In order to construct the operator F for each $d \in \Omega$, we obviously need to have the electric far field pattern for two linearly independent polarization vectors orthogonal to d.

We need to assume the prescribed field $\tilde{E}^s, \tilde{H}^s$ is defined in the exterior of the unknown scatterer. For example, if we have the a priori information that the origin is contained in D, then for actual computations obvious choices for the prescribed scattered field would be the electromagnetic field

$$\tilde{E}^s(x) = \operatorname{curl} a\Phi(x, 0), \quad \tilde{H}^s(x) = \frac{1}{ik} \operatorname{curl} \tilde{E}^s(x)$$

of a magnetic dipole at the origin with electric far field pattern

$$\tilde{E}_\infty(\hat{x}) = \frac{ik}{4\pi}\, \hat{x} \times a,$$

or the electromagnetic field

$$\tilde{E}^s(x) = \operatorname{curl} \operatorname{curl} a\Phi(x, 0), \quad \tilde{H}^s(x) = \frac{1}{ik} \operatorname{curl} \tilde{E}^s(x)$$

of an electric dipole with far field pattern

$$\tilde{E}_\infty(\hat{x}) = \frac{k^2}{4\pi}\, \hat{x} \times (a \times \hat{x}).$$

Another more general possibility is to choose the radiating vector wave functions of Section 6.5 with the far field patterns given in terms of vector spherical harmonics (see Theorem 6.26).

We have already investigated the integral operator F. From Corollary 6.33, we know that F is injective and has dense range if and only if there does not exist an electromagnetic Herglotz pair which satisfies the homogeneous perfect conductor boundary condition on ∂D. Therefore, for the sequel we will make the assumption that k is not a Maxwell eigenvalue for D. This then implies that the inhomogeneous interior Maxwell problem for D is uniquely solvable. The classical approach to solve this boundary value problem is to seek the solution in the form of the electromagnetic field of a magnetic dipole distribution

$$E(x) = \operatorname{curl} \int_{\partial D} a(y)\Phi(x,y)\,ds(y), \quad H(x) = \frac{1}{ik}\operatorname{curl} E(x), \quad x \in D,$$

with a tangential field $a \in T_d^{0,\alpha}(\partial D)$. Then, given $c \in T_d^{0,\alpha}(\partial D)$, by the jump relations of Theorem 6.11 the electric field E satisfies the boundary condition $\nu \times E = c$ on ∂D if the density a solves the integral equation

$$a - Ma = -2c$$

with the magnetic dipole operator M defined by (6.30). The assumption that there exists no nontrivial solution to the homogeneous interior Maxwell problem in D now can be utilized to show with the aid of the jump relations that $I - M$ has a trivial nullspace in $T_d^{0,\alpha}(\partial D)$ (for the details see [32]). Hence, by the Riesz–Fredholm theory $I - M$ has a bounded inverse $(I - M)^{-1}$ from $T_d^{0,\alpha}(\partial D)$ into $T_d^{0,\alpha}(\partial D)$. This implies solvability and well-posedness of the interior Maxwell problem. The proof of the following theorem is now completely analogous to that of Theorem 5.4.

Theorem 7.10 *Assume that k is not a Maxwell eigenvalue for D. Let (E_n, H_n) be a sequence of $C^1(D) \cap C(\bar{D})$ solutions to the Maxwell equations in D such that the boundary data $c_n = \nu \times E_n$ on ∂D are weakly convergent in $T^2(\partial D)$. Then the sequence (E_n, H_n) converges uniformly on compact subsets of D to a solution E, H to the Maxwell equations.*

From now on, we assume that $\mathbb{R}^3 \setminus D$ is contained in the domain of definition of $\tilde{E}^s, \tilde{H}^s$, that is, for the case of the above examples for $\tilde{E}^s, \tilde{H}^s$ with singularities at $x = 0$ we assume the origin to be contained in D. We associate the following uniquely solvable interior Maxwell problem

(7.22) $\operatorname{curl} \tilde{E}^i - ik\tilde{H}^i = 0, \quad \operatorname{curl} \tilde{H}^i + ik\tilde{E}^i = 0 \quad \text{in } D,$

(7.23) $\nu \times (\tilde{E}^i + \tilde{E}^s) = 0 \quad \text{on } \partial D$

to the inverse scattering problem. From Theorem 6.34 we know that the solvability of the integral equation (7.18) is connected to this interior boundary value problem, i.e., (7.18) is solvable for $g \in T^2(\Omega)$ if and only if the solution $\tilde{E}^i, \tilde{H}^i$ to (7.22), (7.23) is an electromagnetic Herglotz pair with kernel ikg.

The Herglotz integral operator $H : T^2(\Omega) \to T^2(\Lambda)$ defined by

$$(7.24) \qquad (Hg)(x) := \nu(x) \times \int_\Omega ik\, e^{ik\,x\cdot d} g(d)\, ds(d), \quad x \in \Lambda,$$

where ν denotes the unit outward normal to the surface Λ represents the tangential component of the electric field on Λ for the Herglotz pair with kernel ikg.

Theorem 7.11 *The Herglotz integral operator H defined by (7.24) is injective and has dense range provided k is not a Maxwell eigenvalue for the interior of Λ.*

Proof. The operator H is related to the adjoint of the far field integral operator given by (7.14). Therefore, the statement of the theorem is equivalent to Theorem 7.4. $\qquad\qquad\square$

We are now ready to reformulate the inverse scattering problem as a nonlinear optimization problem analogous to Definition 5.21 in the acoustic case. We recall the description of the set U of admissable surfaces from the previous section and pick a closed C^2 surface Γ_e such that Λ_e is contained in the interior of Γ_e where we assume that k is not a Maxwell eigenvalue for the interior of Γ_e. We now introduce a cost functional by

$$(7.25)\ \ \mu(g, \Lambda; \alpha) := \|Fg - \tilde{E}_\infty\|^2_{T^2(\Omega)} + \alpha\|Hg\|^2_{T^2(\Gamma_e)} + \gamma\|Hg + \nu \times \tilde{E}^s\|^2_{T^2(\Lambda)}.$$

The operator F in (7.25) is understood in the sense of (7.21) for given functions $E_{\infty,j} \in T^2(\Omega \times \Omega)$, $j = 1, 2, 3$, which satisfy the linear dependence relation (7.20).

Definition 7.12 *Given the (measured) electric far fields $E_{\infty,j} \in T^2(\Omega \times \Omega)$ for $j = 1, 2, 3$ for all incident and observation directions and a regularization parameter $\alpha > 0$, a surface Λ_0 from the compact set U is called* optimal *if*

$$\inf_{g \in T^2(\Omega)} \mu(g, \Lambda_0; \alpha) = m(\alpha)$$

where

$$m(\alpha) := \inf_{g \in T^2(\Omega),\, \Lambda \in U} \mu(\varphi, \Lambda; \alpha).$$

For this electromagnetic optimization problem, we can state the following counterparts to Theorems 5.22–5.24. Variants of these results were first established by Blöhbaum [14].

Theorem 7.13 *For each $\alpha > 0$, there exists an optimal surface $\Lambda \in U$.*

Proof. The proof is analogous to that of Theorem 5.22 with the use of Theorem 7.10 instead of Theorem 5.4. □

Theorem 7.14 *For all incident directions and all polarizations let $E_{\infty,j}(\cdot\,;d) = E_\infty(\cdot\,;d,e_j)$, $j = 1,2,3$, be the exact electric far field pattern of a domain D such that ∂D belongs to U. Then we have convergence of the cost functional $\lim_{\alpha \to 0} m(\alpha) = 0$.*

Proof. The proof is analogous to that of Theorem 5.23. Instead of Theorem 5.20 we use Theorem 7.11 and instead of (5.68) we use the corresponding relation

$$(7.26) \qquad \tilde{E}_\infty - Fg = B(Hg + \nu \times \tilde{E}^s)$$

where $B : T^2(\partial D) \to T^2(\Omega)$ is the bounded injective operator introduced in Theorem 6.37 that maps the electric tangential component of radiating solutions to the Maxwell equations in $\mathbb{R}^3 \setminus D$ onto the electric far field pattern. □

Theorem 7.15 *Let (α_n) be a null sequence and let (Λ_n) be a corresponding sequence of optimal surfaces for the regularization parameter α_n. Then there exists a convergent subsequence of (Λ_n). Assume that for all incident directions and all polarizations $E_\infty(\cdot\,;d,p)$ is the exact electric far field pattern of a domain D such that ∂D belongs to U. Assume further that the solution $\tilde{E}^i, \tilde{H}^i$ to the associated interior Maxwell problem (7.22), (7.23) can be extended as a solution to the Maxwell equations across the boundary ∂D into the interior of Γ_e with continuous boundary values on Γ_e. Then every limit point Λ^* of (Λ_n) represents a surface on which the boundary condition (7.23) is satisfied, i.e., $\nu \times (\tilde{E}^i + \tilde{E}^s) = 0$ on Λ^*.*

Proof. The proof is analogous to that of Theorem 5.24 with the use of Theorems 7.10 and 7.11 instead of Theorems 5.4 and 5.20 and of (7.26) instead of (5.68). □

Using the completeness result of Theorem 6.35, it is possible to design a variant of the above method for which one does not have to assume that k is not a Maxwell eigenvalue for D.

We have not yet numerically tested the electromagnetic version of Colton and Monk's method in the full three-dimensional case. However, Maponi, Misici and Zirilli [124] have implemented the method for the rotationally symmetric case.

8. Acoustic Waves in an Inhomogeneous Medium

Until now, we have only considered the scattering of acoustic and electromagnetic time-harmonic waves in a homogeneous medium in the exterior of an impenetrable obstacle. For the remaining chapters of this book, we shall be considering the scattering of acoustic and electromagnetic waves by an inhomogeneous medium of compact support, and in this chapter we shall consider the direct scattering problem for acoustic waves. We shall content ourselves with the simplest case when the velocity potential has no discontinuities across the boundary of the inhomogeneous medium and shall again use the method of integral equations to investigate the direct scattering problem. However, since boundary conditions are absent, we shall make use of volume potentials instead of surface potentials as in the previous chapters.

We begin the chapter by deriving the linearized equations governing the propagation of small amplitude sound waves in an inhomogeneous medium. We then reformulate the direct scattering problem for such a medium as an integral equation known as the *Lippmann–Schwinger equation*. In order to apply the Riesz–Fredholm theory to this equation, we need to prove a unique continuation principle for second order elliptic partial differential equations. Having used this result to show the existence of a unique solution to the Lippmann–Schwinger equation, we then proceed to investigate the set $\mathcal{F}$ of far field patterns of the scattered fields corresponding to incident time-harmonic plane waves moving in arbitrary directions. By proving a reciprocity relation for far field patterns, we show that the completeness of the set $\mathcal{F}$ is equivalent to the non-existence of eigenvalues to a new type of boundary value problem for the reduced wave equation called the *interior transmission problem*. We then show that if absorption is present there are no eigenvalues whereas if the inhomogeneous medium is non-absorbing and spherically symmetric then there do exist eigenvalues. Continuing in this direction, we present the elements of the theory of operator valued analytic functions and apply this theory to investigate the interior transmission problem for inhomogeneous media that are neither absorbing nor spherically symmetric. We conclude the chapter by presenting some recent results of Kirsch and Monk [97] on the numerical solution of the direct scattering problem by combining the methods of finite elements and integral equations.

8.1 Physical Background

We begin by again considering the propagation of sound waves of small amplitude in $\mathbb{R}^3$ viewed as a problem in fluid dynamics. Let $v(x,t)$, $x \in \mathbb{R}^3$, be the velocity vector of a fluid particle in an inviscid fluid and let $p(x,t)$, $\rho(x,t)$ and $S(x,t)$ denote the pressure, density and specific entropy, respectively, of the fluid. If no external forces are acting on the fluid, then from Section 2.1 we have the equations

(8.1)

$$\frac{\partial v}{\partial t} + (v \cdot \text{grad})\,v + \frac{1}{\rho}\,\text{grad}\,p \;=\; 0 \qquad \text{(Euler's equation)}$$

$$\frac{\partial \rho}{\partial t} + \text{div}(\rho v) \;=\; 0 \qquad \text{(equation of continuity)}$$

$$p \;=\; f(\rho, S) \quad \text{(equation of state)}$$

$$\frac{\partial S}{\partial t} + v \cdot \text{grad}\,S \;=\; 0 \qquad \text{(adiabatic hypothesis)}$$

where f is a function depending on the fluid. Assuming $v(x,t)$, $p(x,t)$, $\rho(x,t)$ and $S(x,t)$ are small, we perturb these quantities around the static state $v = 0$, $p = p_0 = $ constant, $\rho = \rho_0(x)$ and $S = S_0(x)$ with $p_0 = f(\rho_0, S_0)$ and write

(8.2)

$$v(x,t) \;=\; \epsilon v_1(x,t) + \cdots$$

$$p(x,t) \;=\; p_0 + \epsilon p_1(x,t) + \cdots$$

$$\rho(x,t) \;=\; \rho_0(x) + \epsilon \rho_1(x,t) + \cdots$$

$$S(x,t) \;=\; S_0(x) + \epsilon S_1(x,t) + \cdots$$

where $0 < \epsilon \ll 1$ and the dots refer to higher order terms in ϵ. We now substitute (8.2) into (8.1), retaining only the terms of order ϵ. Doing this gives us the linearized equations

$$\frac{\partial v_1}{\partial t} + \frac{1}{\rho_0}\,\text{grad}\,p_1 \;=\; 0$$

$$\frac{\partial \rho_1}{\partial t} + \text{div}(\rho_0 v_1) \;=\; 0$$

$$\frac{\partial p_1}{\partial t} \;=\; c^2(x)\left(\frac{\partial \rho_1}{\partial t} + v_1 \cdot \text{grad}\,\rho_0\right)$$

where the *sound speed* c is defined by

$$c^2(x) = \frac{\partial}{\partial \rho}\,f(\rho_0(x), S_0(x)).$$

From this we deduce that p_1 satisfies

$$\frac{\partial^2 p_1}{\partial t^2} = c^2(x)\rho_0(x) \text{ div}\left(\frac{1}{\rho_0(x)} \text{ grad } p_1\right).$$

If we now assume that terms involving grad ρ_0 are negligible and that p_1 is time harmonic,

$$p_1(x,t) = \text{Re}\left\{u(x)\,e^{-i\omega t}\right\},$$

we see that u satisfies

(8.3)
$$\Delta u + \frac{\omega^2}{c^2(x)}\,u = 0.$$

Equation (8.3) governs the propagation of time harmonic acoustic waves of small amplitude in a slowly varying inhomogeneous medium. We still must prescribe how the wave motion is initiated and what is the boundary of the region containing the fluid. We shall only consider the simplest case when the inhomogeneity is of compact support, the region under consideration is all of $\mathbb{R}^3$ and the wave motion is caused by an incident field u^i satisfying the unperturbed linearized equations being scattered by the inhomogeneous medium. Assuming the inhomogeneous region is contained inside a ball B, i.e., $c(x) = c_0 = \text{constant}$ for $x \in \mathbb{R}^3 \setminus B$, we see that the scattering problem under consideration is now modeled by

(8.4)
$$\Delta u + k^2 n(x)u = 0 \quad \text{in } \mathbb{R}^3,$$

(8.5)
$$u = u^i + u^s,$$

(8.6)
$$\lim_{r\to\infty} r\left(\frac{\partial u^s}{\partial r} - iku^s\right) = 0,$$

where $k = \omega/c_0 > 0$ is the *wave number*,

$$n(x) := \frac{c_0^2}{c^2(x)}$$

is the *refractive index*, u^i is an entire solution of the Helmholtz equation $\Delta u + k^2 u = 0$ and u^s is the scattered field which, as discussed in Section 2.2, satisfies the Sommerfeld radiation condition (8.6) uniformly in all directions. The refractive index is always positive and in our case $n(x) = 1$ for $x \in \mathbb{R}^3 \setminus B$. Occasionally, we would also like to include the possibility that the medium is absorbing, i.e., the refractive index has an imaginary component. This is often modeled in the literature by adding a term that is proportional to v in Euler's equation which implies that n is now of the form

(8.7)
$$n(x) = n_1(x) + i\,\frac{n_2(x)}{k}.$$

8.2 The Lippmann–Schwinger Equation

The aim of this section is to derive an integral equation that is equivalent to the scattering problem (8.4)–(8.6) where $n \in C^1(\mathbb{R}^3)$ has the general form (8.7) such that

$$m := 1 - n$$

has compact support and

$$n_1(x) > 0 \quad \text{and} \quad n_2(x) \geq 0$$

for all $x \in \mathbb{R}^3$. Throughout this chapter, we shall always assume that these assumptions are valid and let $D := \{x \in \mathbb{R}^3 : m(x) \neq 0\}$.

To derive an integral equation equivalent to (8.4)–(8.6), we shall need to consider the *volume potential*

$$(8.8) \qquad u(x) := \int_{\mathbb{R}^3} \varphi(y)\Phi(x,y)\, dy, \quad x \in \mathbb{R}^3,$$

where

$$\Phi(x,y) := \frac{1}{4\pi} \frac{e^{ik|x-y|}}{|x-y|}, \quad x \neq y,$$

is the fundamental solution to the Helmholtz equation and φ is a continuous function in $\mathbb{R}^3$ with compact support, i.e., $\varphi \in C_0(\mathbb{R}^3)$. Extending the definitions given in Section 3.1, for a domain $G \subset \mathbb{R}^3$ the Hölder spaces $C^{p,\alpha}(G)$ are defined as the subspaces of $C^p(G)$ consisting of bounded functions whose p–th order derivates are uniformly Hölder continuous with exponent α. They are Banach spaces with the norms recursively defined by

$$\|\varphi\|_{p,\alpha} := \|\varphi\|_\infty + \|\operatorname{grad}\varphi\|_{p-1,\alpha}.$$

We can now state the following theorem (c.f. [56]).

Theorem 8.1 *The volume potential u given by (8.8) exists as an improper integral for all $x \in \mathbb{R}^3$ and has the following properties. If $\varphi \in C_0(\mathbb{R}^3)$ then $u \in C^{1,\alpha}(\mathbb{R}^3)$ and the orders of differentiation and integration can be interchanged. If $\varphi \in C_0(\mathbb{R}^3) \cap C^{0,\alpha}(\mathbb{R}^3)$ then $u \in C^{2,\alpha}(\mathbb{R}^3)$ and*

$$\Delta u + k^2 u = -\varphi \quad \text{in } \mathbb{R}^3.$$

In addition, we have

$$\|u\|_{2,\alpha,\mathbb{R}^3} \leq C\|\varphi\|_{\alpha,\mathbb{R}^3}$$

for some positive constant C depending only on the support of φ. Furthermore, if $\varphi \in C_0(\mathbb{R}^3) \cap C^{1,\alpha}(\mathbb{R}^3)$ then $u \in C^{3,\alpha}(\mathbb{R}^3)$.

We shall now use Lax's Theorem 3.5 to deduce a mapping property for the volume potential in Sobolev spaces from the classical property in Hölder spaces given above.

Theorem 8.2 *Given two bounded domains D and G, the volume potential*

$$(V\varphi)(x) := \int_D \varphi(y)\Phi(x,y)\,dy, \quad x \in \mathbb{R}^3,$$

defines a bounded operator $V : L^2(D) \to H^2(G)$.

Proof. We choose an open ball B such that $\bar{G} \subset B$ and a nonnegative function $\gamma \in C_0^2(B)$ such that $\gamma(x) = 1$ for all $x \in G$. Consider the spaces $X = C^{0,\alpha}(D)$ and $Y = C^{2,\alpha}(B)$ equipped with the usual Hölder norms. Introduce scalar products on X by the usual L^2 scalar product and on Y by the weighted Sobolev scalar product

$$(u,v)_Y := \int_B \gamma \left\{ u\bar{v} + \sum_{i=1}^3 \frac{\partial u}{\partial x_i} \frac{\partial \bar{v}}{\partial x_i} + \sum_{i,j=1}^3 \frac{\partial^2 u}{\partial x_i \partial x_j} \frac{\partial^2 \bar{v}}{\partial x_i \partial x_j} \right\} dx.$$

We first note that by using $\mathrm{grad}_x \Phi(x,y) = -\mathrm{grad}_y \Phi(x,y)$ and interchanging the order of integration we have

$$(8.9) \qquad \int_B \gamma V\varphi\,\psi\,dx = \int_D \varphi\, V^*(\gamma\psi)\,dx,$$

$$(8.10) \qquad \int_B \gamma \frac{\partial}{\partial x_i} V\varphi \frac{\partial \psi}{\partial x_i}\,dx = -\int_D \varphi \frac{\partial}{\partial x_i} V^*\left(\gamma \frac{\partial \psi}{\partial x_i}\right)dx$$

for $\varphi \in X$ and $\psi \in Y$ where

$$(V^*\psi)(x) := \int_B \psi(y)\Phi(x,y)\,dy, \quad x \in \mathbb{R}^3.$$

Using Gauss' divergence theorem, for $\varphi \in C_0^1(D)$ we have

$$\frac{\partial}{\partial x_i} \int_D \Phi(x,y)\varphi(y)\,dy = \int_D \Phi(x,y)\frac{\partial \varphi}{\partial y_i}(y)\,dy, \quad x \in \mathbb{R}^3,$$

that is,

$$\frac{\partial}{\partial x_i} V\varphi = V \frac{\partial \varphi}{\partial x_i}$$

and consequently, by (8.10) and Gauss' theorem,

$$(8.11) \qquad \int_B \gamma \frac{\partial^2 V\varphi}{\partial x_i \partial x_j} \frac{\partial^2 \psi}{\partial x_i \partial x_j}\,dx = \int_D \varphi \frac{\partial^2}{\partial x_i \partial x_j} V^*\left(\gamma \frac{\partial^2 \psi}{\partial x_i \partial x_j}\right)dx$$

for $\varphi \in C_0^1(D)$ and $\psi \in Y$. Hence, after setting $U = C_0^1(D) \subset X$, from (8.9)–(8.11) we have that the operators $V : U \to Y$ and $W : Y \to X$ given by

$$W\psi := \overline{V^*\bar{\psi}} - \sum_{i=1}^3 \frac{\partial}{\partial x_i} \overline{V^*\left(\gamma \frac{\partial \bar{\psi}}{\partial x_i}\right)} + \sum_{i,j=1}^3 \frac{\partial^2}{\partial x_i \partial x_j} \overline{V^*\left(\gamma \frac{\partial^2 \bar{\psi}}{\partial x_i \partial x_j}\right)}$$

are adjoint, i.e.,

$$(V\varphi, \psi)_X = (\varphi, W\psi)_Y$$

for all $\varphi \in U$ and $\psi \in Y$. By Theorem 8.1, both V and W are bounded with respect to the Hölder norms. Hence, from Lax's Theorem 3.5 and using the fact that the norm on Y dominates the H^2 norm over G, we see that there exists a positive constant c such that

$$\|V\varphi\|_{H^2(G)} \leq c\|\varphi\|_{L^2(D)}$$

for all $\varphi \in C_0^1(D)$. The proof is now finished by observing that $C_0^1(D)$ is dense in $L^2(D)$.   □

We now show that the scattering problem (8.4)–(8.6) is equivalent to the problem of solving the integral equation

$$(8.12) \qquad u(x) = u^i(x) - k^2 \int_{\mathbb{R}^3} \Phi(x,y)m(y)u(y)\,dy, \quad x \in \mathbb{R}^3,$$

for u which is known as the *Lippmann–Schwinger equation*.

Theorem 8.3 *If $u \in C^2(\mathbb{R}^3)$ is a solution of (8.4)–(8.6), then u is a solution of (8.12). Conversely, if $u \in C(\mathbb{R}^3)$ is a solution of (8.12) then $u \in C^2(\mathbb{R}^3)$ and u is a solution of (8.4)–(8.6).*

Proof. Let $u \in C^2(\mathbb{R}^3)$ be a solution of (8.4)–(8.6). Let $x \in \mathbb{R}^3$ be an arbitrary point and choose an open ball B with exterior unit normal ν containing the support of m such that $x \in B$. From Green's formula (2.4) applied to u, we have

$$
\begin{aligned}
u(x) = \int_{\partial B} &\left\{ \frac{\partial u}{\partial \nu}(y)\,\Phi(x,y) - u(y)\frac{\partial \Phi(x,y)}{\partial \nu(y)} \right\} ds(y) \\
&- k^2 \int_B \Phi(x,y)m(y)u(y)\,dy
\end{aligned}
$$

(8.13)

since $\Delta u + k^2 u = mk^2 u$. Note that in the volume integral over B we can integrate over all of $\mathbb{R}^3$ since m has support in B. Green's formula (2.5), applied to u^i, gives

$$(8.14) \qquad u^i(x) = \int_{\partial B} \left\{ \frac{\partial u^i}{\partial \nu}(y)\,\Phi(x,y) - u^i(y)\frac{\partial \Phi(x,y)}{\partial \nu(y)} \right\} ds(y).$$

Finally, from Green's theorem (2.3) and the radiation condition (8.6) we see that

$$(8.15) \qquad \int_{\partial B} \left\{ \frac{\partial u^s}{\partial \nu}(y)\,\Phi(x,y) - u^s(y)\frac{\partial \Phi(x,y)}{\partial \nu(y)} \right\} ds(y) = 0.$$

With the aid of $u = u^i + u^s$ we can now combine (8.13)–(8.15) to conclude that (8.12) is satisfied.

Conversely, let $u \in C(\mathbb{R}^3)$ be a solution of (8.12) and define u^s by

$$u^s(x) := -k^2 \int_{\mathbb{R}^3} \Phi(x,y)m(y)u(y)\,dy, \quad x \in \mathbb{R}^3.$$

Since Φ satisfies the Sommerfeld radiation condition (8.6) uniformly with respect to y on compact sets and m has compact support, it is easily verified that u^s satisfies (8.6). Since $m \in C_0^1(\mathbb{R}^3)$, we can conclude from (8.12) and Theorem 8.1 that first $u \in C^1(\mathbb{R}^3)$ and then that $u^s \in C^2(\mathbb{R}^3)$ with $\Delta u^s + k^2 u^s = k^2 mu$. Finally, since $\Delta u^i + k^2 u^i = 0$, we can conclude from Theorem 8.1 and (8.12) that

$$\Delta u + k^2 u = (\Delta u^i + k^2 u^i) + (\Delta u^s + k^2 u^s) = k^2 mu,$$

that is,

$$\Delta u + k^2 nu = 0$$

in $\mathbb{R}^3$ and the proof is completed. $\square$

We note that in (8.12) we can replace the region of integration by any domain G such that the support of m is contained in $\bar{G}$ and look for solutions in $C(\bar{G})$. Then for $x \in \mathbb{R}^3 \setminus \bar{G}$ we define $u(x)$ by the right hand side of (8.12) and obviously obtain a continuous solution u to the Lippmann–Schwinger equation in all of $\mathbb{R}^3$.

We shall show shortly that (8.12) is uniquely solvable for all values of $k > 0$. This result is nontrivial since it will be based on a unique continuation principle for solutions of (8.4). However, for k sufficiently small we can show the existence of a unique solution to (8.12) by the simple method of successive approximations.

Theorem 8.4 *Suppose that $m(x) = 0$ for $|x| \geq a$ with some $a > 0$ and let $k^2 < 2/Ma^2$ where $M := \max_{|x| \leq a} |m(x)|$. Then there exists a unique solution to the integral equation (8.12).*

Proof. As already pointed out, it suffices to solve (8.12) for $u \in C(\bar{B})$ with the ball $B := \{x \in \mathbb{R}^3 : |x| < a\}$. On the Banach space $C(\bar{B})$, define the operator $T_m : C(\bar{B}) \to C(\bar{B})$ by

$$(8.16) \qquad (T_m u)(x) := \int_B \Phi(x,y)m(y)u(y)\,dy, \quad x \in \bar{B}.$$

By the method of successive approximations, our theorem will be proved if we can show that $\|T_m\|_\infty \leq Ma^2/2$. To this end, we have

$$(8.17) \qquad |(T_m u)(x)| \leq \frac{M\|u\|_\infty}{4\pi} \int_B \frac{dy}{|x-y|}, \quad x \in \bar{B}.$$

To estimate the integral in (8.17), we note that (see Theorem 8.1)

$$h(x) := \int_B \frac{dy}{|x-y|}, \quad x \in \bar{B},$$

is a solution of the Poisson equation $\Delta h = -4\pi$ and is a function only of $r = |x|$. Hence, h solves the differential equation

$$\frac{1}{r^2} \frac{d}{dr} \left(r^2 \frac{dh}{dr} \right) = -4\pi$$

which has the general solution

$$h(r) = -\frac{2}{3} \pi r^2 + \frac{C_1}{r} + C_2$$

where C_1 and C_2 are arbitrary constants. Since h is continuous in a neighborhood of the origin, we must have $C_1 = 0$ and, letting $r \to 0$, we see that

$$C_2 = h(0) = \int_B \frac{dy}{|y|} = 4\pi \int_0^a \rho \, d\rho = 2\pi a^2.$$

Hence, $h(r) = 2\pi(a^2 - r^2/3)$ and thus $\|h\|_\infty = 2\pi a^2$. From (8.17) we can now conclude that

$$|(T_m u)(x)| \le \frac{Ma^2}{2} \|u\|_\infty, \quad x \in \bar{B},$$

i.e., $\|T_m\|_\infty \le Ma^2/2$ and the proof is completed. $\square$

8.3 The Unique Continuation Principle

In order to establish the existence of a unique solution to the scattering problem (8.4)–(8.6) for all positive values of the wave number k, we see from the previous section that it is necessary to establish the existence of a unique solution to the Lippmann–Schwinger equation (8.12). To this end, we would like to apply the Riesz–Fredholm theory since the integral operator (8.16) has a weakly singular kernel and hence is a compact operator $T_m : C(\bar{B}) \to C(\bar{B})$ where B is a ball such that $\bar{B}$ contains the support of m. In order to achieve this aim, we must show that the homogeneous equation has only the trivial solution, or, equivalently, that the only solution of

(8.18) $$\Delta u + k^2 n(x) u = 0 \quad \text{in } \mathbb{R}^3,$$

(8.19) $$\lim_{r \to \infty} r \left(\frac{\partial u}{\partial r} - iku \right) = 0$$

is u identically zero. To prove this, the following unique continuation principle is fundamental. The unique continuation principle for elliptic equations has a long history and we refer the reader to [12] for a historical discussion. The proof of this principle for (8.18) dates back to Müller [141]. Our proof is based on the ideas of Protter [154] and Leis [119].

Lemma 8.5 *Let G be a domain in $\mathbb{R}^3$ and let $u_1, \ldots, u_P \in C^2(G)$ be real valued functions satisfying*

$$(8.20) \qquad |\triangle u_p| \leq c \sum_{q=1}^{P} \{|u_q| + |\operatorname{grad} u_q|\} \quad \text{in } G$$

for $p = 1, \ldots, P$ and some constant c. Assume that u_p vanishes in a neighborhood of some $x_0 \in G$ for $p = 1, \ldots, P$. Then u_p is identically zero in G for $p = 1, \ldots, P$.

Proof. For $0 < R \leq 1$, let $B[x_0; R]$ be the closed ball of radius R centered at x_0. Choose R such that $B[x_0, R] \subset G$. We shall show that $u_p(x) = 0$ for $x \in B[x_0; R/2]$ and $p = 1, \ldots, P$. The theorem follows from this since any other point $x_1 \in G$ can be connected to x_0 by a finite number of overlapping balls. Without loss of generality, we shall assume that $x_0 = 0$ and for convenience we temporarily write $u = u_p$.

For $r = |x|$ and n an arbitrary positive integer, we define $v \in C^2(G)$ by

$$v(x) := \begin{cases} e^{r^{-n}} u(x), & x \neq 0, \\ \\ 0, & x = 0. \end{cases}$$

Then

$$\triangle u = e^{-r^{-n}} \left\{ \triangle v + \frac{2n}{r^{n+1}} \frac{\partial v}{\partial r} + \frac{n}{r^{n+2}} \left(\frac{n}{r^n} - n + 1 \right) v \right\}.$$

Using the inequality $(a+b)^2 \geq 2ab$ and calling the middle term in the above expression in brackets b, we see that

$$(\triangle u)^2 \geq \frac{4n\, e^{-2r^{-n}}}{r^{n+1}} \frac{\partial v}{\partial r} \left\{ \triangle v + \frac{n}{r^{n+2}} \left(\frac{n}{r^n} - n + 1 \right) v \right\}.$$

We now let $\varphi \in C^2(\mathbb{R}^3)$ be such that $\varphi(x) = 1$ for $|x| \leq R/2$ and $\varphi(x) = 0$ for $|x| \geq R$. Then if we define $\hat{u}$ and $\hat{v}$ by $\hat{u} := \varphi u$ and $\hat{v} := \varphi v$ respectively, we see that the above inequality is also valid for u and v replaced by $\hat{u}$ and $\hat{v}$ respectively. In particular, we have the inequality

$$(8.21) \qquad \int_G r^{n+2} e^{2r^{-n}} (\triangle \hat{u})^2 \, dx \geq 4n \int_G r \frac{\partial \hat{v}}{\partial r} \left\{ \triangle \hat{v} + \frac{n}{r^{n+2}} \left(\frac{n}{r^n} - n + 1 \right) \hat{v} \right\} dx.$$

We now proceed to integrate by parts in (8.21), noting that by our choice of φ the boundary terms all vanish. Using the vector identity

$$2 \operatorname{grad}\{x \cdot \operatorname{grad} \hat{v}\} \cdot \operatorname{grad} \hat{v} = \operatorname{div}\{x |\operatorname{grad} \hat{v}|^2\} - |\operatorname{grad} \hat{v}|^2,$$

from Green's theorem and Gauss' divergence theorem we find that

$$\int_G r \frac{\partial \hat{v}}{\partial r} \triangle \hat{v} \, dx = -\int_G \operatorname{grad}\{x \cdot \operatorname{grad} \hat{v}\} \cdot \operatorname{grad} \hat{v} \, dx = \frac{1}{2} \int_G |\operatorname{grad} \hat{v}|^2 \, dx,$$

that is,

(8.22)
$$\int_G r \frac{\partial \hat{v}}{\partial r} \Delta \hat{v} \, dx = \frac{1}{2} \int_G |\operatorname{grad} \hat{v}|^2 \, dx.$$

Furthermore, for m an integer, by partial integration with respect to r we have

$$\int_G \frac{1}{r^m} \hat{v} \frac{\partial \hat{v}}{\partial r} \, dx = - \int_G \hat{v} \frac{\partial}{\partial r} \left(\frac{1}{r^{m-2}} \hat{v} \right) \frac{dx}{r^2}$$

$$= - \int_G \frac{1}{r^m} \hat{v} \frac{\partial \hat{v}}{\partial r} \, dx + (m-2) \int_G \frac{\hat{v}^2}{r^{m+1}} \, dx,$$

that is,

(8.23)
$$\int_G \frac{1}{r^m} \hat{v} \frac{\partial \hat{v}}{\partial r} \, dx = \frac{1}{2} (m-2) \int_G \frac{\hat{v}^2}{r^{m+1}} \, dx.$$

We can now insert (8.22) and (8.23) (for $m = 2n+1$ and $m = n+1$) into the inequality (8.21) to arrive at

(8.24)
$$\int_G r^{n+2} e^{2r^{-n}} (\Delta \hat{u})^2 dx$$

$$\geq 2n \int_G |\operatorname{grad} \hat{v}|^2 dx + 2n^2(n^2+n-1) \int_G \frac{\hat{v}^2}{r^{2n+2}} \, dx.$$

Here we have also used the inequality

$$\int_G \frac{\hat{v}^2}{r^{2n+2}} \, dx \geq \int_G \frac{\hat{v}^2}{r^{n+2}} \, dx$$

which follows from $\hat{v}(x) = 0$ for $r = |x| \geq R$ and $0 < R \leq 1$. From

$$\operatorname{grad} \hat{u} = e^{-r^{-n}} \left\{ \operatorname{grad} \hat{v} + \frac{n}{r^{n+1}} \frac{x}{r} \hat{v} \right\}$$

we can estimate

$$e^{2r^{-n}} |\operatorname{grad} \hat{u}|^2 \leq 2 |\operatorname{grad} \hat{v}|^2 + \frac{2n^2}{r^{2n+2}} |\hat{v}|^2$$

and with this and (8.24) we find that

(8.25)
$$\int_G r^{n+2} e^{2r^{-n}} (\Delta \hat{u})^2 dx \geq n \int_G e^{2r^{-n}} |\operatorname{grad} \hat{u}|^2 dx + n^4 \int_G \frac{e^{2r^{-n}}}{r^{2n+2}} \hat{u}^2 dx.$$

Up to now, we have not used the inequality (8.20). Now we do, relabeling u by u_p. From (8.20) and Schwarz's inequality, we clearly have that

$$|\Delta u_p(x)|^2 \leq 2Pc^2 \sum_{q=1}^{P} \left\{ \frac{|\operatorname{grad} u_q(x)|^2}{r^{n+2}} + \frac{|u_q(x)|^2}{r^{3n+4}} \right\}, \quad |x| \leq \frac{R}{2},$$

since $R \leq 1$. Letting $\| \cdot \|_\infty$ denote the maximum norm over $\bar{G}$, we have

$$| \triangle \hat{u}_p(x) |^2 \leq \| \triangle \hat{u}_p \|_\infty^2 \leq \frac{\| \triangle \hat{u}_p \|_\infty^2}{r^{3n+4}}, \quad \frac{R}{2} \leq |x| \leq R.$$

Observing that $u_p(x) = \hat{u}_p(x)$ for $|x| \leq R/2$, from (8.25) we now have

$$n \int_{|x| \leq R/2} e^{2r^{-n}} |\operatorname{grad} u_p|^2 dx + n^4 \int_{|x| \leq R/2} \frac{e^{2r^{-n}}}{r^{2n+2}} u_p^2 \, dx$$

$$\leq \int_G r^{n+2} \, e^{2r^{-n}} (\triangle \hat{u}_p)^2 dx$$

$$\leq 2Pc^2 \sum_{q=1}^{P} \left\{ \int_{|x| \leq R/2} e^{2r^{-n}} |\operatorname{grad} u_q|^2 dx + \int_{|x| \leq R/2} \frac{e^{2r^{-n}}}{r^{2n+2}} u_p^2 \, dx \right\}$$

$$+ \| \triangle \hat{u}_p \|_\infty^2 \int_{R/2 \leq |x| \leq R} \frac{e^{2r^{-n}}}{r^{2n+2}} \, dx,$$

i.e., for sufficiently large n we have

$$n^4 \int_{|x| \leq R/2} \frac{e^{2r^{-n}}}{r^{2n+2}} u_p^2 \, dx \leq C \int_{R/2 \leq |x| \leq R} \frac{e^{2r^{-n}}}{r^{2n+2}} \, dx, \quad p = 1, \ldots, P,$$

for some constant C. From this, since the function

$$r \mapsto \frac{e^{2r^{-n}}}{r^{2n+2}}, \quad r > 0,$$

is monotonically decreasing, for sufficiently large n we have

$$n^4 \int_{|x| \leq R/2} u_p^2 \, dx \leq C \int_{R/2 \leq |x| \leq R} \, dx, \quad p = 1, \ldots, P.$$

Letting n tend to infinity now shows that $u_p(x) = 0$ for $|x| \leq R/2$ and $p = 1, \ldots, P$ and the theorem is proved. $\qquad \square$

Theorem 8.6 *Let G be a domain in $\mathbb{R}^3$ and suppose $u \in C^2(G)$ is a solution of*

$$\triangle u + k^2 n(x) u = 0$$

in G such that $n \in C(\bar{G})$ and u vanishes in a neighborhood of some $x_0 \in G$. Then u is identically zero in G.

Proof. Apply Lemma 8.5 to $u_1 := \operatorname{Re} u$ and $u_2 := \operatorname{Im} u$. $\qquad \square$

We are now in a position to show that for all $k > 0$ there exists a unique solution to the scattering problem (8.4)–(8.6).

Theorem 8.7 *For each $k > 0$ there exists a unique solution to (8.4)–(8.6) and the solution u depends continuously with respect to the maximum norm on the incident field u^i.*

Proof. As previously discussed, to show existence and uniqueness it suffices to show that the only solution of (8.18), (8.19) is u identically zero. If this is done, by the Riesz–Fredholm theory the integral equation (8.12) can be inverted in $C(\bar{B})$ and the inverse operator is bounded. From this, it follows that u depends continuously on the incident field u^i with respect to the maximum norm. Hence we only must show that the only solution of (8.18), (8.19) is $u = 0$.

Recall that B is chosen to be a ball of radius a centered at the origin such that m vanishes outside of B. As usual ν denotes the exterior unit normal to ∂B. We begin by noting from Green's theorem (2.2) and (8.18) that

$$\int_{|x|=a} u\,\frac{\partial \bar{u}}{\partial \nu}\, ds = \int_{|x| \le a} \{|\operatorname{grad} u|^2 - k^2\,\bar{n}\,|u|^2\}\, dx.$$

From this, since $\operatorname{Im} n \ge 0$, it follows that

$$(8.26) \qquad \operatorname{Im} \int_{|x|=a} u\,\frac{\partial \bar{u}}{\partial \nu}\, ds = k^2 \int_{|x| \le a} \operatorname{Im} n\, |u|^2 dx \ge 0.$$

Theorem 2.12 now shows that $u(x) = 0$ for $|x| \ge a$ and it follows by Theorem 8.6 that $u(x) = 0$ for all $x \in \mathbb{R}^3$. $\qquad\square$

8.4 Far Field Patterns

From (8.12) we see that

$$u^s(x) = -k^2 \int_{\mathbb{R}^3} \Phi(x, y) m(y) u(y)\, dy, \quad x \in \mathbb{R}^3.$$

Hence, letting $|x|$ tend to infinity, with the help of (2.14) we see that

$$u^s(x) = \frac{e^{ik|x|}}{|x|}\, u_\infty(\hat{x}) + O\left(\frac{1}{|x|^2}\right), \quad |x| \to \infty,$$

where the *far field pattern* u_∞ is given by

$$(8.27) \qquad u_\infty(\hat{x}) = -\frac{k^2}{4\pi} \int_{\mathbb{R}^3} e^{-ik\,\hat{x}\cdot y} m(y) u(y)\, dy$$

for $\hat{x} = x/|x|$ on the unit sphere Ω. We note that by Theorem 8.4, for k sufficiently small, u can be obtained by the method of successive approximations. If in (8.27) we replace u by the first term in this iterative process, we obtain the *Born approximation*

$$(8.28) \qquad u_\infty(\hat{x}) = -\frac{k^2}{4\pi} \int_{\mathbb{R}^3} e^{-ik\,\hat{x}\cdot y} m(y) u^i(y)\, dy.$$

We shall briefly return to this approximation in Chapter 10 where it will provide the basis of a linear approach to the inverse scattering problem.

We now consider the case when the incident field u^i is a plane wave, i.e., $u^i(x) = e^{ik\,x\cdot d}$ where d is a unit vector giving the direction of propagation. We denote the dependence of the far field pattern u_∞ on d by writing $u_\infty(\hat{x}) = u_\infty(\hat{x}; d)$ and, similarly, we write $u^s(x) = u^s(x; d)$ and $u(x) = u(x; d)$. Then, analogous to Theorem 3.13, we have the following *reciprocity relation*.

Theorem 8.8 *The far field pattern satisfies the reciprocity relation*

$$u_\infty(\hat{x}; d) = u_\infty(-d; -\hat{x})$$

for all $\hat{x}, d$ on the unit sphere Ω.

Proof. By the relation (3.36) from the proof of Theorem 3.13, we have

$$4\pi\{u_\infty(\hat{x}; d) - u_\infty(-d; -\hat{x})\}$$

$$= \int_{|y|=a} \left\{ u(y; d)\, \frac{\partial}{\partial\nu(y)}\, u(y; -\hat{x}) - u(y; -\hat{x})\, \frac{\partial}{\partial\nu(y)}\, u(y; d) \right\} ds(y)$$

whence the statement follows with the aid of Green's theorem (2.3).   □

As in Chapter 3, we are again concerned with the question if the far field patterns corresponding to all incident plane waves are complete in $L^2(\Omega)$. The reader will recall from Section 3.3 that for the case of obstacle scattering the far field patterns are complete provided k^2 is not a Dirichlet eigenvalue having an eigenfunction that is a Herglotz wave function. In the present case of scattering by an inhomogeneous medium we have a similar result except that the Dirichlet problem is replaced by a new type of boundary value problem introduced by Kirsch in [87] (see also [40]) and called interior transmission problem. This name is motivated by the fact that, as in the classical transmission problem, we have two partial differential equations linked together by their Cauchy data on the boundary but, in this case, the partial differential equations are both defined in the same interior domain instead of in an interior and exterior domain as for the classical transmission problem (c.f. [32]). In particular, let $\{d_n : n = 1, 2, \ldots\}$ be a countable dense set of vectors on the unit sphere Ω and define the class $\mathcal{F}$ of far field patterns by

$$\mathcal{F} := \{u_\infty(\cdot\,; d_n) : n = 1, 2, \ldots\}.$$

Then we have the following theorem. For the rest of this chapter, we shall assume that $D := \{x \in \mathbb{R}^3 : m(x) \neq 0\}$ is connected with a connected C^2 boundary ∂D and D contains the origin.

Theorem 8.9 *The orthogonal complement of $\mathcal{F}$ in $L^2(\Omega)$ consists of the conjugate of those functions $g \in L^2(\Omega)$ for which there exists $w \in C^2(D) \cap C^1(\bar{D})$ and a Herglotz wave function*

$$v(x) = \int_\Omega e^{-ik\,x\cdot d}g(d)\,ds(d), \quad x \in \mathbb{R}^3,$$

such that the pair v, w is a solution to

$$(8.29) \qquad \Delta w + k^2 n(x)w = 0, \quad \Delta v + k^2 v = 0 \quad \text{in } D$$

satisfying

$$(8.30) \qquad w = v, \quad \frac{\partial w}{\partial \nu} = \frac{\partial v}{\partial \nu} \quad \text{on } \partial D.$$

Proof. Let $\mathcal{F}^\perp$ denote the orthogonal complement to $\mathcal{F}$. We will show that $\bar{g} \in \mathcal{F}^\perp$ if and only if g satisfies the assumptions stated in the theorem. From the continuity of u_∞ as a function of d and Theorem 8.8, we have that the property $\bar{g} \in \mathcal{F}^\perp$, i.e.,

$$\int_\Omega u_\infty(\hat{x}; d_n)g(\hat{x})\,ds(\hat{x}) = 0$$

for $n = 1, 2, \ldots$ is equivalent to

$$\int_\Omega u_\infty(-d; -\hat{x})g(\hat{x})\,ds(\hat{x}) = 0$$

for all $d \in \Omega$, i.e.,

$$(8.31) \qquad \int_\Omega u_\infty(\hat{x}; d)g(-d)\,ds(d) = 0$$

for all $\hat{x} \in \Omega$. From the Lippmann–Schwinger equation (8.12) it can be seen that the left hand side of (8.31) is the far field pattern of the scattered field w^s corresponding to the incident field

$$w^i(x) := \int_\Omega e^{ik\,x\cdot d}g(-d)\,ds(d) = \int_\Omega e^{-ik\,x\cdot d}g(d)\,ds(d).$$

But now (8.31) is equivalent to a vanishing far field pattern of w^s and hence by Theorem 2.13 equivalent to $w^s = 0$ in all of $\mathbb{R}^3 \setminus D$, i.e., if $v = w^i$ and $w = w^i + w^s$ then $w = v$ on ∂D and $\partial w / \partial \nu = \partial v / \partial \nu$ on ∂D. Conversely, if v and w satisfy the conditions of the theorem, by setting $w = v$ in $\mathbb{R}^3 \setminus \bar{D}$ and using Green's formula we see that w can be extended into all of $\mathbb{R}^3$ as a C^2 solution of $\Delta w + k^2 n(x)w = 0$. The theorem now follows. □

From the above proof, we see that the boundary conditions (8.30) are equivalent to the condition that $w = v$ in $\mathbb{R}^3 \setminus D$ (In particular, we can impose the boundary conditions (8.30) on the boundary of any domain with C^2 boundary that contains D). We have chosen to impose the boundary conditions on ∂D since it is necessary for the more general scattering problem where the density in D is different from that in $\mathbb{R}^3 \setminus D$ [88]. However, when we later consider weak solutions of (8.29), (8.30), we shall replace (8.30) by the more managable condition $w = v$ in $\mathbb{R}^3 \setminus D$.

Analogous to Theorem 3.19, we also have the following theorem, the proof of which is the same as that of Theorem 8.9 except that w^s is now equal to the spherical wave function $v_p(x) = h_p^{(1)}(k|x|)\, Y_p(\hat{x})$. Note that, in contrast to Theorem 3.19, we are now integrating with respect to $\hat{x}$ instead of d. However, by the reciprocity relation, these two procedures are equivalent.

Theorem 8.10 *Let* $v_p(x) = h_p^{(1)}(k|x|)\, Y_p(\hat{x})$ *be a spherical wave function of order* p. *The integral equation of the first kind*

$$\int_\Omega u_\infty(\hat{x}; d) g_p(\hat{x})\, ds(\hat{x}) = \frac{i^{p-1}}{k}\, Y_p(d), \quad d \in \Omega,$$

has a solution $g_p \in L^2(\Omega)$ *if and only if there exists* $w \in C^2(D) \cap C^1(\bar{D})$ *and a Herglotz wave function*

$$v(x) = \int_\Omega e^{-ik\, x \cdot d} g_p(d)\, ds(d), \quad x \in \mathbb{R}^3,$$

such that v, w *is a solution to*

(8.32) $\triangle w + k^2 n(x) w = 0, \quad \triangle v + k^2 v = 0 \quad \text{in } D$

satisfying

(8.33) $w - v = v_p, \quad \dfrac{\partial w}{\partial \nu} - \dfrac{\partial v}{\partial \nu} = \dfrac{\partial v_p}{\partial \nu} \quad \text{on } \partial D.$

Motivated by Theorems 8.9 and 8.10, we now define the interior transmission problem.

Interior Transmission Problem. *Find functions* $v \in C^2(D) \cap C^1(\bar{D})$ *and* $w \in C^2(D) \cap C^1(\bar{D})$ *such that* (8.32), (8.33) *are satisfied. The boundary value problem* (8.29), (8.30) *is called the homogeneous interior transmission problem.*

In this chapter, we shall only be concerned with the homogeneous interior transmission problem. For information on the inhomogeneous interior transmission problem, we refer the reader to Colton and Kirsch [28], Colton, Kirsch and Päivärinta [30], Kedzierawski [83] and Rynne and Sleeman [164]. Of primary concern to us in this chapter will be the existence of positive values of the wave number k such that nontrivial solutions exist to the homogeneous interior transmission problem since it is only in this case that there is a possibility that $\mathcal{F}$ is not complete in $L^2(\Omega)$. This motivates the following definition.

Definition 8.11 *If $k > 0$ is such that the homogeneous interior transmission problem has a nontrivial solution then k is called a* transmission eigenvalue.

Theorem 8.12 *Suppose $\operatorname{Im} n \neq 0$. Then $k > 0$ is not a transmission eigenvalue, i.e., the set $\mathcal{F}$ of far field patterns is complete in $L^2(\Omega)$ for each $k > 0$.*

Proof. Suppose there exists a nontrivial solution to (8.29), (8.30). Then by Green's theorem (2.3) we have

$$0 = \int_{\partial D} \left(v \frac{\partial \bar{v}}{\partial \nu} - \bar{v} \frac{\partial v}{\partial \nu} \right) ds = \int_{\partial D} \left(w \frac{\partial \bar{w}}{\partial \nu} - \bar{w} \frac{\partial w}{\partial \nu} \right) ds$$

$$= \int_D (w \triangle \bar{w} - \bar{w} \triangle w)\, dx = 2ik^2 \int_D \operatorname{Im} n\, |w|^2\, dx.$$

Hence, w vanishes identically in the open set $D_0 := \{x \in D : \operatorname{Im} n(x) > 0\}$ and by the unique continuation principle (Theorem 8.6) we see that $w = 0$ in D. Hence, v has vanishing Cauchy data $v = \partial v/\partial \nu = 0$ on ∂D and by Theorem 2.1 this implies $v = 0$ in D. This contradicts our assumption and the theorem is proved. □

In the case when $\operatorname{Im} n = 0$, there may exist values of k for which $\mathcal{F}$ is not complete and we shall present partial results in the direction later on in this chapter. In the special case of a spherically stratified medium, i.e., (with a slight abuse of notation) $n(x) = n(r)$, $r = |x|$, Colton and Monk [40] have given a rather complete answer to the question of when the set $\mathcal{F}$ is complete. To motivate the hypothesis of the following theorem, note that the case when $n = 1$ is singular since in this case if $h \in C^2(D) \cap C^1(\bar{D})$ is any solution of the Helmholtz equation in D then $v = w = h$ defines a solution of (8.29), (8.30). The case when $n = 1$ corresponds to the case when the sound speed in the inhomogeneous medium is equal to the sound speed in the host medium. Hence, the hypothesis of the following theorem is equivalent to saying that the sound speed in the inhomogeneous medium is always greater than the sound speed in the host medium. An analogous result is easily seen to hold for the case when $n(x) > 1$ for $x \in D$.

Theorem 8.13 *Suppose that $n(x) = n(r)$, $\operatorname{Im} n = 0$, $0 < n(r) < 1$ for $0 \leq r < a$ and $n(r) = 1$ for $r \geq a$ for some $a > 0$ and, as a function of r, $n \in C^2$. Then transmission eigenvalues exist, and if $k > 0$ is a transmission eigenvalue there exists a solution v of (8.29), (8.30) that is a Herglotz wave function, i.e., the set $\mathcal{F}$ is not complete in $L^2(\Omega)$.*

Proof. We postpone the existence question and assume that there exists a nontrivial solution v, w of the homogeneous interior transmission problem, i.e., $k > 0$ is a transmission eigenvalue. We want to show that v is a Herglotz wave function. To this end, we expand v and w in a series of spherical harmonics

$$v(x) = \sum_{l=0}^{\infty} \sum_{m=-l}^{l} a_l^m \, j_l(kr) \, Y_l^m(\hat{x})$$

and

$$w(x) = \sum_{l=0}^{\infty} \sum_{m=-l}^{l} b_l^m(r) \, Y_l^m(\hat{x})$$

where j_l is the spherical Bessel function of order l (see the proof of Rellich's Lemma 2.11). Then, by the orthogonality of the spherical harmonics, the functions

$$v_l^m(x) := a_l^m \, j_l(kr) \, Y_l^m(\hat{x})$$

and

$$w_l^m(x) := b_l^m(r) \, Y_l^m(\hat{x})$$

also satisfy (8.29), (8.30). By the Funk–Hecke formula (2.44) each of the v_l^m is clearly a Herglotz wave function. At least one of them must be different from zero because otherwise v would vanish identically.

To prove that transmission eigenvalues exist, we confine our attention to a solution of (8.29), (8.30) depending only on $r = |x|$. Then clearly v must be of the form

$$v(x) = a_0 j_0(kr)$$

with a constant a_0. Writing

$$w(x) = b_0 \, \frac{y(r)}{r}$$

with a constant b_0, straightforward calculations show that if y is a solution of

$$y'' + k^2 n(r) \, y = 0$$

satisfying the initial conditions

$$y(0) = 0, \quad y'(0) = 1,$$

then w satifies (8.29). We note that in order for w to satisfy (8.29) at the origin it suffices to construct a solution $y \in C^1[0, a] \cap C^2(0, a]$ to the initial value problem. This can be seen by applying Green's formula (2.4) for w in a domain where we exclude the origin by a small sphere centered at the origin and letting the radius of this sphere tend to zero. Following Erdélyi [54], p. 79, we use the Liouville transformation

$$\xi := \int_0^r [n(\rho)]^{1/2} d\rho, \quad z(\xi) := [n(r)]^{1/4} y(r)$$

to arrive at the initial-value problem for

$$(8.34) \qquad\qquad z'' + [k^2 - p(\xi)] z = 0$$

with initial conditions

(8.35) $$z(0) = 0, \quad z'(0) = [n(0)]^{-1/4}$$

where

$$p(\xi) := \frac{n''(r)}{4\,[n(r)]^2} - \frac{5}{16}\frac{[n'(r)]^2}{[n(r)]^3}.$$

Rewriting (8.34), (8.35) as a Volterra integral equation

$$z(\xi) = \frac{\sin k\xi}{k\,[n(0)]^{1/4}} + \frac{1}{k}\int_0^\xi \sin k(\eta - \xi) z(\eta) p(\eta)\,d\eta$$

and using the method of successive approximations, we see that the solution of (8.34), (8.35) satisfies

$$z(\xi) = \frac{\sin k\xi}{k\,[n(0)]^{1/4}} + O\left(\frac{1}{k^2}\right) \quad \text{and} \quad z'(\xi) = \frac{\cos k\xi}{[n(0)]^{1/4}} + O\left(\frac{1}{k}\right),$$

that is,

$$y(r) = \frac{1}{k\,[n(0)\,n(r)]^{1/4}}\,\sin\left(k\int_0^r [n(\rho)]^{1/2}\,d\rho\right) + O\left(\frac{1}{k^2}\right)$$

and

$$y'(r) = \left[\frac{n(r)}{n(0)}\right]^{1/4}\cos\left(k\int_0^r [n(\rho)]^{1/2}\,d\rho\right) + O\left(\frac{1}{k}\right)$$

uniformly on $[0, a]$.

The boundary condition (8.30) now requires

$$b_0\,\frac{y(a)}{a} \quad - \quad a_0 j_0(ka) \quad = \quad 0$$

$$b_0\,\frac{d}{dr}\left(\frac{y(r)}{r}\right)_{r=a} \quad - \quad a_0 k j_0'(ka) \quad = \quad 0.$$

A nontrivial solution of this system exists if and only if

(8.36) $$d := \det\begin{pmatrix} \dfrac{y(a)}{a} & -j_0(ka) \\[2ex] \dfrac{d}{dr}\left(\dfrac{y(r)}{r}\right)_{r=a} & -k j_0'(ka) \end{pmatrix} = 0.$$

Since $j_0(kr) = \sin kr / kr$, from the above asymptotics for $y(r)$, we find that

(8.37) $$d = \frac{1}{a^2 k\,[n(0)]^{1/4}}\left\{\sin k\left(a - \int_0^a [n(r)]^{1/2}\,dr\right) + O\left(\frac{1}{k}\right)\right\}.$$

Since $0 < n(r) < 1$ for $0 \le r < a$ by hypothesis, we see that

$$a - \int_0^a [n(r)]^{1/2}\,dr \neq 0.$$

Hence, from (8.37) we see that for k sufficiently large there exists an infinite set of values of k such that (8.36) is true. Each such k is a transmission eigenvalue and this completes the proof of the theorem.   □

8.5 The Analytic Fredholm Theory

Our aim, in the next section of this chapter, is to show that under suitable conditions on the refractive index the transmission eigenvalues form at most a discrete set. Our proof will be based on the theory of operator valued analytic functions. Hence, in this section we shall present the rudiments of the theory.

Definition 8.14 *Let D be a domain in the complex plane $\mathbb{C}$ and $f : D \to X$ a function from D into the (complex) Banach space X. f is said to be* strongly holomorphic *in D if for every $z \in D$ the limit*

$$\lim_{h \to 0} \frac{f(z+h) - f(z)}{h}$$

exists in X. f is said to be weakly holomorphic D *if for every bounded linear functional ℓ in the dual space X^* we have that $z \mapsto \ell(f(z))$ is a holomorphic function of z for $z \in D$.*

Strongly holomorphic functions are obviously continuous. As we shall see in the next section of this chapter, it is often easier to verify that a function is weakly holomorphic than that it is strongly holomorphic. What is surprising is that these two definitions of holomorphic functions are in fact equivalent. Note that strongly holomorphic functions are clearly weakly holomorphic.

Theorem 8.15 *Every weakly holomorphic function is strongly holomorphic.*

Proof. Let $f : D \to X$ be weakly holomorphic in D. Let $z_0 \in D$ and let Γ be a circle of radius $r > 0$ centered at z_0 with counterclockwise orientation whose closed interior is contained in D. Then if $\ell \in X^*$, the function $z \mapsto \ell(f(z))$ is holomorphic in D. Since $\ell(f)$ is continuous on Γ, we have that

$$(8.38) \qquad |\ell(f(\zeta))| \leq C(\ell)$$

for all $\zeta \in \Gamma$ and some positive number $C(\ell)$ depending on ℓ. Now, for each $\zeta \in \Gamma$ let $\Lambda(\zeta)$ be the linear functional on X^* which assigns to each $\ell \in X^*$ the number $\ell(f(\zeta))$. From (8.38), for each $\ell \in X^*$ we have that $|\Lambda(\zeta)(\ell)| \leq C(\ell)$ for all $\zeta \in \Gamma$ and hence by the uniform boundedness principle we have that $\|\Lambda(\zeta)\| \leq C$ for all $\zeta \in \Gamma$ for some positive constant C. From this, using the Hahn–Banach theorem, we conclude that

$$(8.39) \qquad \|f(\zeta)\| = \sup_{\|\ell\|=1} |\ell(f(\zeta))| = \sup_{\|\ell\|=1} |\Lambda(\zeta)(\ell)| = \|\Lambda(\zeta)\| \leq C$$

for all $\zeta \in \Gamma$.

For $|h| \leq r/2$, by Cauchy's integral formula, we have

$$\ell\left(\frac{f(z_0 + h) - f(z_0)}{h}\right) = \frac{1}{2\pi i} \int_\Gamma \frac{1}{h}\left(\frac{1}{\zeta - (z_0 + h)} - \frac{1}{\zeta - z_0}\right) \ell(f(\zeta))\, d\zeta.$$

Since for $\zeta \in \Gamma$ and $|h_1|, |h_2| \leq r/2$ we have

$$\left| \frac{1}{h_1} \left(\frac{1}{\zeta - (z_0 + h_1)} - \frac{1}{\zeta - z_0} \right) - \frac{1}{h_2} \left(\frac{1}{\zeta - (z_0 + h_2)} - \frac{1}{\zeta - z_0} \right) \right|$$

$$= \left| \frac{h_1 - h_2}{(\zeta - z_0)(\zeta - z_0 - h_1)(\zeta - z_0 - h_2)} \right| \leq \frac{4|h_1 - h_2|}{r^3},$$

using (8.39) and Cauchy's integral formula we can estimate

$$\left| \ell \left(\frac{f(z_0 + h_1) - f(z_0)}{h_1} \right) - \ell \left(\frac{f(z_0 + h_2) - f(z_0)}{h_2} \right) \right| \leq \frac{4C}{r^2} |h_1 - h_2|$$

for all $\ell \in X^*$ with $\|\ell\| \leq 1$. Again by the Hahn–Banach theorem, this implies that

(8.40) $$\left\| \frac{f(z_0 + h_1) - f(z_0)}{h_1} - \frac{f(z_0 + h_2) - f(z_0)}{h_2} \right\| \leq \frac{4C}{r^2} |h_1 - h_2|$$

for all h_1, h_2 with $|h_1|, |h_2| \leq r/2$. Therefore,

$$\lim_{h \to 0} \frac{f(z_0 + h) - f(z_0)}{h}$$

exists since the Banach space X is complete. Since z_0 was an arbitrary point of D, the theorem follows. $\qquad \square$

Corollary 8.16 *Let X and Y be two Banach spaces and denote by $\mathcal{L}(X, Y)$ the Banach space of bounded linear operators mapping X into Y. Let D be a domain in $\mathbb{C}$ and let $A : D \to \mathcal{L}(X, Y)$ be an operator valued function such that for each $\varphi \in X$ the function $A\varphi : D \to Y$ is weakly holomorphic. Then A is strongly holomorphic.*

Proof. For each $\varphi \in X$, we apply the analysis of the previous proof to the weakly holomorphic function $z \mapsto f(z) := A(z)\varphi$. By (8.39), we have

$$\|A(\zeta)\varphi\| = \|f(\zeta)\| \leq C_\varphi$$

for all $\zeta \in \Gamma$ and some positive constant C_φ depending on φ. This, again by the uniform boundedness principle, implies that

$$\|A(\zeta)\| \leq C$$

for all $\zeta \in \Gamma$ and some constant $C > 0$. Hence, we can estimate in Cauchy's formula for $z \mapsto \ell(A(z)\varphi)$ with the aid of

$$\|\ell(A(\zeta)\varphi)\| \leq C$$

for all $\zeta \in \Gamma$, all $\ell \in Y^*$ with $\|\ell\| \leq 1$ and all $\varphi \in X$ with $\|\varphi\| \leq 1$ and obtain the inequality (8.40) for A in the operator norm. This concludes the proof. $\qquad \square$

Definition 8.17 *Let D be a domain in the complex plane $\mathbb{C}$ and $f : D \to X$ a function from D into the (complex) Banach space X. f is said to be* analytic *in D if for every $z_0 \in D$ there exists a power series expansion*

$$f(z) = \sum_{m=0}^{\infty} a_m(z - z_0)^m$$

that converges in the norm on X uniformly for all z in a neighborhood of z_0 and where the cofficients a_m are elements from X.

As in classical complex function theory, the concepts of holomorphic and analytic functions coincide as stated in the following theorem. Therefore, we can synonymously talk about (weakly and strongly) holomorphic and analytic functions.

Theorem 8.18 *Every analytic function is holomorphic and vice versa.*

Proof. Let $f : D \to X$ be analytic. Then, for each ℓ in the dual space X^*, ·the function $z \mapsto \ell(f(z))$, by the continuity of ℓ, is a complex valued analytic function. Therefore, by classical function theory it is a holomorphic complex valued function. Hence, f is weakly holomorphic and thus by Theorem 8.15 it is strongly holomorhic.

Conversely, let $f : D \to X$ be holomorphic. Then, by Definition 8.14, for each $z \in D$ the derivative

$$f'(z) := \lim_{h \to 0} \frac{f(z + h) - f(z)}{h}$$

exists. By continuity, for every $\ell \in X^*$ the function $z \mapsto \ell(f'(z))$ clearly represents the derivative of the complex valued function $z \mapsto \ell(f(z))$. Classical function theory again implies that $z \mapsto f'(z)$ is weakly holomorphic and hence by Theorem 8.15 strongly holomorphic. Therefore, by induction the derivatives $f^{(m)}$ of order m exist and for each $\ell \in X^*$ the m–th derivative of $\ell(f)$ is given by $\ell(f^{(m)})$. Then, using the notation of the proof of Theorem 8.15, by Cauchy's integral formula we have

$$\ell\left(f(z) - \sum_{m=0}^{n} \frac{1}{m!} f^{(m)}(z_0)(z - z_0)^m \right)$$

$$= \frac{1}{2\pi i} \int_\Gamma \frac{\ell(f(\zeta))}{\zeta - z} \, d\zeta - \frac{1}{2\pi i} \sum_{m=0}^{n} (z - z_0)^m \int_\Gamma \frac{\ell(f(\zeta))}{(\zeta - z_0)^{m+1}} \, d\zeta$$

$$= \frac{1}{2\pi i} \int_\Gamma \frac{\ell(f(\zeta))}{\zeta - z} \left(\frac{z - z_0}{\zeta - z_0} \right)^{n+1} d\zeta.$$

From this we have the estimate

$$\left| \ell\left(f(z) - \sum_{m=0}^{n} \frac{1}{m!} f^{(m)}(z_0)(z - z_0)^m \right) \right| \leq \sup_{\zeta \in \Gamma} |\ell(f(\zeta))| \frac{1}{2^n}$$

for all $\ell \in X^*$ and all $z \in D$ with $|z - z_0| \leq r/2$. Using the uniform boundedness principle and the Hahn–Banach theorem as in the proof of Theorem 8.15, we can now conclude uniform convergence of the series

$$f(z) = \sum_{m=0}^{\infty} \frac{1}{m!} \, f^{(m)}(z_0)(z - z_0)^m$$

for all $z \in D$ with $|z - z_0| \leq r/2$ and the proof is finished.   □

We now want to establish the analytic Riesz–Fredholm theory for compact operators in a Banach space. For this we recall that for a single compact linear operator $A : X \to X$ mapping a Banach space X into itself either the inverse operator $(I - A)^{-1} : X \to X$ exists and is bounded or the operator $I - A$ has a nontrivial nullspace of finite dimension (see [106]). In the latter case, it can be proved (see Theorem 1.21 in [32] or Theorem 3.9 in [106]) that there exists a bounded operator P on X with finite dimensional range such that the inverse of $I - A - P : X \to X$ exists and is bounded. Actually, P can be chosen to be the projection of X onto the generalized nullspace of $I - A$.

We can now prove the following theorem, where $\mathcal{L}(X)$ denotes the Banach space of bounded linear operators mapping the Banach space X into itself.

Theorem 8.19 *Let D be a domain in $\mathbb{C}$ and let $A : D \to \mathcal{L}(X)$ be an operator valued analytic function such that $A(z)$ is compact for each $z \in D$. Then either*

a) $(I - A(z))^{-1}$ does not exist for any $z \in D$

or

b) $(I - A(z))^{-1}$ exists for all $z \in D \setminus S$ where S is a discrete subset of D.

Proof. Given an arbitrary $z_0 \in D$, we shall show that for z in a neighborhood of z_0 either a) or b) holds. The theorem will then follow by a straightforward connectedness argument. As mentioned above, for fixed z_0 either the inverse of $I - A(z_0)$ exists and is bounded or the operator $I - A(z_0)$ has a nontrivial nullspace of finite dimension.

In the case where $I - A(z_0)$ has a bounded inverse, since A is continuous we can choose $r > 0$ such that

$$\|A(z) - A(z_0)\| < \frac{1}{\|(I - A(z_0))^{-1}\|}$$

for all $z \in B_r := \{z \in \mathbb{C} : |z - z_0| < r\}$. Then the Neumann series for $[I - (I - A(z_0))^{-1}(A(z) - A(z_0))]^{-1}$ converges and we can conclude that the inverse operator $(I - A(z))^{-1}$ exists for all $z \in B_r$, is bounded and depends continuously on z. Hence, in B_r property b) holds. In particular, from

$$\frac{1}{h} \left\{ (I - A(z + h))^{-1} - (I - A(z))^{-1} \right\}$$

$$= \frac{1}{h} \, (I - A(z + h))^{-1} \, (A(z + h) - A(z)) \, (I - A(z))^{-1}$$

we observe that $z \mapsto (I - A(z))^{-1}$ is holomorphic and hence analytic in B_r.

In the case where $I - A(z_0)$ has a nontrivial nullspace, by the above remark there exists a bounded linear operator of the form

$$P\varphi = \sum_{j=1}^{n} \ell_j(\varphi)\psi_j$$

with linearly independent elements $\psi_1, \ldots, \psi_n \in X$ and bounded linear functionals $\ell_1, \ldots, \ell_n \in X^*$ such that $I - A(z_0) - P : X \to X$ has a bounded inverse. We now choose $r > 0$ such that

$$\|A(z) - A(z_0)\| < \frac{1}{\|(I - A(z_0) - P)^{-1}\|}$$

for all $z \in B_r := \{z \in \mathbb{C} : |z - z_0| < r\}$. Then as above we have that the inverse $T(z) := (I - A(z) - P)^{-1}$ exists for all $z \in B_r$, is bounded and depends analytically on z. Now define

$$B(z) := P(I - A(z) - P)^{-1}$$

Then

(8.41)
$$B(z)\varphi = \sum_{j=1}^{n} \ell_j(T(z)\varphi)\psi_j$$

and since

$$I - A(z) = (I + B(z))(I - A(z) - P)$$

we see that for $z \in B_r$ the operator $I - A(z)$ is invertible if and only if $I + B(z)$ is invertible.

Since $B(z)$ is an operator with finite dimensional range, the invertibility of $I + B(z)$ depends on whether or not the homogeneous equation $\varphi + B(z)\varphi = 0$ has a nontrivial solution. Given $\psi \in X$, let φ be a solution of

(8.42)
$$\varphi + B(z)\varphi = \psi.$$

Then from (8.41) we see that φ must be of the form

(8.43)
$$\varphi = \psi - \sum_{j=1}^{n} \beta_j \psi_j$$

where the coefficients $\beta_j := \ell_j(T(z)\varphi)$ satisfy

(8.44)
$$\beta_j + \sum_{i=1}^{n} \ell_j(T(z)\psi_i)\beta_i = \ell_j(T(z)\psi), \quad j = 1, \ldots n.$$

Conversely, if (8.44) has a solution $\beta_1, \beta_2, \ldots, \beta_n$ then φ defined by (8.43) is easily seen to be a solution of (8.42). Hence, $I + B(z)$ is invertible if and only if the linear system (8.44) is uniquely solvable for each right hand side, i.e., if and only if

$$d(z) := \det \{\delta_{ij} + \ell_j(T(z)\psi_i)\} \neq 0.$$

The analyticity of $z \mapsto T(z)$ implies analyticity of the functions $z \mapsto \ell_j(T(z)\psi_i)$ in B_r. Therefore, d also is analytic, i.e., either $S_r := \{z \in B_r : d(z) = 0\}$ is a discrete set in B_r or $S_r = B_r$. Hence, in the case where $I - A(z_0)$ has a nontrivial nullspace we have also established existence of a neighborhood where either a) or b) holds. This completes the proof of the theorem.   □

8.6 Transmission Eigenvalues

From Theorem 8.12 we see that if $\operatorname{Im} n > 0$ then transmission eigenvalues do not exist whereas from Theorem 8.13 they do exist if $\operatorname{Im} n = 0$ and $n(x) = n(r)$ is spherically stratified. In this section we shall remove the condition of spherical stratification and give sufficient conditions on n such that there exist at most a countable number of transmission eigenvalues. By Theorem 8.9 this implies that the set $\mathcal{F}$ of far field patterns is complete in $L^2(\Omega)$ except for possibly a discrete set of values of the wave number. Throughout this section, we shall always assume that $\operatorname{Im} n(x) = 0$ and $m(x) > 0$ for $x \in D$ where again $m := 1-n$. Analogous results are easily seen to hold for the case when $m(x) < 0$ for $x \in D$.

We begin our analysis by introducing the linear space W by

$$W := \left\{ u \in H^2(\mathbb{R}^3) : u = 0 \text{ in } \mathbb{R}^3 \setminus D, \int_D \frac{1}{m} \left(|u|^2 + |\triangle u|^2\right) dx < \infty \right\}$$

where $H^2(\mathbb{R}^3)$ is the usual Sobolev space and on W we define the scalar product

$$(u, v) := \int_D \frac{1}{m} \left(u\bar{v} + \triangle u \triangle \bar{v}\right) dx, \quad u, v \in W.$$

Lemma 8.20 *The space W is a Hilbert space.*

Proof. We first note that there exists a positive constant c such that

$$(8.45) \qquad \int_D \left(|u|^2 + |\triangle u|^2\right) dx \leq c \int_D \frac{1}{m} \left(|u|^2 + |\triangle u|^2\right) dx$$

for all $u \in W$. Furthermore, from the representation

$$u(x) = -\frac{1}{4\pi} \int_{\mathbb{R}^3} \frac{\triangle u(y)}{|x - y|} \, dy, \quad x \in \mathbb{R}^3,$$

for $u \in C_0^2(\mathbb{R}^3)$ and Theorem 8.2 we see that

$$(8.46) \qquad \|u\|_{H^2(\mathbb{R}^3)} \leq C\|\triangle u\|_{L^2(D)}$$

for functions $u \in W$ and some positive constant C (Here we have used the fact that $u = 0$ in $\mathbb{R}^3 \setminus D$). From (8.45) and (8.46) we have that for functions $u \in W$

the norm in W dominates the norm in $H^2(\mathbb{R}^3)$. Now let (u_n) be a Cauchy sequence in W. Then (u_n) is a Cauchy sequence in $H^2(\mathbb{R}^3)$ and hence converges to a function $u \in H^2(\mathbb{R}^3)$ with respect to the norm in $H^2(\mathbb{R}^3)$. Since, by the Sobolev imbedding theorem, the H^2 norm dominates the maximum norm, and each $u_n = 0$ in $\mathbb{R}^3 \setminus D$, we can conclude that $u = 0$ in $\mathbb{R}^3 \setminus D$. Since $(u_n/\sqrt{m})$ is a Cauchy sequence in $L^2(D)$ we have that (u_n) converges in $L^2(D)$ to a function of the form $\sqrt{m}\, v$ for $v \in L^2(D)$. Similarly, (Δu_n) converges in $L^2(D)$ to a function of the form $\sqrt{m}\, w$ for $w \in L^2(D)$. Since (u_n) converges to u in $H^2(\mathbb{R}^3)$, we have that $\sqrt{m}\, v = u$ and $\sqrt{m}\, v = \Delta u$. Hence $u \in W$ and (u_n) converges to u with respect to the norm in W. $\qquad\square$

Now let G be Green's function for the Laplacian in D and make the assumption that

$$(8.47) \qquad \int_D \int_D [G(x,y)]^2 \, \frac{m(y)}{m(x)} \, dx\, dy < \infty.$$

The condition (8.47) is not vacuous since if m is continuously differentiable, D is a ball, m is spherically stratified near ∂D and there exists a positive constant c such that for $y \in \partial D$, $m(x) \geq c|x-y|^\alpha$, $1 \leq \alpha < 2$, as $x \to y$, then it can be shown that (8.47) is satisfied.

Lemma 8.21 *Assume that $m(x) > 0$ for $x \in D$ and (8.47) is valid. For $x \in D$, let $d(x, \partial D)$ denote the distance between x and ∂D and for $\delta > 0$ sufficiently small define the set $U_\delta := \{x \in D : d(x, \partial D) < \delta\}$. Then for all $u \in W$ we have*

$$\int_{U_\delta} \frac{1}{m} \, |u|^2 \, dx \leq C(\delta) \, \|u\|^2$$

where $\lim\limits_{\delta \to 0} C(\delta) = 0$.

Proof. Applying Green's theorem over D to $G(x, \cdot)$ and functions $u \in C_0^2(\mathbb{R}^3)$ and then using a limiting argument shows that

$$(8.48) \qquad u(x) = -\int_D G(x,y)\, \Delta\, u(y)\, dy, \quad x \in D,$$

for functions $u \in W$. By Schwarz's inequality, we have

$$(8.49) \quad |u(x)|^2 \leq \int_D m(y)\, [G(x,y)]^2 dy \int_D \frac{1}{m(y)}\, |\Delta\, u(y)|^2 dy, \quad x \in D.$$

It now follows that

$$\int_{U_\delta} \frac{1}{m(x)}\, |u(x)|^2 dx \leq \int_{U_\delta} \frac{1}{m(x)} \int_D m(y)\, [G(x,y)]^2 dy\, dx\, \|u\|^2.$$

If we define $C(\delta)$ by

$$C(\delta) := \int_D \int_{U_\delta} [G(x,y)]^2 \, \frac{m(y)}{m(x)} \, dx\, dy,$$

the lemma follows. $\qquad\square$

Using Lemma 8.21, we can now prove a version of Rellich's selection theorem for the weighted function spaces W and

$$L^2_{1/m}(D) := \left\{ u : D \to \mathbb{C} : u \text{ measurable}, \int_D \frac{1}{m}|u|^2 dx < \infty \right\}.$$

Theorem 8.22 *Assume that $m(x) > 0$ for $x \in D$ and (8.47) is valid. Then the imbedding from W into $L^2_{1/m}(D)$ is compact.*

Proof. For $u \in W$, by Green's theorem and Schwarz's inequality we have

$$(8.50) \qquad \|\operatorname{grad} u\|^2_{L^2(D)} = -\int_D \bar{u} \, \triangle u \, dx \leq \|u\|^2_{L^2(D)} \|\triangle u\|^2_{L^2(D)}.$$

Suppose now that (u_n) is a bounded sequence from W, i.e., $\|u_n\| \leq M$ for $n = 1, 2, \ldots$ and some positive constant M. Then, from the fact that the norm in $L^2_{1/m}(D)$ dominates the norm in $L^2(D)$ and (8.50), we see that each u_n is in the Sobolev space $H^1(D)$ and there exists a positive constant, which we again designate by M, such that $\|u_n\|_{H^1(D)} \leq M$ for $n = 1, 2, \ldots$. Hence, by Rellich's selection theorem, there exists a subsequence, again denoted by (u_n), such that (u_n) is convergent to u in $L^2(D)$. We now must show that in fact $u_n \to u$, $n \to \infty$, in $L^2_{1/m}(D)$. To this end, let U_δ be as in Lemma 8.21. Let $\varepsilon > 0$ and choose δ such that $C(\delta) \leq \varepsilon/8M^2$ and n_0 such that

$$\int_{D \setminus U_\delta} \frac{1}{m}|u_n - u_l|^2 dx < \frac{\varepsilon}{2}$$

for $n, l \geq n_0$. Then for $n, l \geq n_0$ we have

$$\int_D \frac{1}{m}|u_n - u_l|^2 dx = \int_{U_\delta} \frac{1}{m}|u_n - u_l|^2 dx + \int_{D \setminus U_\delta} \frac{1}{m}|u_n - u_l|^2 dx$$

$$\leq C(\delta) \|u_n - u_l\|^2 + \frac{\varepsilon}{2} \leq \frac{\varepsilon}{2} + \frac{\varepsilon}{2} = \varepsilon.$$

Hence, (u_n) is a Cauchy sequence in $L^2_{1/m}(D)$ and thus $u_n \to u \in L^2_{1/m}(D)$ for $n \to \infty$. The theorem is now proved. $\qquad\square$

To prove that there exist at most a countable number of transmission eigenvalues, we could now use Theorem 8.22 and proceed along the lines of Rynne and Sleeman [164]. However, we choose an alternate route based on analytic projection operators since these operators are of interest in their own right. We begin with two lemmas.

Lemma 8.23 *Assume that $m(x) > 0$ for $x \in D$ and (8.47) is valid. Then for all k there exists a positive constant $\gamma = \gamma(k)$ such that for $u \in W$*

$$\|u\|^2 \leq \gamma(k) \int_D \frac{1}{m}|\triangle u + k^2 u|^2 dx.$$

Proof. We first choose $k = 0$. Integrating (8.49) we obtain

$$\int_D \frac{1}{m(x)} |u(x)|^2 dx \leq \int_D \int_D \frac{m(y)}{m(x)} [G(x,y)]^2 dy\, dx \int_D \frac{1}{m(y)} |\triangle u(y)|^2 dy,$$

and hence the lemma is true for $k = 0$.

Now assume that $k \neq 0$. From the above analysis for $k = 0$, we see that $\triangle : W \to L^2_{1/m}(D)$ is injective and has closed range. In particular, $\triangle$ is a semi-Fredholm operator (c.f. [166], p. 125). Since compact perturbations of semi-Fredholm operators are semi-Fredholm ([166], p. 128), we have from Theorem 8.22 that $\triangle + k^2$ is also semi-Fredholm. Applying Green's formula (2.4) to functions $u \in C^2_0(\mathbb{R}^3)$ and then using a limiting argument shows that

$$u(x) = - \int_D \varPhi(x,y)\{\triangle u(y) + k^2 u(y)\}\, dy, \quad x \in \mathbb{R}^3,$$

for functions $u \in W$. Hence, by Schwarz's inequality, there exists a positive constant C such that

$$\|u\|_{L^2(D)} \leq C \|\triangle u + k^2 u\|_{L^2_{1/m}(D)}$$

for all $u \in W$. From this we can now conclude that $\triangle + k^2 : W \to L^2_{1/m}(D)$ is injective and, since the range of a semi-Fredholm operator is closed, by the bounded inverse theorem the lemma is now seen to be true for $k \neq 0$.   □

Now for $k \geq 0$ and for $u, v \in W$ define the scalar product

$$(u,v)_k = \int_D \frac{1}{m} (\triangle u + k^2 u)(\triangle \bar{v} + k^2 \bar{v})\, dx$$

with norm $\|u\|_k = \sqrt{(u,u)_k}$. By Lemma 8.23 and Minkowski's inequality, the norm $\|\cdot\|_k$ is equivalent to $\|\cdot\|$ for any $k \geq 0$. For arbitrary complex k, we define the sesquilinear form B on W by

$$B(u,v;k) = \int_D \frac{1}{m} (\triangle u + k^2 u)(\triangle \bar{v} + k^2 \bar{v})\, dx.$$

We then have the following result.

Lemma 8.24 *Assume that $m(x) > 0$ for $x \in D$ and (8.47) is valid. Then for every $k_0 \geq 0$ there exists $\varepsilon > 0$ such that if $|k - k_0| \leq \varepsilon$ then*

$$|B(u,v;k) - B(u,v;k_0)| \leq C\|u\|_{k_0}\|v\|_{k_0}$$

for all $u, v \in W$ where C is a constant satisfying $0 < C < 1$.

Proof. We have

$$B(u,v;k) - B(u,v;k_0) = (k^2 - k_0^2) \int_D \frac{1}{m} (u\, \triangle\, \bar{v} + \bar{v}\, \triangle\, u)\, dx$$

$$+ (k^4 - k_0^4) \int_D \frac{1}{m} u\bar{v}\, dx$$

and hence by Schwarz's inequality

$$
\begin{aligned}
|B(u,v;k) - B(u,v;k_0)| \ &\leq\ |k^2 - k_0^2| \left(\int_D \frac{1}{m} |u|^2 dx \right)^{\frac{1}{2}} \left(\int_D \frac{1}{m} |\triangle v|^2 dx \right)^{\frac{1}{2}} \\
&+\ |k^2 - k_0^2| \left(\int_D \frac{1}{m} |v|^2 dx \right)^{\frac{1}{2}} \left(\int_D \frac{1}{m} |\triangle u|^2 dx \right)^{\frac{1}{2}} \\
&+\ |k^4 - k_0^4| \left(\int_D \frac{1}{m} |u|^2 dx \right)^{\frac{1}{2}} \left(\int_D \frac{1}{m} |v|^2 dx \right)^{\frac{1}{2}}.
\end{aligned}
$$

From Lemma 8.23, we now have that

$$
|B(u,v;k) - B(u,v;k_0)| \leq (2|k^2 - k_0^2| + |k^4 - k_0^4|)\gamma(k_0)\|u\|_{k_0}\|v\|_{k_0}.
$$

Hence, if $|k - k_0|$ is sufficiently small, then $(2|k^2 - k_0^2| + |k^4 - k_0^4|)\gamma(k_0)$ is less than one and the lemma follows. $\qquad\square$

From Lemma 8.24, we see that for $|k - k_0| \leq \varepsilon$ we have

$$
|B(u,v;k)| \leq (1 + C)\|u\|_{k_0}\|v\|_{k_0}
$$

for all $u, v \in W$, i.e., the sesquilinear form B is bounded, and

$$
\mathrm{Re}\, B(u,u;k) \geq \mathrm{Re}\, B(u,u;k_0) - |B(u,u;k) - B(u,u;k_0)| \geq (1 - C)\|u\|_{k_0}^2
$$

for all $u \in W$, i.e., B is strictly coercive. Hence, by the Lax–Milgram theorem, for each $k \in \mathbb{C}$ with $|k - k_0| \leq \varepsilon$ where ε is defined as in Lemma 8.24 there exists a bounded linear operator $S(k) : W \to W$ with a bounded inverse $S^{-1}(k)$ such that

$$
(8.51) \qquad\qquad B(u,v;k) = (S(k)u, v)_{k_0}
$$

holds for all $u, v \in W$. From (8.51), we see that for each $u \in W$ the function $k \mapsto S(k)u$ is weakly analytic and hence from Corollary 8.16 we conclude that $k \mapsto S(k)$ is strongly analytic. Then, in particular, the inverse $S^{-1}(k)$ is also strongly analytic in k. We are now in a position to prove the main result of this section.

Theorem 8.25 *Assume that $m(x) > 0$ for $x \in D$ and (8.47) is valid. Then the set of transmission eigenvalues is either empty or forms a discrete set.*

Proof. Our first aim is to define a projection operator P_k in $L_m^2(D)$ which depends on k in a neighborhood of the positive real axis. To this end, let $f \in L_m^2(D)$ and for $k \in \mathbb{C}$ define the antilinear functional ℓ_f on W by

$$
(8.52) \qquad\qquad \ell_f(\varphi) = \int_D f(\triangle\bar{\varphi} + k^2\bar{\varphi})\, dx, \quad \varphi \in W.
$$

Then by Minkowski's inequality, the functional ℓ_f is bounded on W. Therefore, by the Riesz representation theorem, there exists $p_f \in W$ such that for all $\varphi \in W$ and fixed $k_0 \geq 0$ we have

$$\ell_f(\varphi) = (p_f, \varphi)_{k_0}$$

where p_f depends on k and the linear mapping $p(k) : L_m^2(D) \to W$ with $p(k)f := p_f$ is bounded from $L_m^2(D)$ into W. From (8.52) we see that for each $f \in L_m^2(D)$ the mapping $k \mapsto p(k)f = p_f$ is weakly analytic. Hence, by Corollary 8.16, the mapping $k \mapsto p(k)$ is strongly analytic.

For all $k \in \mathbb{C}$ with $|k - k_0| \leq \varepsilon$ where ε is defined as in Lemma 8.24, we now introduce the analytic operator $P_k : L_m^2(D) \to L_m^2(D)$ by

$$P_k f := \frac{1}{m}(\Delta + k^2)S^{-1}(k)p_f, \quad f \in L_m^2(D).$$

Note that $P_k f \in L_m^2(D)$ since, with the aid of Minkowski's inequality and the boundedness of $S^{-1}(k)$ and $p(k)$, we have

$$\int_D m|P_k f|^2 dx = \int_D \frac{1}{m}|(\Delta + k^2)S^{-1}(k)p_f|^2 dx \leq C_1 \|S^{-1}(k)p_f\|_{k_0}^2 \leq C_2 \|f\|_{L_m^2(D)}^2$$

for some positive constants C_1 and C_2.

We now want to show that P_k is an orthogonal projection operator. To this end, let H be the linear space

$$H := \{u \in C^2(\mathbb{R}^3) : \Delta u + k^2 u = 0 \text{ in } \mathbb{R}^3\},$$

$\bar{H}$ the closure of H in $L_m^2(D)$ and $H^\perp$ the orthogonal complement of H in $L_m^2(D)$. Then for $f \in H$ and $\varphi \in W$, by Green's theorem (2.3) (using our by now familiar limiting argument), we have

$$\ell_f(\varphi) = \int_D f(\Delta \bar{\varphi} + k^2 \bar{\varphi})\, dx = \int_D \bar{\varphi}(\Delta f + k^2 f)\, dx = 0.$$

Hence, $p_f = 0$ for $f \in H$ and since $f \mapsto p_f$ is bounded from $L_m^2(D)$ into W we also have $p_f = 0$ for $f \in \bar{H}$ and consequently $P_k f = 0$ for $f \in \bar{H}$.

Recall now the definition (8.16) of the operator

$$(T_m f)(x) := \int_D \Phi(x, y)m(y)f(y)\, dy, \quad x \in \mathbb{R}^3.$$

Note that we can write

$$T_m f = \tilde{T}_m(\sqrt{m}\, f)$$

where $\tilde{T}_m$ has a weakly singular kernel, i.e., $\tilde{T}_m : L^2(D) \to L^2(D)$ is a compact operator. Therefore, since the L^2 norm dominates the L_m^2 norm, we have that $T_m : L_m^2(D) \to L_m^2(D)$ is also compact. Furthermore, from Theorems 8.1 and 8.2 and a limiting argument, we see that

$$(8.53) \qquad\qquad mf = -(\Delta + k^2)T_m f.$$

For $f \in H^\perp$, we clearly have $(T_m f)(x) = 0$ for $x \in \mathbb{R}^3 \setminus \bar{D}$ since $\Phi(x, \cdot) \in H$ for $x \in \mathbb{R}^3 \setminus \bar{D}$. Therefore, from Lemma 8.23 and (8.53) we see that $T_m f$ is in W for $f \in H^\perp$. From (8.51)–(8.53) we have

$$-\ell_f(\varphi) = \int_D \frac{1}{m} (\Delta + k^2) \bar{\varphi} (\Delta + k^2) T_m f \, dx = B(T_m f, \varphi; k) = (S(k) T_m f, \varphi)_{k_0}$$

for all $\varphi \in W$ and $f \in H^\perp$. Since $T_m f \in W$, from the definition of p_f we conclude that

$$p_f = -S(k) T_m f$$

for $f \in H^\perp$ and consequently

$$P_k f = -\frac{1}{m} (\Delta + k^2) T_m f = f$$

for $f \in H^\perp$, i.e., P_k is indeed an orthogonal projection operator (depending analytically on the parameter k).

Having defined the projection operator P_k, we now turn to the homogeneous interior transmission problem (8.29), (8.30) and note that k is real. From the above analysis, we see that if v, w is a solution of (8.29), (8.30) then

$$P_k v = 0$$

since clearly $v \in H$. From (8.29), that is, $(\Delta + k^2)(w - v) = k^2 m w$, and the homogeneous Cauchy data (8.30), by Green's theorem (2.3), we obtain

$$k^2 \int_D m w \bar{u} \, dx = \int_D \bar{u} (\Delta + k^2)(w - v) \, dx = \int_D \{\bar{u} \Delta (w - v) - (w - v) \Delta \bar{u}\} \, dx = 0$$

for all $u \in H$, that is, $w \in H^\perp$ and consequently

$$P_k w = w.$$

If we now use Green's formula (2.4) to rewrite (8.29), (8.30) in the form

$$w - v = -k^2 T_m w$$

and apply the operator P_k to both sides of this equation, we arrive at the operator equation

(8.54) $$w + k^2 P_k T_m w = 0.$$

Now consider (8.54) defined for $w \in L^2_m(D)$. Since T_m is compact and P_k is bounded, $P_k T_m$ is compact. Since $P_k T_m$ is an operator valued analytic function of k, we can apply Theorem 8.19 to conclude that $(I + k^2 P_k T_m)^{-1}$ exist for all k in a neighborhood of the positive real axis with the possible exception of a discrete set (We note that for k sufficiently small, $(I + k^2 P_k T_m)^{-1}$ exists by the contraction mapping principle). Hence, we can conclude from (8.54) that $w = 0$ except for possibly a discrete set of values of $k > 0$. From (8.29), (8.30) this now implies that $v = 0$ by Green's formula (2.5) and the theorem follows. $\square$

Corollary 8.26 *Assume that $m(x) > 0$ for $x \in D$ and (8.47) is valid. Then, except possibly for a discrete set of values of $k > 0$, the set $\mathcal{F}$ of far field patterns is complete in $L^2(\Omega)$.*

Proof. This follows from Theorem 8.25 and Theorem 8.9.   □

We remind the reader that Theorem 8.25 and Corollary 8.26 remain valid if the condition $m(x) > 0$ is replaced by $m(x) < 0$.

8.7 Numerical Methods

In this final section we shall make some brief remarks on the numerical solution of the scattering problem (8.4)–(8.6) for the inhomogeneous medium with particular emphasis on a recent approach proposed by Kirsch and Monk [97]. The principle problem associated with the numerical solution of (8.4)–(8.6) is that the domain is unbounded. A variety of methods have been proposed for the numerical solution. Broadly speaking, these methods can be grouped into three categories: 1) volume integral equations, 2) expanding grid methods and 3) coupled finite element and boundary element methods.

The volume integral equation method seeks to numerically solve the Lippmann–Schwinger equation (8.12). The advantage of this method is that the problem of an unbounded domain is handled in a simple and natural way. A disadvantage is that care must be used to approximate the three-dimensional singular integral appearing in (8.12). Furthermore, the discrete problem derived from (8.12) has a non-sparse matrix and hence a suitable iteration scheme must be used to obtain a solution, preferably a multi-grid method. However, at this time, a numerical analysis comparable to the analysis for boundary integral equations as described in Sections 3.5 and 3.6 has yet to be done.

The expanding grid method seeks a solution of (8.4)–(8.6) in a ball B_R of radius R centered at the origin where on the boundary ∂B_R the scattered field u^s is required to satisfy

$$(8.55) \qquad \frac{\partial u^s}{\partial r} - iku^s = 0 \quad \text{on } \partial B_R.$$

This boundary condition is clearly motivated by the Sommerfeld radiation condition (8.6) where it is understood that $R \gg a$ with $n(x) = 1$ for $|x| \geq a$. Having posed the boundary condition (8.55), the resulting interior problem is solved by finite element methods. A complete analysis of this method has been provided by Goldstein [59] who has shown how to choose R as well as how the mesh size must be graded in order to obtain optimal convergence. The expanding grid method has the advantage that any standard code for solving the Helmholtz equation in an interior domain can be used to approximate the infinite domain problem (8.4)–(8.6). A disadvantage of this approach for solving (8.4)–(8.6) is that R must be taken to be large and hence computations must be made over a very large (but bounded) domain.

In order to avoid computing on a large domain, various numerical analysts have suggested combining a finite element method inside the inhomogeneity with an appropriate boundary integral equation outside the inhomogeneity. This leads to a coupled finite element and boundary element method for solving (8.4)–(8.6). In this method, a domain D is chosen which contains the support of m and then u is approximated inside D by finite element methods and outside D by a boundary integral representation of u such that u and its normal derivative are continuous across ∂D. For a survey of such coupled methods, we refer the reader to Hsiao [70]. We shall now present a version of this method due to Kirsch and Monk [97] which uses Nyström's method to approximately solve the boundary integral equation. The advantage of this approach is that Nyström's method is exponentially convergent for analytic boundaries and easy to implement (compare Section 3.5). A difficulty is that Nyström's method is defined pointwise whereas the finite element solution for the interior domain is defined variationally. However, as we shall see, this difficulty can be overcome.

For the sake of simplicity, we shall only present the method of Kirsch and Monk for the nonabsorbing two-dimensional case, i.e., we want to construct a solution to the scattering problem

$$(8.56) \qquad \triangle u + k^2 n(x) u = 0 \quad \text{in } \mathbb{R}^2,$$

$$(8.57) \qquad u(x) = u^i(x) + u^s(x),$$

$$(8.58) \qquad \lim_{r \to \infty} \sqrt{r} \left(\frac{\partial u^s}{\partial r} - iku^s \right) = 0,$$

where we assume that $k > 0$ and $n \in C^1(\mathbb{R}^3)$ is real valued such that $m := 1 - n$ has compact support. We choose a simply connected bounded domain D with analytic boundary ∂D that contains the support of m. We shall use the standard notation $L^2(D)$ and $L^2(\partial D)$ for the spaces of square integrable functions defined on D and ∂D, respectively, and the corresponding Sobolev spaces will be denoted by $H^s(D)$ and $H^s(\partial D)$. The inner product on $L^2(D)$ will be denoted by $(\,\cdot\,,\cdot\,)$ and on $L^2(\partial D)$ (or the dual pairing between $H^{-1/2}(\partial D)$ and $H^{1/2}(\partial D)$) by $\langle\,\cdot\,,\cdot\,\rangle$. Finally, we recall the Sobolev space $H^1_{\text{loc}}(\mathbb{R}^2 \setminus \bar{D})$ of all functions u for which the restriction onto $D_R := \{x \in \mathbb{R}^2 \setminus \bar{D} : |x| < R\}$ belongs to $H^1(D_R)$ for all sufficiently large R.

To describe the method due to Kirsch and Monk for numerically solving (8.56)–(8.58), we begin by defining two operators $G_i : H^{-1/2}(\partial D) \to H^1(D)$ and $G_e : H^{-1/2}(\partial D) \to H^1_{\text{loc}}(\mathbb{R}^2 \setminus \bar{D})$ in terms of the following boundary value problems. Given $\psi \in H^{-1/2}(\partial D)$, define $G_i \psi := w \in H^1(D)$ as the weak solution of

$$(8.59) \qquad \triangle w + k^2 n(x) w = 0 \quad \text{in } D,$$

$$(8.60) \qquad \frac{\partial w}{\partial \nu} + ikw = \psi \quad \text{on } \partial D,$$

where ν is the unit outward normal to ∂D. Similarly, $G_e \psi := w \in H^1_{\text{loc}}(\mathbb{R}^2 \setminus \bar{D})$ is the weak solution of

$$(8.61) \qquad \triangle w + k^2 w = 0 \quad \text{in } \mathbb{R}^2 \setminus \bar{D},$$

$$(8.62) \qquad \frac{\partial w}{\partial \nu} + ikw = \psi \quad \text{on } \partial D,$$

$$(8.63) \qquad \lim_{r \to \infty} \sqrt{r} \left(\frac{\partial w}{\partial r} - ikw \right) = 0,$$

where (8.63) holds uniformly in all directions. Now define $u(\psi)$ by

$$u(\psi) := \begin{cases} G_e \psi + u^i & \text{in } \mathbb{R}^2 \setminus \bar{D}, \\[2mm] G_i \psi + G_i \left(\dfrac{\partial u^i}{\partial \nu} + iku^i \right) & \text{in } D. \end{cases}$$

Note that $\partial u/\partial \nu + iku$ has the same limiting values on both sides of ∂D. Furthermore, we see that if $\psi \in H^{-1/2}(\partial D)$ can be chosen such that

$$(8.64) \qquad (G_i - G_e)\, \psi = u^i - G_i \left(\frac{\partial u^i}{\partial \nu} + iku^i \right) \quad \text{on } \partial D,$$

then u has the same limiting values on both sides of ∂D. From these facts it can be deduced that u solves the scattering problem (8.56)–(8.58) (c.f. the proof of Theorem 5.7). Hence, we need to construct an approximate solution to the operator equation (8.64).

To solve (8.64), we first choose a finite element space $S_h \subset H^1(D)$ and define $G_i^h \psi$ to be the usual finite element approximation of (8.59), (8.60). In particular, if $\psi \in H^{-1/2}(\partial D)$ then $G_i^h \psi \in S_h$ satisfies (we do not distinguish between $G_i^h \psi$ defined on D and its trace defined on ∂D)

$$(8.65) \quad (\text{grad } G_i^h \psi, \text{grad } \varphi_h) - k^2 (n G_i^h \psi, \varphi_h) + ik \langle G_i^h \psi, \varphi_h \rangle - \langle \psi, \varphi_h \rangle = 0$$

for all $\varphi_h \in S_h$. For h sufficiently small, the existence and uniqueness of a solution to (8.65) is well known (Schatz [165]). Having defined $G_i^h \psi$, we next define $G_e^M \psi$ to be the approximate solution of (8.61)–(8.63) obtained by numerically solving an appropriate boundary integral equation using Nyström's method with M knots (c.f. Section 3.5).

We now need to discretize $H^{-1/2}(\partial D)$. To this end, we parameterize ∂D by

$$x = (x_1(t), x_2(t)), \quad 0 \le t \le 2\pi,$$

and define S_N by

$$S_N := \left\{ g : \partial D \to \mathbb{C} : g(x) = \sum_{j=-N+1}^{N} a_j e^{ijt}, \; a_j \in \mathbb{C}, \; x = (x_1(t), x_2(t)) \right\}.$$

Note that the indices on the sum in this definition are chosen such that S_N has an even number of degrees of freedom which is convenient for using fast Fourier transforms. We now want to define a projection $P_N : L^2(\partial D) \to S_N$. For $g \in L^2(\partial D)$, this is done by defining $P_N g \in S_N$ to be the unique solution of

$$\langle g - P_N g, \varphi \rangle = 0$$

for every $\varphi \in S_N$. We next define a projection P_N^M from discrete functions defined on the Nyström's points into functions in S_N. To do this, let x_i for $i = 1, \dots, M$ be the Nyström points on ∂D and let g be a discrete function on ∂D so that $g(x_i) = g_i$, $i = 1, \dots, M$. Then, provided $M \geq 2N$, we define $P_N^M g \in S_N$ by requiring that

$$\langle P_N^M g, \varphi \rangle_M = \langle g, \varphi \rangle_M$$

for every $\varphi \in S_N$ where

$$\langle u, v \rangle_M := \frac{1}{M} \sum_{i=1}^{M} u(x_i) \overline{v(x_i)}.$$

Note that $P_N^M g$ is the uniquely determined element in S_N that is the closest to g with respect to the norm $\| \cdot \|_M$ assoziated with $\langle \cdot, \cdot \rangle_M$.

Following Kirsch and Monk, we can now easily define a discrete method for solving the scattering problem (8.56)–(8.58). We first seek $\varphi_N \in S_N$ such that

$$(P_N G_i^h - P_N^M G_e^M)\varphi_N = P_N \left(u^i - G_i^h \left(\frac{\partial u^i}{\partial \nu} + iku^i \right) \right).$$

Then, having found φ_N, we approximate the solution u of (8.56)–(8.58) by

$$u_N^{h,M} := \begin{cases} G_e^M \varphi_N + u^i & \text{in } \mathbb{R}^2 \setminus \bar{D}, \\[2ex] G_i^h \varphi_N + G_i^h \left(\dfrac{\partial u^i}{\partial \nu} + iku^i \right) & \text{in } D. \end{cases}$$

Error estimates and numerical examples of the implementation of this scheme for solving the scattering problem (8.56)–(8.58) can be found in Kirsch and Monk [97].

9. Electromagnetic Waves in an Inhomogeneous Medium

In the previous chapter, we considered the direct scattering problem for acoustic waves in an inhomogeneous medium. We now consider the case of electromagnetic waves. However, our aim is not to simply prove the electromagnetic analogue of each theorem in Chapter 8, but rather to select the basic ideas of the previous chapter, extend them to the electromagnetic case, and then consider some themes that were not considered in Chapter 8, but ones that are particularly relevant to the case of electromagnetic waves. In particular, we shall consider two simple problems, one in which the electromagnetic field has no discontinuities across the boundary of the medium and the second where the medium is an imperfect conductor such that the electromagnetic field does not penetrate deeply into the body. This last problem is an approximation to the more complicated transmission problem for a piecewise constant medium and leads to what is called the exterior impedance problem for electromagnetic waves.

After a brief discussion of the physical background to electromagnetic wave propagation in an inhomogeneous medium, we show existence and uniqueness of a solution to the direct scattering problem for electromagnetic waves in an inhomogeneous medium. By means of a reciprocity relation for electromagnetic waves in an inhomogeneous medium, we then show that, for a conducting medium, the set of electric far field patterns corresponding to incident time-harmonic plane waves moving in arbitrary directions is complete in the space of square integrable tangential vector fields on the unit sphere. However, we show that this set of far field patterns is in general not complete for a dielectric medium. Finally, we establish the existence and uniqueness of a solution to the exterior impedance problem and show that the set of electric far field patterns is again complete in the space of square integrable tangential vector fields on the unit sphere. These results for the exterior impedance problem will be used in the next chapter when we discuss the inverse scattering problem for electromagnetic waves in an inhomogeneous medium. We note, as in the case of acoustic waves, that our ideas and methods can be extended to more complicated scattering problems involving discontinuous fields, piecewise dielectric media, etc. but, for the sake of clarity and brevity, we do not consider these more general problems in this book.

9.1 Physical Background

We consider electromagnetic wave propagation in an inhomogeneous isotropic medium in $\mathbb{R}^3$ with electric permittivity $\varepsilon = \varepsilon(x) > 0$, magnetic permeability $\mu = \mu_0$ and electric conductivity $\sigma = \sigma(x)$ where μ_0 is a positive constant. We assume that $\varepsilon(x) = \varepsilon_0$ and $\sigma(x) = 0$ for all x outside some sufficiently large ball where ε_0 is a constant. Then if J is the current density, the electric field $\mathcal{E}$ and magnetic field $\mathcal{H}$ satisfy the *Maxwell equations*, namely

$$(9.1) \qquad \operatorname{curl} \mathcal{E} + \mu_0 \frac{\partial \mathcal{H}}{\partial t} = 0, \quad \operatorname{curl} \mathcal{H} - \varepsilon(x) \frac{\partial \mathcal{E}}{\partial t} = J.$$

Furthermore, in an isotropic conductor, the current density is related to the electric field by *Ohm's law*

$$(9.2) \qquad J = \sigma \mathcal{E}.$$

For most metals, σ is very large and hence it is often reasonable in many theoretical investigations to approximate a metal by a fictitious *perfect conductor* in which σ is taken to be infinite. However, in this chapter, we shall assume that the inhomogeneous medium is not a perfect conductor, i.e., σ is finite. If σ is nonzero, the medium is called a *conductor*, whereas if $\sigma = 0$ the medium is referred to as a *dielectric*.

We now assume that the electromagnetic field is time-harmonic, i.e., of the form

$$\mathcal{E}(x,t) = \frac{1}{\sqrt{\varepsilon_0}} E(x) e^{-i\omega t}, \quad \mathcal{H}(x,t) = \frac{1}{\sqrt{\mu_0}} H(x) e^{-i\omega t}$$

where ω is the frequency. Then from (9.1) and (9.2) we see that E and H satisfy the time-harmonic Maxwell equations

$$(9.3) \qquad \operatorname{curl} E - ikH = 0, \quad \operatorname{curl} H + ikn(x)E = 0$$

in $\mathbb{R}^3$ where the (positive) wave number k is defined by $k^2 = \varepsilon_0 \mu_0 \omega^2$ and the *refractive index* $n = n(x)$ is given by

$$n(x) := \frac{1}{\varepsilon_0} \left(\varepsilon(x) + i \frac{\sigma(x)}{\omega} \right).$$

We shall assume that $n \in C^{1,\alpha}(\mathbb{R}^3)$ for some $0 < \alpha < 1$ and that $m := 1 - n$ has compact support. As usual, we define $D := \{x \in \mathbb{R}^3 : m(x) \neq 0\}$.

We consider the following scattering problem for (9.3). Let $E^i, H^i \in C^1(\mathbb{R}^3)$ be a solution of the Maxwell equations for a homogeneous medium

$$(9.4) \qquad \operatorname{curl} E^i - ikH^i = 0, \quad \operatorname{curl} H^i + ikE^i = 0$$

in all of $\mathbb{R}^3$. We then want to find a solution $E, H \in C^1(\mathbb{R}^3)$ of (9.3) in $\mathbb{R}^3$

such that if

(9.5)
$$E = E^i + E^s, \quad H = H^i + H^s$$

the scattered field E^s, H^s satisfies the *Silver–Müller radiation condition*

(9.6)
$$\lim_{r \to \infty} (H^s \times x - rE^s) = 0$$

uniformly for all directions $x/|x|$ where $r = |x|$.

For the next three sections of this chapter, we shall be concerned with the scattering problem (9.3)–(9.6). The existence and uniqueness of a solution to this problem were first given by Müller [142] for the more general case when $\mu = \mu(x)$. The proof simplifies considerably for the case we are considering, i.e., $\mu = \mu_0$, and we shall present this proof in the next section.

9.2 Existence and Uniqueness

Under the assumptions given in the previous section for the refractive index n, we shall show in this section that there exists a unique solution to the scattering problem (9.3)–(9.6). Our analysis follows that of Colton and Kress [32] and is based on reformulating (9.3)–(9.6) as an integral equation. We first prove the following theorem, where

$$\Phi(x, y) := \frac{1}{4\pi} \frac{e^{ik|x-y|}}{|x-y|}, \quad x \neq y,$$

as usual, denotes the fundamental solution to the Helmholtz equation and

$$m := 1 - n.$$

Theorem 9.1 *Let $E, H \in C^1(\mathbb{R}^3)$ be a solution of the scattering problem (9.3)–(9.6). Then E satisfies the integral equation*

(9.7)
$$E(x) = E^i(x) - k^2 \int_{\mathbb{R}^3} \Phi(x, y)m(y)E(y)\, dy$$
$$+ \operatorname{grad} \int_{\mathbb{R}^3} \frac{1}{n(y)} \operatorname{grad} n(y) \cdot E(y)\, \Phi(x, y)\, dy, \quad x \in \mathbb{R}^3.$$

Proof. Let $x \in \mathbb{R}^3$ be an arbitrary point and choose an open ball B with unit outward normal ν such that B contains the support of m and $x \in B$. From the

Stratton–Chu formula (6.5) applied to E, H, we have

$$E(x) = -\operatorname{curl} \int_{\partial B} \nu(y) \times E(y) \, \Phi(x,y) \, ds(y)$$

$$+ \operatorname{grad} \int_{\partial B} \nu(y) \cdot E(y) \, \Phi(x,y) \, ds(y)$$

(9.8)
$$-ik \int_{\partial B} \nu(y) \times H(y) \, \Phi(x,y) \, ds(y)$$

$$+ \operatorname{grad} \int_{B} \frac{1}{n(y)} \operatorname{grad} n(y) \cdot E(y) \, \Phi(x,y) \, dy$$

$$-k^2 \int_{B} m(y) E(y) \, \Phi(x,y) \, dy$$

since $\operatorname{curl} H + ik E = ikmE$ and $n \operatorname{div} E = -\operatorname{grad} n \cdot E$. Note that in the volume integrals over B we can integrate over all of $\mathbb{R}^3$ since m has support in B. The Stratton–Chu formula applied to E^i, H^i gives

$$E^i(x) = -\operatorname{curl} \int_{\partial B} \nu(y) \times E^i(y) \, \Phi(x,y) \, ds(y)$$

(9.9)
$$+ \operatorname{grad} \int_{\partial B} \nu(y) \cdot E^i(y) \, \Phi(x,y) \, ds(y)$$

$$-ik \int_{\partial B} \nu(y) \times H^i(y) \, \Phi(x,y) \, ds(y).$$

Finally, from the version of the Stratton–Chu formula corresponding to Theorem 6.6, we see that

$$-\operatorname{curl} \int_{\partial B} \nu(y) \times E^s(y) \, \Phi(x,y) \, ds(y)$$

(9.10)
$$+ \operatorname{grad} \int_{\partial B} \nu(y) \cdot E^s(y) \, \Phi(x,y) \, ds(y)$$

$$-ik \int_{\partial B} \nu(y) \times H^s(y) \, \Phi(x,y) \, ds(y) = 0.$$

With the aid of $E = E^i + E^s$, $H = H^i + H^s$ we can now combine (9.8)–(9.10) to conclude that (9.7) is satisfied. $\qquad\square$

We now want to show that every solution of the integral equation (9.7) is also a solution to (9.3)–(9.6).

Theorem 9.2 *Let $E \in C(\mathbb{R}^3)$ be a solution of the integral equation (9.7). Then E and $H := \operatorname{curl} E/ik$ are a solution of (9.3)–(9.6).*

Proof. Since m has compact support, from Theorem 8.1 we can conclude that if $E \in C(\mathbb{R}^3)$ is a solution of (9.7) then $E \in C^{1,\alpha}(\mathbb{R}^3)$. Hence, by the relation $\mathrm{grad}_x\, \Phi(x,y) = -\mathrm{grad}_y\, \Phi(x,y)$, Gauss' divergence theorem and Theorem 8.1, we have

$$(9.11) \quad \mathrm{div} \int_{\mathbb{R}^3} \Phi(x,y)m(y)E(y)\, dy = \int_{\mathbb{R}^3} \mathrm{div}\{m(y)E(y)\}\Phi(x,y)\, dy$$

and

$$(9.12) \quad (\Delta+k^2) \int_{\mathbb{R}^3} \frac{1}{n(y)}\, \mathrm{grad}\, n(y)\cdot E(y)\, \Phi(x,y)\, dy = -\frac{1}{n(x)}\, \mathrm{grad}\, n(x)\cdot E(x)$$

for $x \in \mathbb{R}^3$. Taking the divergence of (9.7) and using (9.11) and (9.12), we see that

$$u := \frac{1}{n}\, \mathrm{div}(nE)$$

satisfies the integral equation

$$u(x) + k^2 \int_{\mathbb{R}^3} \Phi(x,y)m(y)u(y)\, dy = 0, \quad x \in \mathbb{R}^3.$$

Hence, from Theorems 8.3 and 8.7 we can conclude that $u(x) = 0$ for $x \in \mathbb{R}^3$, that is,

$$(9.13) \qquad\qquad \mathrm{div}(nE) = 0 \quad \text{in } \mathbb{R}^3.$$

Therefore, the integral equation (9.7) can be written in the form

$$E(x) = E^i(x) - k^2 \int_{\mathbb{R}^3} \Phi(x,y)m(y)E(y)\, dy$$

$$(9.14)$$

$$- \mathrm{grad} \int_{\mathbb{R}^3} \Phi(x,y)\, \mathrm{div}\, E(y)\, dy, \quad x \in \mathbb{R}^3,$$

and thus for $H := \mathrm{curl}\, E/ik$ we have

$$(9.15) \quad H(x) = H^i(x) + ik\, \mathrm{curl} \int_{\mathbb{R}^3} \Phi(x,y)m(y)E(y)\, dy, \quad x \in \mathbb{R}^3.$$

In particular, by Theorem 8.1 this implies $H \in C^{1,\alpha}(\mathbb{R}^3)$ since $E \in C^{1,\alpha}(\mathbb{R}^3)$. We now use the vector identity (6.4), the Maxwell equations (9.4), and (9.11),

(9.13)–(9.15) to deduce that

$$\operatorname{curl} H(x) + ikE(x) \;=\; ik(\operatorname{curl}\operatorname{curl} - k^2) \int_{\mathbb{R}^3} \Phi(x,y)m(y)E(y)\,dy$$

$$-ik\operatorname{grad} \int_{\mathbb{R}^3} \Phi(x,y)\operatorname{div} E(y)\,dy$$

$$=\; -ik(\triangle + k^2) \int_{\mathbb{R}^3} \Phi(x,y)m(y)E(y)\,dy$$

$$-ik\operatorname{grad} \int_{\mathbb{R}^3} \operatorname{div}\{n(y)E(y)\}\Phi(x,y)\,dy$$

$$=\; ikm(x)E(x)$$

for $x \in \mathbb{R}^3$. Therefore E, H satisfy (9.3). Finally, the decomposition (9.5) and the radiation condition (9.6) follow readily from (9.7) and (9.15) with the aid of (2.14) and (6.25). $\qquad\square$

We note that in (9.7) we can replace the region of integration by any domain G such that the support of m is contained in $\bar G$ and look for solutions in $C(\bar G)$. Then for $x \in \mathbb{R}^3 \setminus \bar G$ we define $E(x)$ by the right hand side of (9.7) and obviously obtain a continuous solution to (9.7) in all of $\mathbb{R}^3$.

In order to show that (9.7) is uniquely solvable we need to establish the following unique continuation principle for the Maxwell equations.

Theorem 9.3 *Let G be a domain in $\mathbb{R}^3$ and let $E, H \in C^1(G)$ be a solution of*

(9.16) $\qquad\qquad \operatorname{curl} E - ikH = 0, \quad \operatorname{curl} H + ikn(x)E = 0$

in G such that $n \in C^{1,\alpha}(G)$. Suppose E, H vanishes in a neighborhood of some $x_0 \in G$. Then E, H is identically zero in G.

Proof. From the representation formula (9.8) and Theorem 8.1, since by assumption $n \in C^{1,\alpha}(G)$, we first can conclude that $E \in C^{1,\alpha}(B)$ for any ball B with $\bar B \subset G$. Then, using $\operatorname{curl} E = ikH$ from (9.8) we have $H \in C^{2,\alpha}(B)$ whence, in particular, $H \in C^2(G)$ follows.

Using the vector identity (6.8), we deduce from (9.16) that

$$\triangle H + \frac{1}{n(x)} \operatorname{grad} n(x) \times \operatorname{curl} H + k^2 n(x)H = 0 \quad \text{in } G$$

and the proof is completed by applying Lemma 8.5 to the real and imaginary parts of the cartesian components of H. $\qquad\square$

Theorem 9.4 *The scattering problem (9.3)–(9.6) has at most one solution $E, H \in C^1(\mathbb{R}^3)$.*

Proof. Let E, H denote the difference between two solutions. Then E, H clearly satisfy the radiation condition (9.6) and the Maxwell equations for a homogeneous medium outside some ball B containing the support of m. From Gauss' divergence theorem and the Maxwell equations (9.3), denoting as usual by ν the exterior unit normal to B, we have that

$$\int_{\partial B} \nu \times E \cdot \bar{H} \, ds = \int_B (\operatorname{curl} E \cdot \bar{H} - E \cdot \operatorname{curl} \bar{H}) \, dx$$

(9.17)

$$= ik \int_B (|H|^2 - \bar{n} |E|^2) \, dx$$

and hence

$$\operatorname{Re} \int_{\partial B} \nu \times E \cdot \bar{H} \, ds = -k \int_B \operatorname{Im} n \, |E|^2 dx \leq 0.$$

Hence, by Theorem 6.10, we can conclude that $E(x) = H(x) = 0$ for $x \in \mathbb{R}^3 \setminus \bar{B}$. By Theorem 9.3 the proof is complete. $\qquad \square$

We are now in a position to show that there exists a unique solution to the electromagnetic scattering problem.

Theorem 9.5 *The scattering problem (9.3)–(9.6) for an inhomogeneous medium has a unique solution and the solution E, H depends continuously on the incident field E^i, H^i with respect to the maximum norm.*

Proof. By Theorems 9.2 and 9.4, it suffices to prove the existence of a solution $E \in C(\mathbb{R}^3)$ to (9.7). As in the proof of Theorem 8.7, it suffices to look for solutions of (9.7) in an open ball B containing the support of m. We define an electromagnetic operator $T_e : C(\bar{B}) \to C(\bar{B})$ on the Banach space of continuous vector fields in $\bar{B}$ by

$$(T_e E)(x) := -k^2 \int_B \Phi(x, y) m(y) E(y) \, dy$$

(9.18)

$$+ \operatorname{grad} \int_B \frac{1}{n(y)} \operatorname{grad} n(y) \cdot E(y) \, \Phi(x, y) \, dy, \quad x \in \bar{B}.$$

Since T_e has a weakly singular kernel it is a compact operator. Hence, we can apply the Riesz–Fredholm theory and must show that the homogeneous equation corresponding to (9.7) has only the trivial solution. If this is done, equation (9.7) can be solved and the inverse operator $(I - T_e)^{-1}$ is bounded. From this it follows that E, H depend continuously on the incident field with respect to the maximum norm.

By Theorem 9.2, a continuous solution E of $E - T_e E = 0$ solves the homogeneous scattering problem (9.3)–(9.6) with $E^i = 0$ and hence, by Theorem 9.4, it follows that $E = 0$. The theorem is now proved. $\qquad \square$

9.3 Far Field Patterns

We now want to examine the set of far field patterns of the scattering problem (9.3)–(9.6) where the refractive index $n = n(x)$ again satisfies the assumptions of Section 9.1 and the incident electromagnetic field is given by the plane wave

$$E^i(x; d, p) = \frac{i}{k} \operatorname{curl} \operatorname{curl} p \, e^{ik\,x\cdot d} = ik \, (d \times p) \times d \, e^{ik\,x\cdot d},$$

(9.19)

$$H^i(x; d, p) = \operatorname{curl} p \, e^{ik\,x\cdot d} = ik \, d \times p \, e^{ik\,x\cdot d},$$

where d is a unit vector giving the direction of propagation and $p \in \mathbb{R}^3$ is a constant vector giving the polarization. From Theorem 6.8, we see that

$$E^s(x; d, p) = \frac{e^{ik|x|}}{|x|} \, E_\infty(\hat{x}; d, p) + O\left(\frac{1}{|x|^2}\right), \quad |x| \to \infty,$$

(9.20)

$$H^s(x; d, p) = \frac{e^{ik|x|}}{|x|} \, \hat{x} \times E_\infty(\hat{x}; d, p) + O\left(\frac{1}{|x|^2}\right), \quad |x| \to \infty,$$

where E_∞ is the electric far field pattern. Furthermore, from (6.78) and Green's vector theorem (6.3), we can immediately deduce the following reciprocity relation.

Theorem 9.6 *Let E_∞ be the electric far field pattern of the scattering problem (9.3)–(9.6) and (9.19). Then for all vectors $\hat{x}, d \in \Omega$ and $p, q \in \mathbb{R}^3$ we have*

$$q \cdot E_\infty(\hat{x}; d, p) = p \cdot E_\infty(-d; -\hat{x}, q).$$

Motivated by our study of acoustic waves in Chapter 8, we now want to use this reciprocity relation to show the equivalence of the completeness of the set of electric far field patterns and the uniqueness of the solution to an electromagnetic interior transmission problem. In this chapter, we shall only be concerned with the homogeneous problem, defined as follows.

Homogeneous Electromagnetic Interior Transmission Problem. *Find a solution $E_0, E_1, H_0, H_1 \in C^1(D) \cap C(\bar{D})$ of*

$$\operatorname{curl} E_1 - ikH_1 = 0, \quad \operatorname{curl} H_1 + ikn(x)E_1 = 0 \quad \text{in } D,$$

(9.21)

$$\operatorname{curl} E_0 - ikH_0 = 0, \quad \operatorname{curl} H_0 + ikE_0 = 0 \quad \text{in } D,$$

satisfying the boundary condition

(9.22) $$\nu \times (E_1 - E_0) = 0, \quad \nu \times (H_1 - H_0) = 0 \quad \text{on } \partial D,$$

where again $D := \{x \in \mathbb{R}^3 : m(x) \neq 0\}$ and where we assume that D is connected with a connected C^2 boundary.

In order to establish the connection between electric far field patterns and the electromagnetic interior transmission problem, we now recall the definition of the Hilbert space

$$T^2(\Omega) := \{g : \Omega \to \mathbb{C}^3 : g \in L^2(\Omega),\ \nu \cdot g = 0 \text{ on } \Omega\}$$

of square integrable tangential fields on the unit sphere. Let $\{d_n : n = 1, 2, \ldots\}$ be a countable dense set of unit vectors on Ω and consider the set $\mathcal{F}$ of electric far field patterns defined by

$$\mathcal{F} := \{E_\infty(\cdot\,; d_n, e_j) : n = 1, 2, \ldots, \ j = 1, 2, 3\}$$

where e_1, e_2, e_3 are the cartesian unit coordinate vectors in $\mathbb{R}^3$. Recalling the definition of an electromagnetic Herglotz pair and Herglotz kernel given in Section 6.6, we can now prove the following theorem due to Colton and Päivärinta [46].

Theorem 9.7 *A tangential vector field g is in the orthogonal complement $\mathcal{F}^\perp$ of $\mathcal{F}$ if and only if there exists a solution of the homogeneous electromagnetic interior transmission problem such that E_0, H_0 is an electromagnetic Herglotz pair with Herglotz kernel h given by $h(d) = ik\, g(-d)$.*

Proof. Suppose that $g \in T^2(\Omega)$ satisfies

$$\int_\Omega E_\infty(\hat{x}; d_n, e_j) \cdot \overline{g(\hat{x})}\, ds(\hat{x}) = 0$$

for $n = 1, 2, \ldots$ and $j = 1, 2, 3$. Then by continuity and superposition we have that

$$\int_\Omega E_\infty(\hat{x}; d, p) \cdot \overline{g(\hat{x})}\, ds(\hat{x}) = 0$$

for all $d \in \Omega$ and $p \in \mathbb{R}^3$. By the reciprocity relation, this is equivalent to

$$p \cdot \int_\Omega E_\infty(-d; -\hat{x}, \overline{g(\hat{x})})\, ds(\hat{x}) = 0$$

for all $d \in \Omega$ and $p \in \mathbb{R}^3$, i.e.,

$$(9.23) \qquad \int_\Omega E_\infty(\hat{x}; d, h(d))\, ds(d) = 0$$

for all $\hat{x} \in \Omega$ where $h(d) = \overline{g(-d)}$. Analogous to Lemma 6.31, from the integral equation (9.7) it can be seen that the left hand side of (9.23) represents the electric far field pattern of the scattered wave E_0^s, H_0^s corresponding to the incident wave E_0^i, H_0^i given by the the electromagnetic Herglotz pair

$$E_0^i(x) = \int_\Omega E^i(x; d, h(d))\, ds(d) = ik \int_\Omega h(d)\, e^{ik\,x\cdot d}\, ds(d),$$

$$H_0^i(x) = \int_\Omega H^i(x; d, h(d))\, ds(d) = \operatorname{curl} \int_\Omega h(d)\, e^{ik\,x\cdot d}\, ds(d).$$

Hence, (9.23) is equivalent to a vanishing far field pattern of E_0^s, H_0^s and thus, by Theorem 6.9, equivalent to $E_0^s = H_0^s = 0$ in $\mathbb{R}^3 \backslash B$, i.e., with $E_0 := E_0^i$, $H_0 := H_0^i$ and $E_1 := E_0^i + E_0^s$, $H_1 := H_0^i + H_0^s$ we have solutions to (9.21) satisfying the boundary condition (9.22). □

In the case of a conducting medium, i.e., $\operatorname{Im} n \neq 0$, we can use Theorem 9.7 to deduce the following result [46].

Theorem 9.8 *In a conducting medium, the set $\mathcal{F}$ of electric far field patterns is complete in $T^2(\Omega)$.*

Proof. Recalling that an electromagnetic Herglotz pair vanishes if and only if its Herglotz kernel vanishes (Theorem 3.15 and Definition 6.29), we see from Theorem 9.7 that it suffices to show that the only solution of the homogeneous electromagnetic interior transmission problem (9.21), (9.22) is $E_0 = E_1 = H_0 = H_1 = 0$. However, analogous to (9.17), from Gauss' divergence theorem and the Maxwell equations (9.21) we have

$$\int_{\partial D} \nu \cdot E_1 \times \bar{H}_1 \, ds = ik \int_D (|H_1|^2 - \bar{n} \, |E_1|^2) \, dx,$$

$$\int_{\partial D} \nu \cdot E_0 \times \bar{H}_0 \, ds = ik \int_D (|H_0|^2 - |E_0|^2) \, dx.$$

From these two equations, using the transmission conditions (9.22) we obtain

$$\int_D (|H_1|^2 - \bar{n} \, |E_1|^2) \, dx = \int_D (|H_0|^2 - |E_0|^2) \, dx$$

and taking the imaginary part of both sides gives

$$\int_D \operatorname{Im} n \, |E_1|^2 dx = 0.$$

From this, we conclude by unique continuation that $E_1 = H_1 = 0$ in D. From (9.22) we now have vanishing tangential components of E_0 and H_0 on the boundary ∂D whence $E_0 = H_0 = 0$ in D follows from the Stratton–Chu formulas (6.8) and (6.9). □

In contrast to Theorem 9.8, the set $\mathcal{F}$ of electric far field patterns is not in general complete for a dielectric medium. We shall show this for a spherically stratified medium in the next section.

9.4 The Spherically Stratified Dielectric Medium

In this section, we shall consider the class $\mathcal{F}$ of electric far field patterns for a spherically stratified dielectric medium. Our aim is to show that in this case there exist wave numbers k such that $\mathcal{F}$ is not complete in $T^2(\Omega)$. It suffices to show that when $n(x) = n(r)$, $r = |x|$, $\operatorname{Im} n = 0$ and, as a function of r, $n \in C^2$, there exist values of k such that there exists a nontrivial solution to the homogeneous electromagnetic interior transmission problem

$$\operatorname{curl} E_1 - ik H_1 = 0, \quad \operatorname{curl} H_1 + ikn(r)E_1 = 0 \quad \text{in } B,$$

(9.24)

$$\operatorname{curl} E_0 - ik H_0 = 0, \quad \operatorname{curl} H_0 + ik E_0 = 0 \quad \text{in } B,$$

with the boundary condition

(9.25) $\nu \times (E_1 - E_0) = 0, \quad \nu \times (H_1 - H_0) = 0 \quad \text{on } \partial B,$

where E_0, H_0 is an electromagnetic Herglotz pair, where now B is an open ball of radius a with exterior unit normal ν and where $\operatorname{Im} n = 0$. Analogous to the construction of the spherical vector wave functions in Theorem 6.24 from the scalar spherical wave functions, we will develop special solutions to the electromagnetic transmission problem (9.24), (9.25) from solutions to the acoustic interior transmission problem

(9.26) $$\Delta w + k^2 n(r) w = 0, \quad \Delta v + k^2 v = 0 \quad \text{in } B,$$

(9.27) $$w - v = 0, \quad \frac{\partial w}{\partial \nu} - \frac{\partial v}{\partial \nu} = 0 \quad \text{on } \partial B.$$

Assume that w, v are solutions of (9.26), (9.27), are three times continuously differentiable and define

$$E_1(x) := \operatorname{curl}\{x w(x)\}, \quad H_1(x) := \frac{1}{ik} \operatorname{curl} E_1(x),$$

(9.28)

$$E_0(x) := \operatorname{curl}\{x v(x)\}, \quad H_0(x) := \frac{1}{ik} \operatorname{curl} E_0(x).$$

Then, from the identity (6.4) together with

$$\Delta\{x w(x)\} = x \, \Delta \, w(x) + 2 \operatorname{grad} w(x)$$

and (9.26) we have that

$$ik \operatorname{curl} H_1(x) = \operatorname{curl} \operatorname{curl} \operatorname{curl}\{x w(x)\} = -\operatorname{curl} \Delta\{x w(x)\}$$

$$= k^2 \operatorname{curl}\{x n(r) w(x)\} = k^2 n(r) \operatorname{curl}\{x w(x)\} = k^2 n(r) E_1(x),$$

that is,

$$\operatorname{curl} H_1 + ikn(r)E_1 = 0,$$

and similarly

$$\operatorname{curl} H_0 + ik E_0 = 0.$$

Hence, E_1, H_1 and E_0, H_0 satisfy (9.24). From $w - v = 0$ on ∂B we have that

$$x \times \{E_1(x) - E_0(x)\} = x \times \{\operatorname{grad}[w(x) - v(x)] \times x\} = 0, \quad x \in \partial B,$$

that is,

$$\nu \times (E_1 - E_0) = 0 \quad \text{on } \partial B.$$

Finally, setting $u = w - v$ in the relation

$$\operatorname{curl}\operatorname{curl}\{xu(x)\} = -\Delta\{xu(x)\} + \operatorname{grad}\operatorname{div}\{xu(x)\}$$

$$= -x\,\Delta\,u(x) + \operatorname{grad}\left\{u(x) + \frac{\partial u}{\partial r}(x)\right\}$$

and using the boundary condition (9.27), we deduce that

$$\nu \times (H_1 - H_0) = 0 \quad \text{on } \partial B$$

is also valid. Hence, from a three times continuously differentiable solution w, v to the scalar transmission problem (9.26), (9.27), via (9.28) we obtain a solution E_1, H_1 and E_0, H_0 to the electromagnetic transmission problem (9.24), (9.25). Note, however, that in order to obtain a nontrivial solution through (9.28) we have to insist that w and v are not spherically symmetric.

We proceed as in Section 8.4 and, after introducing spherical coordinates (r, θ, φ), look for solutions to (9.26), (9.27) of the form

$$v(r, \theta) = a_l j_l(kr)\, P_l(\cos\theta),$$

(9.29)

$$w(r, \theta) = b_l \frac{y_l(r)}{r}\, P_l(\cos\theta),$$

where P_l is Legendre's polynomial, j_l is a spherical Bessel function, a_l and b_l are constants to be determined and the function y_l is a solution of

(9.30)
$$y_l'' + \left(k^2 n(r) - \frac{l(l+1)}{r^2}\right) y_l = 0$$

for $r > 0$ such that y_l is continuous for $r \geq 0$. However, in contrast to the analysis of Section 8.4, we are only interested in solutions which are dependent on θ, i.e., in solutions for $l \geq 1$. In particular, the ordinary differential equation (9.30) now has singular coefficients. We shall show that if $n(r) > 1$ for $0 \leq r < a$ or $0 < n(r) < 1$ for $0 \leq r < a$, then for each $l \geq 1$ there exist an infinite set of values of k and constants $a_l = a_l(k)$, $b_l = b_l(k)$, such that (9.29) is a nontrivial solution of (9.26), (9.27). From Section 6.6 we know that E_0, H_0, given by (9.28), is an electromagnetic Herglotz pair. Hence, by Theorem 9.7, for such values of k the set of electric far field patterns is not complete.

To show the existence of values of k such that (9.29) yields a nontrivial solution of (9.26), (9.27), we need to examine the asymptotic behavior of solutions to (9.30). To this end, we use the Liouville transformation

(9.31)
$$\xi := \int_0^r [n(\rho)]^{1/2}\,d\rho, \quad z(\xi) := [n(r)]^{1/4} y_l(r)$$

to transform (9.30) to

(9.32) $$z'' + [k^2 - p(\xi)]z = 0$$

where

$$p(\xi) := \frac{n''(r)}{4\,[n(r)]^2} - \frac{5}{16}\frac{[n'(r)]^2}{[n(r)]^3} + \frac{l(l+1)}{r^2 n(r)} \ .$$

Note that since $n(r) > 0$ for $r \geq 0$ and n is in C^2, the transformation (9.31) is invertible and p is well defined and continuous for $r > 0$. In order to deduce the required asymptotic estimates, we rewrite (9.32) in the form

(9.33) $$z'' + \left(k^2 - \frac{l(l+1)}{n(0)r^2} - g(r)\right)z = 0$$

where

(9.34) $$g(r) := \frac{l(l+1)}{r^2 n(r)} - \frac{l(l+1)}{r^2 n(0)} + \frac{n''(r)}{4\,[n(r)]^2} - \frac{5}{16}\frac{[n'(r)]^2}{[n(r)]^3}$$

and note that since $n(r) = 1$ for $r \geq a$ we have

$$\int_1^\infty |g(r)|\,dr < \infty \quad \text{and} \quad \int_0^1 r|g(r)|\,dr < \infty.$$

For $\lambda > 0$ we now define the functions E_λ and M_λ by

$$E_\lambda(r) := \begin{cases} \left[-\dfrac{Y_\lambda(r)}{J_\lambda(r)}\right]^{1/2}, & 0 < r < r_\lambda, \\[2mm] 1, & r_\lambda \leq r < \infty, \end{cases}$$

and

$$M_\lambda(r) := \begin{cases} [2\,|Y_\lambda(r)|\,J_\lambda(r)]^{1/2}, & 0 < r < r_\lambda, \\[2mm] [J_\lambda^2(r) + Y_\lambda^2(r)]^{1/2}, & r_\lambda \leq r < \infty, \end{cases}$$

where J_λ is the Bessel function, Y_λ the Neumann function and r_λ is the smallest positive root of the equation

$$J_\lambda(r) + Y_\lambda(r) = 0.$$

Note that r_λ is less than the first positive zero of J_λ. For the necessary information on Bessel and Neumann functions of non-integral order we refer the reader to [25] and [117]. We further define G_λ by

$$G_\lambda(k,r) := \frac{\pi}{2}\int_0^r tM_\lambda^2(k\rho)\,|g(\rho)|\,d\rho$$

where g is given by (9.34). Noting that for $k > 0$ and $\lambda \geq 0$ we have that G_λ is finite when r is finite, we can now state the following result from Olver ([150], p. 450).

Theorem 9.9 *Let $k > 0$ and $l \geq 0$. Then (9.33) has a solution z which, as a function of ξ, is continuous in $[0, \infty)$, twice continuously differentiable in $(0, \infty)$, and is given by*

$$(9.35) \qquad z(\xi) = \sqrt{\frac{\pi \xi}{2k}} \, \{J_\lambda(k\xi) + \varepsilon_l(k, \xi)\}$$

where

$$\lambda = \sqrt{\frac{l(l+1)}{n(0)} + \frac{1}{4}}$$

and

$$|\varepsilon_l(k, \xi)| \leq \frac{M_\lambda(k\xi)}{E_\lambda(k\xi)} \, \{e^{G_\lambda(k\xi)} - 1\} .$$

In order to apply Theorem 9.9 to obtain an asymptotic estimate for a continuous solution y_l of (9.30), we fix $\xi > 0$ and let k be large. Then for $\lambda > 0$ we have that there exist constants C_1 and C_2, both independent of k, such that

$$(9.36) \qquad
\begin{aligned}
|G_\lambda(k, \xi)| &\leq C \left\{ \int_0^1 M_\lambda^2(kr)\, dr + \frac{1}{k} \int_1^\infty |g(r)|\, dr \right\} \\
&\leq C_1 \left\{ \frac{1}{k} \int_{1/k}^1 \frac{dr}{r} + \frac{1}{k} \right\} = C_1 \left\{ \frac{\ln k}{k} + \frac{1}{k} \right\} .
\end{aligned}$$

Hence, for z defined by (9.35) we have from Theorem 9.9, (9.36) and the asymptotics for the Bessel function J_λ that

$$(9.37) \qquad
\begin{aligned}
z(\xi) &= \sqrt{\frac{\pi \xi}{2k}} \left\{ J_\lambda(k\xi) + O\left(\frac{\ln k}{k^{3/2}}\right) \right\} \\
&= \frac{1}{k} \cos\left(k\xi - \frac{\lambda\pi}{2} - \frac{\pi}{4} \right) + O\left(\frac{\ln k}{k^2}\right)
\end{aligned}$$

for fixed $\xi > 0$ and λ as defined in Theorem 9.9. Furthermore, it can be shown that the asymptotic expansion (9.37) can be differentiated with respect to ξ, the error estimate being $O(\ln k / k)$. Hence, from (9.31) and (9.37) we can finally conclude that if y_l is defined by (9.31) then

$$(9.38) \quad y_l(r) = \frac{1}{k[n(r)]^{1/4}} \cos\left(k \int_0^r [n(\rho)]^{1/2} d\rho - \frac{\lambda\pi}{2} - \frac{\pi}{4} \right) + O\left(\frac{\ln k}{k^2}\right)$$

where the asymptotic expansion for $[n(r)]^{1/4} y_l(r)$ can be differentiated with respect to r, the error estimate again being $O(\ln k / k)$.

We now note that, from the above estimates, w, as defined by (9.29), is a C^2 solution of $\Delta w + k^2 n(r) w = 0$ in $B \setminus \{0\}$ and is continuous in B. Hence, by the removable singularity theorem for elliptic differential equations (c.f. [155],p.

104) we have that $w \in C^2(B)$. Since $n \in C^{1,\alpha}(\mathbb{R}^3)$, we can conclude from Green's formula (8.13) and Theorem 8.1 that $w \in C^3(B)$ and hence E_1 and H_1 are continuously differentiable in B.

We now return to the scalar interior transmission problem (9.26), (9.27) and note that (9.29) will be a nontrivial solution provided there exists a nontrivial solution a_l, b_l of the homogeneous linear system

$$b_l \frac{y_l(a)}{a} \quad - \quad a_l j_l(ka) \quad = \quad 0$$

(9.39)

$$b_l \frac{d}{dr} \left(\frac{y_l(r)}{r} \right)_{r=a} \quad - \quad a_l k j_l'(ka) \quad = \quad 0.$$

The system (9.39) will have a nontrivial solution provided the determinant of the coefficients vanishes, that is,

(9.40) $$d := \det \begin{pmatrix} \dfrac{y_l(a)}{a} & -j_l(ka) \\ \dfrac{d}{dr}\left(\dfrac{y_l(r)}{r}\right)_{r=a} & -k j_l'(ka) \end{pmatrix} = 0.$$

Recalling the asymptotic expansions (2.38) for the spherical Bessel functions, i.e.,

$$j_l(kr) = \frac{1}{kr} \cos\left(kr - \frac{l\pi}{2} - \frac{\pi}{2}\right) + O\left(\frac{1}{k^2}\right), \quad k \to \infty,$$

(9.41)

$$j_l'(kr) = \frac{1}{kr} \sin\left(kr - \frac{l\pi}{2} + \frac{\pi}{2}\right) + O\left(\frac{1}{k^2}\right), \quad k \to \infty,$$

we see from (9.38), (9.41) and the addition formula for the sine function that

$$d = \frac{1}{a^2 k}\left\{ \sin\left(k \int_0^a [n(r)]^{1/2} dr - ka - \frac{\lambda\pi}{2} + \frac{l\pi}{2} + \frac{\pi}{4}\right) + O\left(\frac{\ln k}{k}\right)\right\}.$$

Therefore, a sufficient condition for (9.40) to be valid for a discrete set of values of k is that either $n(r) > 1$ for $0 \le r < a$ or $n(r) < 1$ for $0 \le r < a$. Hence we have the following theorem [46].

Theorem 9.10 *Assume that $\operatorname{Im} n = 0$ and that $n(x) = n(r)$ is spherically stratified, $n(r) = 1$ for $r \ge a$, $n(r) > 1$ or $0 < n(r) < 1$ for $0 \le r < a$ and, as a function of r, $n \in C^2$. Then there exists an infinite set of wave numbers k such that the set $\mathcal{F}$ of electric far field patterns is not complete in $T^2(\Omega)$.*

9.5 The Exterior Impedance Boundary Value Problem

The mathematical treatment of the scattering of time harmonic electromagnetic waves by a body which is not perfectly conducting but which does not allow the electric and magnetic field to penetrate deeply into the body leads to what is called an exterior impedance boundary value problem for electromagnetic waves (c.f. [78], p. 511, [178], p. 304). In particular, such a model is frequently used for coated media instead of the more complicated transmission problem. In addition to being an appropriate theme for this chapter, we shall also need to make use of the mathematical theory of the exterior impedance boundary value problem in our later treatment of the inverse scattering problem for electromagnetic waves. The first rigorous proof of the existence of a unique solution to the exterior impedance boundary value problem for electromagnetic waves was given by Colton and Kress in [31]. Here we shall provide a simpler proof of this result by basing our ideas on those developed for a perfect conductor in Chapter 6. We first define the problem under consideration where for the rest of this section D is a bounded domain in $\mathbb{R}^3$ with connected C^2 boundary ∂D with unit outward normal ν.

Exterior Impedance Problem. *Given a Hölder continuous tangential field c on ∂D and a positive constant λ, find a solution $E, H \in C^1(\mathbb{R}^3 \setminus \bar{D}) \cap C(\mathbb{R}^3 \setminus D)$ of the Maxwell equations*

$$\text{(9.42)} \qquad \operatorname{curl} E - ikH = 0, \quad \operatorname{curl} H + ikE = 0 \quad in \ \mathbb{R}^3 \setminus \bar{D}$$

satisfying the impedance boundary condition

$$\text{(9.43)} \qquad \nu \times \operatorname{curl} E - i\lambda \, (\nu \times E) \times \nu = c \quad on \ \partial D$$

and the Silver–Müller radiation condition

$$\text{(9.44)} \qquad \lim_{r \to \infty} (H \times x - rE) = 0$$

uniformly for all directions $\hat{x} = x/|x|$.

The uniqueness of a solution to (9.42)–(9.44) is easy to prove.

Theorem 9.11 *The exterior impedance problem has at most one solution provided $\lambda > 0$.*

Proof. If $c = 0$, then from (9.43) and the fact that $\lambda > 0$ we have that

$$\operatorname{Re} k \int_{\partial D} \nu \times E \cdot \bar{H} \, ds = -\lambda \int_{\partial D} |\nu \times E|^2 ds \leq 0.$$

We can now conclude from Theorem 6.10 that $E = H = 0$ in $\mathbb{R}^3 \setminus \bar{D}$.   □

We now turn to the existence of a solution to the exterior impedance problem, always assuming that $\lambda > 0$. To this end, we recall the definition of the space $C^{0,\alpha}(\partial D)$ of Hölder continuous functions defined on ∂D from Section 3.1 and the the space $T^{0,\alpha}(\partial D)$ of Hölder continuous tangential fields defined on

∂D from Section 6.3. We also recall from Theorems 3.2, 3.4 and 6.13 that the single-layer operator $S : C^{0,\alpha}(\partial D) \to C^{0,\alpha}(\partial D)$ defined by

$$(S\varphi)(x) := 2 \int_{\partial D} \Phi(x,y)\varphi(y)\,ds(y), \quad x \in \partial D,$$

the double-layer operator $K : C^{0,\alpha}(\partial D) \to C^{0,\alpha}(\partial D)$ defined by

$$(K\varphi)(x) := 2 \int \frac{\partial \Phi(x,y)}{\partial \nu(y)}\,\varphi(y)\,ds(y), \quad x \in \partial D,$$

and the magnetic dipole operator $M : T^{0,\alpha}(\partial D) \to T^{0,\alpha}(\partial D)$ defined by

$$(Ma)(x) := 2 \int_{\partial D} \nu(x) \times \mathrm{curl}_x \{a(y)\Phi(x,y)\}\,ds(y), \quad x \in \partial D,$$

are all compact. Furthermore, with the spaces

$$T_d^{0,\alpha}(\partial D) = \{a \in T^{0,\alpha}(\partial D) : \mathrm{Div}\,a \in C^{0,\alpha}(\partial D)\}$$

and

$$T_r^{0,\alpha}(\partial D) = \{b \in T^{0,\alpha}(\partial D) : \nu \times b \in T_d^{0,\alpha}(\partial D)\}$$

which were also introduced in Section 6.3, the electric dipole operator $N : T_r^{0,\alpha}(\partial D) \to T_d^{0,\alpha}(\partial D)$ defined by

$$(Na)(x) := 2\nu(x) \times \mathrm{curl}\,\mathrm{curl} \int_{\partial D} \nu(y) \times a(y)\,\Phi(x,y)\,ds(y), \quad x \in \partial D,$$

is bounded by Theorem 6.17.

With these definitions and facts recalled, following Hähner [64], we now look for a solution of the exterior impedance problem in the form

$$E(x) = \int_{\partial D} b(y)\Phi(x,y)\,ds(y)$$

$$+ i\lambda\,\mathrm{curl} \int_{\partial D} \nu(y) \times (S_0^2 b)(y)\,\Phi(x,y)\,ds(y)$$

(9.45)

$$+ \mathrm{grad} \int_{\partial D} \varphi(y)\Phi(x,y)\,ds(y)$$

$$+ i\lambda \int_{\partial D} \nu(y)\varphi(y)\Phi(x,y)\,ds(y),$$

$$H(x) = \frac{1}{ik}\,\mathrm{curl}\,E(x), \quad x \in \mathrm{I\!R}^3 \setminus \bar{D},$$

where S_0 is the single-layer operator in the potential theoretic limit $k = 0$ and the densities $b \in T^{0,\alpha}(\partial D)$ and $\varphi \in C^{0,\alpha}(\partial D)$ are to be determined. The vector field E clearly satisfies the vector Helmholtz equation and its cartesian components satisfy the (scalar) Sommerfeld radiation condition. Hence, if we insist that $\operatorname{div} E = 0$ in $\mathbb{R}^3 \setminus \bar{D}$, then by Theorems 6.4 and 6.7 we have that E, H satisfy the Maxwell equations and the Silver–Müller radiation condition. Since $\operatorname{div} E$ satisfies the scalar Helmholtz equation and the Sommerfeld radiation condition, by the uniqueness for the exterior Dirichlet problem it suffices to impose $\operatorname{div} E = 0$ only on the boundary ∂D. From the jump and regularity conditions of Theorems 3.1, 3.3, 6.11 and 6.12, we can now conclude that (9.45) for $b \in T^{0,\alpha}(\partial D)$ and $\varphi \in C^{0,\alpha}(\partial D)$ ensures the regularity $E, H \in C^{0,\alpha}(\mathbb{R}^3 \setminus D)$ up to the boundary and that it solves the exterior impedance problem provided b and φ satisfy the integral equations

$$b + M_{11}b + M_{12}\varphi = 2c$$

(9.46)

$$-i\lambda\varphi + M_{21}b + M_{22}\varphi = 0$$

where

$$M_{11}b := Mb + i\lambda N P S_0^2 b - i\lambda P S b + \lambda^2 \{M(\nu \times S_0^2 b)\} \times \nu + \lambda^2 P S_0^2 b,$$

$$(M_{12}\varphi)(x) := 2i\lambda\,\nu(x) \times \int_{\partial D} \operatorname{grad}_x \Phi(x,y) \times \{\nu(y) - \nu(x)\}\varphi(y)\, ds(y)$$

$$+ \lambda^2 (PS\nu\varphi)(x), \quad x \in \partial D,$$

$$(M_{21}b)(x) := -2 \int_{\partial D} \operatorname{grad}_x \Phi(x,y) \cdot b(y)\, ds(y), \quad x \in \partial D,$$

$$M_{22}\varphi := k^2 S\varphi + i\lambda K\varphi,$$

and where P stands for the orthogonal projection of a vector field defined on ∂D onto the tangent plane, that is, $Pa := (\nu \times a) \times \nu$. Noting the smoothing property $S_0 : C^{0,\alpha}(\partial D) \to C^{1,\alpha}(\partial D)$ from Theorem 3.4, as in the proof of Theorem 6.19 it is not difficult to verify that $M_{11} : T^{0,\alpha}(\partial D) \to T^{0,\alpha}(\partial D)$ is compact. Compactness of the operator $M_{12} : C^{0,\alpha}(\partial D) \to T^{0,\alpha}(\partial D)$ follows by applying Corollary 2.9 from [32] to the first term in the definition of M_{12}. Loosely speaking, compactness of M_{12} rests on the fact that the factor $\nu(x) - \nu(y)$ makes the kernel weakly singular. Finally, $M_{22} : C^{0,\alpha}(\partial D) \to C^{0,\alpha}(\partial D)$ clearly is compact, whereas $M_{21} : T^{0,\alpha}(\partial D) \to C^{0,\alpha}(\partial D)$ is merely bounded. Writing the system (9.46) in the form

$$\begin{pmatrix} I & 0 \\ M_{21} & -i\lambda I \end{pmatrix} \begin{pmatrix} b \\ \varphi \end{pmatrix} + \begin{pmatrix} M_{11} & M_{12} \\ 0 & M_{22} \end{pmatrix} \begin{pmatrix} b \\ \varphi \end{pmatrix} = \begin{pmatrix} 2c \\ 0 \end{pmatrix},$$

we now see that the first of the two matrix operators has a bounded inverse because of its triangular form and the second is compact. Hence, we can apply the Riesz–Fredholm theory to (9.46).

For this purpose, suppose b and φ are a solution to the homogeneous equation corresponding to (9.46) (i.e., $c = 0$). Then the field E, H defined by (9.45) satisfies the homogeneous exterior impedance problem in $\mathbb{R}^3 \setminus \bar{D}$. Since $\lambda > 0$, we can conclude from Theorem 9.11 that $E = H = 0$ in $\mathbb{R}^3 \setminus D$. Viewing (9.45) as defining a solution of the vector Helmholtz equation in D, from the jump relations of Theorems 3.1, 3.3, 6.11 and 6.12 we see that

$$(9.47) \qquad -\nu \times E_- = i\lambda\nu \times S_0^2 b, \quad -\nu \times \operatorname{curl} E_- = b \quad \text{on } \partial D,$$

$$(9.48) \qquad -\operatorname{div} E_- = -i\lambda\varphi, \quad -\nu \cdot E_- = -\varphi \quad \text{on } \partial D.$$

Hence, with the aid of Green's vector theorem (6.2), we derive from (9.47) and (9.48) that

$$\int_D \{|\operatorname{curl} E|^2 + |\operatorname{div} E|^2 - k^2|E|^2\}\,dx = i\lambda \int_{\partial D} \{|S_0 b|^2 + |\varphi|^2\}\,ds.$$

Taking the imaginary part of the last equation and recalling that $\lambda > 0$ now shows that $S_0 b = 0$ and $\varphi = 0$ on ∂D. Since S_0 is injective (see the proof of Theorem 3.10), we have that $b = 0$ on ∂D. The Riesz–Fredholm theory now implies the following theorem. The statement on the boundedness of the operator A follows from the fact that by the Riesz–Fredholm theory the inverse operator for (9.46) is bounded from $T^{0,\alpha}(\partial D) \times C^{0,\alpha}(\partial D)$ into itself and by applying the mapping properties of Theorems 3.3 and 6.12 to the solution (9.45).

Theorem 9.12 *Suppose $\lambda > 0$. Then for each $c \in T^{0,\alpha}(\partial D)$ there exists a unique solution to the exterior impedance problem. The operator A mapping the boundary data c onto the tangential component $\nu \times E$ of the solution is a bounded operator $A : T^{0,\alpha}(\partial D) \to T_d^{0,\alpha}(\partial D)$.*

For technical reasons, we shall need in Chapter 10 sufficient conditions for the invertibility of the operator $NR - i\lambda R(I + M) : T_d^{0,\alpha}(\partial D) \to T^{0,\alpha}(\partial D)$ where the operator $R : T^{0,\alpha}(\partial D) \to T^{0,\alpha}(\partial D)$ is given by

$$Ra := a \times \nu.$$

To this end, we first try to express the solution of the exterior impedance problem in the form

$$E(x) = \operatorname{curl} \int_{\partial D} a(y)\Phi(x,y)\,ds(y), \quad x \in \mathbb{R}^3 \setminus \bar{D},$$

where $a \in T_d^{0,\alpha}(\partial D)$. From the jump conditions of Theorems 6.11 and 6.12, this leads to the integral equation

$$(9.49) \qquad NRa - i\lambda RMa - i\lambda Ra = 2c$$

for the unknown density a. However, we can interpret the solution of the exterior impedance problem as the solution of the exterior Maxwell problem with boundary condition

$$\nu \times E = Ac \quad \text{on } \partial D,$$

and hence a also is required to satisfy the integral equation

$$a + Ma = 2Ac.$$

The last equation turns out to be a special case of equation (6.49) with $\eta = 0$ (and a different right hand side). From the proof of Theorem 6.19, it can be seen that if k is not a Maxwell eigenvalue for D then $I + M$ has a trivial nullspace. Hence, since by Theorem 6.16 the operator $M : T_d^{0,\alpha}(\partial D) \to T_d^{0,\alpha}(\partial D)$ is compact, by the Riesz–Fredholm theory $(I + M)^{-1} : T_d^{0,\alpha}(\partial D) \to T_d^{0,\alpha}(\partial D)$ exists and is bounded. Hence, $(I + M)^{-1}A : T^{0,\alpha}(\partial D) \to T_d^{0,\alpha}(\partial D)$ is the bounded inverse of $NR - i\lambda R(I + M)$ and we have proven the following theorem.

Theorem 9.13 *Assume that $\lambda > 0$ and that k is not a Maxwell eigenvalue for D. Then the operator $NR - i\lambda R(I + M) : T_d^{0,\alpha}(\partial D) \to T^{0,\alpha}(\partial D)$ has a bounded inverse.*

We shall now conclude this chapter by briefly considering the electric far field patterns corresponding to the exterior impedance problem (9.42)–(9.44) with c given by

$$c := -\nu \times \operatorname{curl} E^i + i\lambda \left(\nu \times E^i\right) \times \nu \quad \text{on } \partial D$$

where E^i and H^i are given by (9.19). This corresponds to the scattering of the incident field (9.19) by the imperfectly conducting obstacle D where the total electric field $E = E^i + E^s$ satisfies the impedance boundary condition

$$(9.50) \qquad \nu \times \operatorname{curl} E - i\lambda \left(\nu \times E\right) \times \nu = 0 \quad \text{on } \partial D$$

and E^s is the scattered electric field. From Theorem 6.8 we see that E^s has the asymptotic behavior

$$E^s(x; d, p) = \frac{e^{ik|x|}}{|x|} E_\infty^\lambda(\hat{x}; d, p) + O\left(\frac{1}{|x|^2}\right), \quad |x| \to \infty,$$

where E_∞^λ is the electric far field pattern. From (6.78) and (9.50) we can easily deduce the following reciprocity relation [3].

Theorem 9.14 *For all vectors $\hat{x}, d \in \Omega$ and $p, q \in \mathbb{R}^3$ we have*

$$q \cdot E_\infty^\lambda(\hat{x}; d, p) = p \cdot E_\infty^\lambda(-d; -\hat{x}, q).$$

We are now in a position to prove the analogue of Theorem 9.8 for the exterior impedance problem. In particular, recall the Hilbert space $T^2(\Omega)$ of tangential L^2 vector fields on the unit sphere, let $\{d_n : n = 1, 2, \ldots\}$ be a countable dense set of unit vectors on Ω and denote by e_1, e_2, e_3 the cartesian unit coordinate vectors in $\mathbb{R}^3$. For the electric far field patterns we now have the following theorem due to Angell, Colton and Kress [3].

Theorem 9.15 *Assume* $\lambda > 0$. *Then the set*

$$\mathcal{F}_\lambda = \{E_\infty^\lambda(\,\cdot\,; d_n, e_j) : n = 1, 2, \ldots, j = 1, 2, 3\}$$

of electric far field patterns for the exterior impedance problem is complete in $T^2(\Omega)$.

Proof. Suppose that $g \in T^2(\Omega)$ satisfies

$$\int_\Omega E_\infty^\lambda(\hat{x}; d_n, e_j) \cdot g(\hat{x})\, ds(\hat{x}) = 0$$

for $n = 1, 2, \ldots$ and $j = 1, 2, 3$. We must show that $g = 0$. As in the proof of Theorem 9.7, by continuity and superposition we have that

$$\int_\Omega E_\infty^\lambda(\hat{x}; d, p) \cdot g(\hat{x})\, ds(\hat{x}) = 0$$

for all $d \in \Omega$ and $p \in \mathbb{R}^3$. By the reciprocity Theorem 9.14, we have

$$p \cdot \int_\Omega E_\infty^\lambda(-d; -\hat{x}, g(\hat{x}))\, ds(\hat{x}) = 0$$

for all $d \in \Omega$ and $p \in \mathbb{R}^3$, i.e.,

$$(9.51) \qquad \int_\Omega E_\infty^\lambda(\hat{x}; d, h(d))\, ds(d) = 0$$

for all $\hat{x} \in \Omega$ where $h(d) = g(-d)$.

Now define the electromagnetic Herglotz pair E_0^i, H_0^i by

$$E_0^i(x) = \int_\Omega E^i(x; d, h(d))\, ds(d) = ik \int_\Omega h(d)\, e^{ik\, x \cdot d}\, ds(d),$$

$$H_0^i(x) = \int_\Omega H^i(x; d, h(d))\, ds(d) = \operatorname{curl} \int_\Omega h(d)\, e^{ik\, x \cdot d}\, ds(d).$$

Analogous to Lemma 6.31 it can be seen that the left hand side of (9.51) represents the electric far field pattern of the scattered field E_0^s, H_0^s corresponding to the incident field E_0^i, H_0^i. Then from (9.51) we see that the electric far field pattern of E_0^s vanishes and hence, from Theorem 6.9, both E_0^s and H_0^s are identically zero in $\mathbb{R}^3 \setminus D$. We can now conclude that E_0^i, H_0^i satisfies the impedance boundary condition

$$(9.52) \qquad \nu \times \operatorname{curl} E_0^i - i\lambda\, (\nu \times E_0^i) \times \nu = 0 \quad \text{on } \partial D.$$

Gauss' theorem and the Maxwell equations (compare (9.17)) now imply that

$$\int_{\partial D} \nu \times E_0^i \cdot \bar{H}_0^i\, ds = ik \int_D \{|H_0^i|^2 - |E_0^i|^2\}\, dx$$

and hence from (9.52) we have that

$$\lambda \int_{\partial D} |\nu \times E_0^i|^2 ds = ik^2 \int_D \{|E_0^i|^2 - |H_0^i|^2\} \, dx$$

whence $\nu \times E_0^i = 0$ on ∂D follows since $\lambda > 0$. From (9.52) we now see that $\nu \times H_0^i = 0$ on ∂D and hence from the Stratton–Chu formulas (6.8) and (6.9) we have that $E_0^i = H_0^i = 0$ in D and by analyticity (Theorem 6.3) $E_0^i = H_0^i = 0$ in $\mathbb{R}^3$. But now from Theorem 3.15 we conclude that $h = 0$ and consequently $g = 0$. $\qquad\square$

10. The Inverse Medium Problem

We now turn our attention to the problem of reconstructing the refractive index from a knowledge of the far field pattern of the scattered acoustic or electromagnetic wave. We shall call this problem the *inverse medium problem*. Of particular interest to us will be the use of a dual space method to determine the refractive index. This method has the numerical advantage of being able to increase the number of probing waves with a minimum amount of extra cost and, in addition, leads to a number of mathematical problems which are of interest in their own right. Our aim in this chapter is to develop the theory of the inverse medium problem to the point where an optimization scheme can be formulated for the solution such that under appropriate conditions the infimum of the cost functional is zero. However, since similar optimization schemes were analyzed in depth in Chapters 5 and 7, we shall not dwell on the specific optimization scheme itself, except in Section 10.6 where we present some numerical examples.

We first consider the case of acoustic waves and the use of the Lippmann–Schwinger equation to reformulate the acoustic inverse medium problem as a problem in constrained optimization. Included here is a brief discussion of the use of the Born approximation to linearize the problem. We then proceed to the proof of a uniqueness theorem for the acoustic inverse medium problem. The next two sections are then devoted to the formulation of two versions of the dual space method to solve the acoustic inverse medium problem. Although the first version is numerically simpler, it has the disadvantage of only being applicable to absorbing media. The second version overcomes this disadvantage at the cost of needing to construct a solution to an exterior impedance boundary value problem for the Helmholtz equation. We then proceed to extend these results for acoustic waves to the case of electromagnetic waves and, in the last section of this chapter, we give some numerical examples illustrating the use of our methods.

10.1 The Inverse Medium Problem for Acoustic Waves

We now consider the inverse scattering problem for time-harmonic acoustic waves in an inhomogeneous medium which we shall from now on refer to as the acoustic *inverse medium problem*. Recall from Chapter 8 that the direct scattering problem we are now concerned with is, given the refractive index

$$n(x) = n_1(x) + i\,\frac{n_2(x)}{k}$$

where $k > 0$, $n \in C^1(\mathbb{R}^3)$ and

$$m := 1 - n$$

has compact support to determine u such that

(10.1) $$\Delta u + k^2 n(x)u = 0 \quad \text{in } \mathbb{R}^3,$$

(10.2) $$u(x) = e^{ik\,x\cdot d} + u^s(x),$$

(10.3) $$\lim_{r\to\infty} r\left(\frac{\partial u^s}{\partial r} - iku^s\right) = 0$$

uniformly for all directions. As in Chapter 8, we shall in addition always assume that $n_1(x) > 0$ and $n_2(x) \geq 0$ for $x \in \mathbb{R}^3$. The existence of a unique solution to (10.1)–(10.3) was established in Chapter 8 where it was also shown that u^s has the asymptotic behavior

$$u^s(x) = \frac{e^{ik|x|}}{|x|}\,u_\infty(\hat{x}; d) + O\left(\frac{1}{|x|^2}\right), \quad |x| \to \infty,$$

with the far field pattern u_∞ given by

(10.4) $$u_\infty(\hat{x}; d) = -\frac{k^2}{4\pi}\int_{\mathbb{R}^3} e^{-ik\,\hat{x}\cdot y}m(y)u(y)\,dy, \quad \hat{x} \in \Omega.$$

The inverse medium problem for acoustic waves is to determine n from $u_\infty(\hat{x}; d)$ for $\hat{x} \in \Omega$ and a finite number of directions d. We shall also often consider data for different values of k and in this case we shall write $u_\infty(\hat{x}; d) = u_\infty(\hat{x}; d, k)$.

The most obvious method for solving the acoustic inverse medium problem is to rewrite (10.1)–(10.3) as the Lippmann–Schwinger equation

(10.5) $$u(x) = e^{ik\,x\cdot d} - k^2\int_{\mathbb{R}^3} \Phi(x, y)m(y)u(y)\,dy, \quad x \in \mathbb{R}^3,$$

which we discussed in detail in Section 8.2 and look for a solution of (10.5) such that m and u satisfy the constraint (10.4). In particular, assuming the far field pattern $u_\infty(\hat{x}; d, k)$ is known for $\hat{x} \in \Omega$, $d \in \{d_1, \ldots, d_P\}$ and $k \in \{k_1, \ldots, k_Q\}$ and defining the operators $F : L^2(B) \to L^2(\Omega)$ and $T_m : L^2(B) \to L^2(B)$ by

$$(Fu)(\hat{x}) := -\frac{k^2}{4\pi}\int_B e^{-ik\,\hat{x}\cdot y}m(y)u(y)\,dy, \quad \hat{x} \in \Omega,$$

and

(10.6) $$(T_m u)(x) := \int_D \Phi(x, y)m(y)u(y)\,dy, \quad x \in B,$$

where B is a ball containing the support of m, we seek a solution to the optimization problem

$$(10.7) \quad \min_{\substack{u_{pq} \in U_1 \\ m \in U_2}} \sum_{p=1}^{P} \sum_{q=1}^{Q} \left\{ \|u_{\infty,pq} - Fu_{pq}\|_{L^2(\Omega)}^2 + \|u_{pq} - u_{pq}^i + k^2 T_m u_{pq}\|_{L^2(B)}^2 \right\}$$

where $u_{pq}^i(x) := e^{ik_q x \cdot d_p}$, $u_{\infty,pq} := u_\infty(\cdot; d_p, k_q)$ and U_1, U_2 are appropriate compact sets. The formulation of (10.4), (10.5) as the optimization problem (10.7) is similar to that used in Chapter 5 for the inverse obstacle problem and hence we shall not dwell further on the details of this procedure. Indeed, throughout this chapter we shall proceed in a similar fashion, i.e., having presented the overdetermined set of equations for the determination of the unknown refractive index we shall only briefly discuss how this set can be reformulated as a problem in constrained optimization. The precise norms and constraints used are, of course, important, as is the numerical method used to solve the resulting optimization problem. However, having done this in Chapter 5 for the case of obstacle scattering, we do not want to repeat a similar analysis here. Variations of the optimization scheme (10.7) to solve the inverse medium problem have been used with considerable success by Johnson and Tracy [77, 176], Kleinman and van den Berg [98], Wang and Chew [181] and Weston [187, 188] among others.

The solution of the inverse medium problem becomes particularly simple when use is made of the Born approximation (8.28). In this case, instead of the nonlinear system (10.4), (10.5) we have the linear integral equations

$$(10.8) \quad u_\infty(\hat{x}; d_p, k_q) = -\frac{k_q^2}{4\pi} \int_{\mathbb{R}^3} e^{-ik_q \hat{x} \cdot y} m(y) u_{pq}^i(y) \, dy, \quad \hat{x} \in \Omega,$$

for $p = 1, \ldots, P$ and $q = 1, \ldots, Q$ to solve for the unknown function m and any one of the linear methods described in Chapter 4 can be used to do this. (Note that since the kernel of each equation is analytic, this problem is severely ill-posed and regularization methods must be used.) The obvious advantage to the Born approximation approach is that the nonlinear inverse medium problem is reduced to considering a set of linear integral equations (albeit of the first kind). The equally obvious disadvantage is that the approach is only valid if $k_q^2 \|m\|_\infty \ll 1$, a condition that is often not satisfied in applications. For a further discussion of the Born approximation approach to the inverse medium problem, the reader is referred to Bleistein [13], Chew [22] and Langenberg [114] where additional references may be found.

10.2 A Uniqueness Theorem

In this section we shall prove a uniqueness theorem for the inverse acoustic medium problem. Although our result is not the best possible, it is easy to prove and will suffice for our purposes. We begin with the following theorem due to Calderón [20] (see also Ramm [157]).

Theorem 10.1 *Let Y_n^m, $n = 0, 1, 2, \ldots$, $m = -n, \ldots, n$, denote a system of spherical harmonics and let H_n^m be the homogeneous harmonic polynomials*

$$H_n^m(x) = |x|^n Y_n^m(\hat{x})$$

where $\hat{x} := x/|x|$. Then the set

$$\{H_n^m H_p^q : n, p = 0, 1, 2, \ldots, \ m = -n, \ldots, n, \ q = -p, \ldots, p\}$$

of products of homogeneous harmonic polynomials is complete in $L^2(D)$ for any bounded domain $D \subset \mathbb{R}^3$.

Proof. From the discussion preceeding the proof of Theorem 6.23, we know that the Fourier expansions with respect to spherical harmonics for twice continuously differentiable functions converge uniformly. Therefore, with the aid of the maximum-minimum principle for harmonic functions, we conclude that for any entire harmonic function h we have an expansion

$$h(x) = \sum_{n=0}^{\infty} \sum_{m=-n}^{n} a_n^m |x|^n Y_n^m(\hat{x}), \quad x \in \mathbb{R}^3,$$

that converges uniformly on compact subsets.

Let $\rho \in \mathbb{C}^3$ be such that $\rho \cdot \rho = 0$ and consider the harmonic functions

$$h_1(x) := e^{x \cdot \rho}, \quad h_2(x) := e^{-x \cdot \bar{\rho}}$$

for $x \in \mathbb{R}^3$. Note that for $\rho = \alpha + i\beta$ with $\alpha, \beta \in \mathbb{R}^3$, we have $\rho \cdot \rho = 0$ if and only if $\alpha \cdot \beta = 0$ and $|\alpha| = |\beta|$. From the above expansion, applied to h_1 and h_2, it follows that the property

$$\int_D \varphi \, H_n^m H_p^q \, dx = 0$$

for all $n, p = 0, 1, 2, \ldots$, $m = -n, \ldots, n$, $q = -p, \ldots, p$, implies that

(10.9)
$$\int_D \varphi h_1 h_2 \, dx = 0$$

for all $\rho \in \mathbb{C}^3$ such that $\rho \cdot \rho = 0$. But for $\rho = \alpha + i\beta$, equation (10.9) becomes

$$\int_D \varphi(x) \, e^{2i \, x \cdot \beta} \, dx = 0$$

for $\beta \in \mathbb{R}^3$ and we can now conclude by the Fourier integral theorem that $\varphi = 0$ almost everywhere in D. $\qquad \square$

Theorem 10.1 now enables us to prove the following uniqueness theorem for the inverse acoustic medium problem.

Theorem 10.2 *The refractive index n is uniquely determined by a knowledge of the far field pattern $u_\infty(\hat{x}; d, k)$ for $\hat{x}, d \in \Omega$ and an interval of values of k.*

Proof. Suppose there exist two refractive indices n_1 and n_2 with the same far field pattern. We want to show that these indices must be the same. We set $m_1 := 1 - n_1$ and $m_2 := 1 - n_2$ and choose a ball B containing the support of m_1 and m_2. By assumption, we have

$$(10.10) \quad \int_B e^{-ik\,\hat{x}\cdot y} m_1(y) u_1(y; d, k)\, dy = \int_B e^{-ik\,\hat{x}\cdot y} m_2(y) u_2(y; d, k)\, dy$$

for $\hat{x}, d \in \Omega$ and for an interval I of positive k values. We now want to show that $m_1 = m_2$. To this end, we see from the Funk–Hecke formula (2.44) and (10.10) that

$$(10.11) \quad \int_B j_n(k|y|)\, Y_n(\hat{y})\, [m_1(y) u_1(y; d, k) - m_2(y) u_2(y; d, k)]\, dy = 0$$

for all spherical harmonics Y_n of order $n = 0, 1, \ldots,$ all k in the interval I and all $d \in \Omega$. After defining

$$v_{p,j}(x; k) := \int_\Omega u_j(x; d, k)\, Y_p(d)\, ds(d)$$

for $p = 0, 1, \ldots,$ and $j = 1, 2$, by multiplying (10.11) by $Y_p(d)$ and integrating with respect to d we see that

$$(10.12) \quad \int_B j_n(k|y|)\, Y_n(\hat{y})\, [m_1(y) v_{p,1}(y; k) - m_2(y) v_{p,2}(y; k)]\, dy = 0$$

for all $k \in I$ and all $n, p = 0, 1, \ldots.$ Multiplying the Lippmann–Schwinger equation by $Y_p(d)$ and integrating with respect to d we obtain

$$v_{p,j} = u_p - k^2 T_{m_j} v_{p,j}$$

where u_p is the spherical wave function

$$u_p(x) := \int_\Omega e^{ik\,x\cdot d}\, Y_p(d)\, ds(d) = 4\pi i^p j_p(k|x|)\, Y_p(\hat{x})$$

and the operators T_{m_j} are defined by

$$(T_{m_j} u)(x) := \int_B \Phi(x, y) m_j(y) u(y)\, dy, \quad x \in B, \ j = 1, 2.$$

We now recall from Chapter 8 that $I + k^2 T_{m_j}$ is invertible in $C(\bar{B})$ for each $k > 0$ and is an operator valued analytic function of k. Since u_p is an analytic function of k, we can now conclude from a Neumann series argument (see in particular the proof of Theorem 8.19) that $v_{p,j}$ is an analytic function of k for

$k > 0$ and hence (10.12) is valid for $k > 0$ and not only for $k \in I$. From the Neumann series, we see that

$$\|(I + k^2 T_{m_j})^{-1} - I\|_{C(\bar{B})} = O(k^2), \quad k \to 0,$$

and hence

$$\lim_{k \to 0} \frac{v_{p,j}(x; k)}{k^p} = \lim_{k \to 0} \frac{u_p(x)}{k^p} = 4\pi i^p Y_p(\hat{x}) \lim_{k \to 0} \frac{j_p(k|x|)}{k^p}$$

uniformly with respect to x. From the power series (2.31) for the spherical Bessel functions, we now see that

$$\lim_{k \to 0} \frac{v_{p,j}(x; k)}{k^p} = \frac{4\pi i^p |x|^p Y_p(\hat{x})}{1 \cdot 3 \cdots (2p+1)}$$

uniformly with respect to x. Dividing (10.12) by k^{n+p} and letting k tend to zero now shows that

$$\int_B [m_1(y) - m_2(y)] |y|^n Y_n(\hat{y}) |y|^p Y_p(\hat{y}) \, dy = 0$$

for all spherical harmonics of order $n, p = 0, 1, \ldots$. From Theorem 10.1 we can now conclude that $m_1 = m_2$ almost everywhere in $\mathbb{R}^3$ and, since $m_j \in C^1(\mathbb{R}^3)$ by assumption, the theorem follows.   □

In Theorem 10.2, the assumption that u_∞ is known for an interval of values of k can be replaced by only knowing u_∞ for a single value of k (see [144, 148, 158]). The proof of this proceeds in three steps. In the first step, it is shown that the set of solutions to (10.1)–(10.3) corresponding to all $d \in \Omega$ is complete in the closure in $L^2(B)$ of all solutions to (10.1) in B where $B \supset \operatorname{supp} m$. It is then shown that if there exist two refractive indices n_1 and n_2 having the same far field pattern then

$$(10.13) \qquad \int_{\mathbb{R}^3} u_1(m_1 - m_2)u_2 \, dx = 0$$

where u_j is any solution of (10.1) with $n = n_j$, $j = 1, 2$. Finally, a special solution of (10.1) is constructed in the form

$$(10.14) \qquad u(x) = e^{i\zeta \cdot x}[1 + R_\zeta(x)]$$

where $\zeta \in \mathbb{C}^3$ such that $\zeta \cdot \zeta = k^2$ and $R_\zeta = O(1/|\zeta|)$. Choosing u_j to be of the form (10.14) where $\zeta = \zeta_j$ with $\zeta_1 + \zeta_2 = \xi \in \mathbb{R}^3$, and substituting u_j, $j = 1, 2$, into (10.13) and then letting $|\zeta_1|$ and $|\zeta_2|$ tend to infinity now shows that the Fourier transform of $m_1 - m_2$ vanishes, i.e., $m_1 = m_2$. For more details, we refer the reader to the above cited references.

Although of obvious theoretical interest, a uniqueness theorem such as Theorem 10.2 is often of limited practical interest, other than suggesting the amount of information that is necessary in order to reconstruct the refractive index. The reason for this is that in order to numerically solve the inverse medium problem one usually reduces the problem to a constrained nonlinear optimization problem involving inexact far field data. Hence, the uniqueness question of primary interest is whether or not there exist local minima to this optimization problem and, more specifically, whether or not there exists a unique global minimum. In general, these questions are still unanswered for the optimization problems thus far introduced. However, one can also ask the simpler question if, for exact far field data, the minimum value of the cost functional is equal to zero. This is clearly the case for the optimization problem (10.7), provided the compact sets U_1 and U_2 are chosen sufficiently large. In the following sections, we shall introduce two other optimization schemes which also have this property.

10.3 A Dual Space Method

In this section we shall discuss an optimization method for solving the inverse medium problem for acoustic waves due to Colton and Monk [40] (see also Colton and Kirsch [28]) that is based on an application of Theorem 8.10. This method (as well as the method to be discussed in Section 10.4) has the advantage of being able to increase the number of incident fields without increasing the cost of solving the inverse problem. We shall call this approach a *dual space method* since it requires the determination of a function $g_{pq} \in L^2(\Omega)$ such that

$$(10.15) \qquad \int_\Omega u_\infty(\hat{x}; d) g_{pq}(\hat{x}) \, ds(\hat{x}) = \frac{i^{p-1}}{k} Y_p^q(d), \quad d \in \Omega,$$

i.e., the determination of a linear functional in the dual space of $L^2(\Omega)$ having prescribed values on the class $\mathcal{F} := \{u_\infty(\cdot; d_n) : n = 1, 2, 3, \ldots\}$ of far field patterns where $\{d_n : n = 1, 2, 3, \ldots\}$ is a countable dense set of vectors on the unit sphere Ω. We shall assume throughout this section that $\operatorname{Im} n(x) > 0$ for all $x \in D := \{x \in \mathbb{R}^3 : m(x) \neq 0\}$ where again $m := 1 - n$. From Theorem 8.12 we see that this implies that, if a solution exists to the integral equation (10.15), then this solution is unique. As in Chapter 8, we shall, for the sake of simplicity, always assume that D is connected with a connected C^2 boundary ∂D and D contains the origin.

We shall begin our analysis by giving a different proof of the "if" part of Theorem 8.10. In particular, assume that there exist functions $v \in C^2(D) \cap C^1(\bar{D})$ and $w \in C^2(D) \cap C^1(\bar{D})$ which satisfy the interior transmission problem

$$(10.16) \qquad \Delta w + k^2 n(x) w = 0, \quad \Delta v + k^2 v = 0 \quad \text{in } D,$$

$$(10.17) \qquad w - v = u_p^q, \quad \frac{\partial w}{\partial \nu} - \frac{\partial v}{\partial \nu} = \frac{\partial u_p^q}{\partial \nu} \quad \text{on } \partial D$$

where

$$u_p^q(x) := h_p^{(1)}(k|x|) Y_p^q(\hat{x})$$

denotes a radiating spherical wave function and where ν is the unit outward normal to ∂D. If we further assume that v is a Herglotz wave function

(10.18)
$$v(x) = \int_\Omega e^{-ik\,x\cdot d}\, g_{pq}(d)\, ds(d), \quad x \in \mathbb{R}^3,$$

where $g_{pq} \in L^2(\Omega)$, then from the representation (2.13) for the far field pattern, Green's theorem and the radiation condition we have for fixed k and every $d \in \Omega$ that

$$\int_\Omega u_\infty(\hat{x}; d) g_{pq}(\hat{x})\, ds(\hat{x}) = \frac{1}{4\pi} \int_{\partial D} \left(u\, \frac{\partial v}{\partial \nu} - v\, \frac{\partial u}{\partial \nu} \right) ds$$

(10.19)
$$= \frac{1}{4\pi} \int_{\partial D} \left(u\, \frac{\partial w}{\partial \nu} - w\, \frac{\partial u}{\partial \nu} \right) ds - \frac{1}{4\pi} \int_{\partial D} \left(u\, \frac{\partial u_p^q}{\partial \nu} - u_p^q\, \frac{\partial u}{\partial \nu} \right) ds$$

$$= -\frac{1}{4\pi} \int_{\partial D} \left(e^{ik\,x\cdot d}\, \frac{\partial u_p^q}{\partial \nu}(x) - u_p^q(x)\, \frac{\partial e^{ik\,x\cdot d}}{\partial \nu(x)} \right) ds(x) = \frac{i^{p-1}}{k}\, Y_p^q(d).$$

From (10.19) we see that the identity (10.15) is approximately satisfied if there exists a Herglotz wave function v such that the Cauchy data (10.17) for w is approximately satisfied in $L^2(\partial D)$.

The dual space method for solving the inverse acoustic medium problem is to determine $g_{pq} \in L^2(\Omega)$ such that (10.15) is satisfied and, given v defined by (10.18), to determine w and n from the overdetermined boundary value problem (10.16), (10.17). This is done for a finite set of values of k and integers p and q with $q = -p, \ldots, p$. To reformulate this scheme as an optimization problem, we define the operator T_m as in (10.6) and the operator $F_k : L^2(\Omega) \to L^2(\Omega)$ by

(10.20)
$$(F_k g)(d) := \int_\Omega u_\infty(\hat{x}; d, k) g(\hat{x})\, ds(\hat{x}), \quad d \in \Omega.$$

Then, from Green's formula (2.4) we can rewrite the boundary value problem (10.16), (10.17) in the form

$$w_{pq} = v_{pq} - k^2 T_m w_{pq} \quad \text{in } B,$$

(10.21)
$$-k^2 T_m w_{pq} = u_p^q \quad \text{on } \partial B$$

where $v = v_{pq}$ and $w = w_{pq}$. Note, that by the uniqueness for the exterior Dirichlet problem for the Helmholtz equation the boundary condition in (10.21) ensures that $k^2 T_m w_{pq} + u_p^q = 0$ first in $\mathbb{R}^3 \setminus B$ and then, by unique continuation $k^2 T_m w_{pq} + u_p^q = 0$ in $\mathbb{R}^3 \setminus \bar{D}$. This implies that both boundary conditions in (10.17) are satisfied.

The dual space method for solving the inverse medium problem can then be formulated as the optimization problem

$$
\min_{\substack{g_{pq} \in W \\ w_{pq} \in U_1 \\ m \in U_2}} \left\{ \sum_{p=1}^{P} \sum_{q=-p}^{p} \sum_{r=1}^{R} \sum_{s=1}^{S} \left| (F_{k_s} g_{pq})(d_r) - \frac{i^{p-1}}{k_s} Y_p^q(d_r) \right|^2 \right.
$$

(10.22)
$$
+ \sum_{p=1}^{P} \sum_{q=-p}^{p} \sum_{s=1}^{S} \| w_{pq} + k_s^2 T_m w_{pq} - v_{pq} \|_{L^2(B)}^2
$$

$$
\left. + \sum_{p=1}^{P} \sum_{q=-p}^{p} \sum_{s=1}^{S} \| k_s^2 T_m w_{pq} + u_p^q \|_{L^2(\partial B)}^2 \right\},
$$

noting that v_{pq}, w_{pq}, u_p^q and the operator T_m all depend on $k = k_s$, $s = 1, \ldots S$. Here, as in (10.7), W, U_1 and U_2 are appropriate (possibly weakly) compact sets. We shall again not dwell on the details of the optimization scheme (10.22) except to note that if exact far field data is used and W, U_1 and U_2 are large enough then the minimum value of the cost functional in (10.22) will be zero, provided the approximation property stated after (10.19) is valid. We now turn our attention to showing that this approximation property is indeed true. In the analysis which follows, we shall occasionally apply Green's theorem to functions in the Sobolev space $H^2(D)$ or to functions in $C^2(D)$ having L^2 Cauchy data. When doing so, we shall always be implicitly appealing to a limiting argument involving smooth functions or parallel surfaces (c.f. [90]).

For the Herglotz wave function

(10.23)
$$
v_g(y) = \int_\Omega e^{-ik\, y \cdot \hat{x}} g(\hat{x}) \, ds(\hat{x}), \quad y \in \mathbb{R}^3,
$$

with kernel g and

$$
V(D) := \{ w \in H^2(D) : \Delta w + k^2 n(x) w = 0 \text{ in } D \},
$$

we define the subspace $W \subset L^2(\partial D) \times L^2(\partial D)$ by

$$
W := \left\{ \left(v_g - w, \frac{\partial}{\partial \nu}(v_g - w) \right) : g \in L^2(\Omega),\ w \in V(D) \right\}.
$$

The desired approximation property will be valid provided W is dense in $L^2(\partial D) \times L^2(\partial D)$. To this end, we have the following theorem due to Colton and Kirsch [28].

Theorem 10.3 *Suppose* $\operatorname{Im} n(x) > 0$ *for* $x \in D$. *Then the subspace* W *is dense in* $L^2(\partial D) \times L^2(\partial D)$.

Proof. Let $\varphi, \psi \in L^2(\partial D)$ be such that

(10.24)
$$
\int_{\partial D} \left\{ \varphi(v_g - w) + \psi \frac{\partial}{\partial \nu}(v_g - w) \right\} ds = 0
$$

for all $g \in L^2(\Omega)$, $w \in V(D)$. We first set $w = 0$ in (10.24). Then from (10.23) and (10.24) we have that

$$\int_\Omega g(\hat{x}) \int_{\partial D} \left\{ \varphi(y) e^{-ik\, y \cdot \hat{x}} + \psi(y) \frac{\partial e^{-ik\, y \cdot \hat{x}}}{\partial \nu(y)} \right\} ds(y)\, ds(\hat{x}) = 0$$

for all $g \in L^2(\Omega)$ and hence

$$\int_{\partial D} \left\{ \varphi(y) e^{-ik\, y \cdot \hat{x}} + \psi(y) \frac{\partial e^{-ik\, y \cdot \hat{x}}}{\partial \nu(y)} \right\} ds(y) = 0$$

for $\hat{x} \in \Omega$. Therefore, the far field pattern of the combined single- and double-layer potential

$$u(x) := \int_{\partial D} \left\{ \varphi(y) \Phi(x,y) + \psi(y) \frac{\partial \Phi(x,y)}{\partial \nu(y)} \right\} ds(y), \quad x \in \mathbb{R}^3 \setminus \partial D,$$

vanishes, i.e., from Theorem 2.13 we can conclude that $u(x) = 0$ for $x \in \mathbb{R}^3 \setminus \bar{D}$. Since $\varphi, \psi \in L^2(\partial D)$, we can apply the generalized jump relations (3.19)–(3.22) to conclude that

$$(10.25) \qquad \varphi = \frac{\partial u_-}{\partial \nu}, \quad \psi = -u_- \quad \text{on } \partial D$$

and hence $u \in H^{3/2}(D)$ (c.f. [90]). If we now set $g = 0$ in (10.24), we see from Green's theorem that

$$(10.26) \qquad k^2 \int_D muw\, dx = \int_{\partial D} \left(u \frac{\partial w}{\partial \nu} - w \frac{\partial u}{\partial \nu} \right) ds = 0$$

for all $w \in V(D)$.

Now consider the boundary value problem

$$(10.27) \qquad \begin{aligned} \Delta v + k^2 n(x) v &= k^2 m(x) u \quad \text{in } D, \\ v &= 0 \quad \text{on } \partial D. \end{aligned}$$

Since $\operatorname{Im} n(x) > 0$ for $x \in D$, this problem has a unique solution $v \in H^2(D)$ (c.f. [196]). Then from Green's theorem and (10.26) we have that

$$(10.28) \qquad \int_{\partial D} w \frac{\partial v}{\partial \nu} ds = \int_D w(\Delta v + k^2 nv)\, dx = k^2 \int_D muw\, dx = 0$$

for all $w \in V(D)$. Note that from the trace theorem we have that the boundary integral in (10.28) is well defined. Since $\operatorname{Im} n(x) > 0$ for $x \in D$, the boundary values of functions $w \in V(D)$ are dense in $L^2(\partial D)$ and hence we can conclude from (10.28) that $\partial v / \partial \nu(x) = 0$ for $x \in \partial D$. From (10.27) and Green's theorems we now see that, since $u \in H^{3/2}(D)$ and $\Delta u + k^2 u = 0$ in D, we have

$$0 = \int_D u(\Delta \bar{v} + k^2 \bar{v}) \, dx = \frac{1}{k^2} \int_D \frac{1}{m}(\Delta v + k^2 n v)(\Delta \bar{v} + k^2 \bar{v}) \, dx$$

$$= \frac{1}{k^2} \int_D \left(\frac{1}{m} |\Delta v + k^2 v|^2 - k^2 v(\Delta \bar{v} + k^2 \bar{v}) \right) dx$$

$$= \frac{1}{k^2} \int_D \left(\frac{1}{m} |\Delta v + k^2 v|^2 - k^4 |v|^2 + k^2 |\operatorname{grad} v|^2 \right) dx$$

and, taking the imaginary part, we see that

$$\int_D \frac{\operatorname{Im} n}{|m|^2} |\Delta v + k^2 v|^2 \, dx = 0.$$

Hence, $\Delta v + k^2 v = 0$ in D and since the Cauchy data for v vanish on ∂D we have from Theorem 2.1 (using a straightforward limiting argument) that $v(x) = 0$ for $x \in D$. Hence, $u(x) = 0$ for $x \in D$ and thus, from (10.25), $\varphi = \psi = 0$. This completes the proof of the theorem. $\square$

In order for the dual space method presented in this section to work, we have required that $\operatorname{Im} n(x) > 0$ for $x \in D$. In particular, if this is not the case, or $\operatorname{Im} n$ is small, the presence of transmission eigenvalues can contaminate the method to the extent of destroying its ability to reconstruct the refractive index. Numerical examples using this method for solving the inverse medium problem can be found in [40, 43] and Section 10.6 of this book.

10.4 A Modified Dual Space Method

As mentioned above, a disadvantage of the dual space method presented in the previous section is that the presence of transmission eigenvalues can lead to numerical instabilities and poor reconstructions of the refractive index. We shall now introduce a modified version of the identity (10.15) that leads to a method for solving the acoustic inverse medium problem that avoids this difficulty.

We begin by considering the following auxiliary problem. Let $\lambda \geq 0$ and let $h \in C^2(\mathbb{R}^3 \setminus \bar{B}) \cap C^1(\mathbb{R}^3 \setminus B)$ be the solution of the exterior impedance boundary value problem

(10.29) $$\Delta h + k^2 h = 0 \quad \text{in } \mathbb{R}^3 \setminus \bar{B},$$

(10.30) $$h(x) = e^{ik\,x \cdot d} + h^s(x),$$

(10.31) $$\frac{\partial h}{\partial \nu} + i\lambda h = 0 \quad \text{on } \partial B,$$

(10.32) $$\lim_{r \to \infty} r\left(\frac{\partial h^s}{\partial r} - ikh^s \right) = 0$$

where again B is an open ball centered at the origin containing the support of m and where ν denotes the exterior normal to ∂B. (Domains other than balls could also be used provided they contain the support of m.) The uniqueness of a solution to (10.29)–(10.32) follows from Theorem 2.12 whereas the existence of a solution follows by seeking a solution in the form (3.27) and imitating the proof of Theorem 3.10 (See also the remarks after Theorem 3.10). We note that by the Analytic Fredholm Theorem 8.19 the integral equation obtained from the use of (3.27) is uniquely solvable for all $\lambda \in \mathbb{C}$ with the possible exception of a countable set of values of λ, i.e., there exists a solution to (10.29)–(10.32) for a range of λ where the condition $\lambda \geq 0$ is violated. Finally, from the representation (3.27), we see that h^s has the asymptotic behavior

$$h^s(x) = \frac{e^{ik|x|}}{|x|}\, h_\infty(\hat{x}; d) + O\left(\frac{1}{|x|^2}\right), \quad |x| \to \infty,$$

and from (3.36) we see that the far field pattern h_∞ satisfies the reciprocity relation

(10.33) $$h_\infty(\hat{x}; d) = h_\infty(-d; -\hat{x}), \quad \hat{x}, d \in \Omega.$$

Now let u_∞ be the far field pattern of the scattering problem (10.1)–(10.3) for acoustic waves in an inhomogeneous medium and consider the problem of when there exists a function $g_{pq} \in L^2(\Omega)$ such that

(10.34) $$\int_\Omega [u_\infty(\hat{x}; d) - h_\infty(\hat{x}; d)]\, g_{pq}(\hat{x})\, ds(\hat{x}) = \frac{i^{p-1}}{k}\, Y_p^q(d)$$

for all d in a countable dense set of vectors on the unit sphere, i.e., by continuity, for all $d \in \Omega$. By the Reciprocity Theorem 8.8 and (10.33) we see that (10.34) is equivalent to the identity

(10.35) $$\int_\Omega [u_\infty(\hat{x}; d) - h_\infty(\hat{x}; d)]\, g_{pq}(-d)\, ds(d) = \frac{(-i)^{p+1}}{k}\, Y_p^q(\hat{x})$$

for all $\hat{x} \in \Omega$. If we define the Herglotz wave function w^i by

(10.36) $$w^i(x) := \int_\Omega e^{-ik\, x \cdot d}\, g_{pq}(d)\, ds(d) = \int_\Omega e^{ik\, x \cdot d}\, g_{pq}(-d)\, ds(d), \quad x \in \mathbb{R}^3,$$

then

$$w_\infty(\hat{x}) := \int_\Omega u_\infty(\hat{x}; d)\, g_{pq}(-d)\, ds(d), \quad \hat{x} \in \Omega,$$

is the far field pattern corresponding to the scattering problem

$$\Delta w + k^2 n(x) w = 0 \quad \text{in } \mathbb{R}^3,$$

$$w(x) = w^i(x) + w^s(x),$$

$$\lim_{r \to \infty} r\left(\frac{\partial w^s}{\partial r} - ik w^s\right) = 0,$$

and

$$v_\infty(\hat{x}) := \int_\Omega h_\infty(\hat{x}; d)\, g_{pq}(-d)\, ds(d), \quad \hat{x} \in \Omega,$$

is the far field pattern corresponding to the scattering problem

$$\Delta v + k^2 v = 0 \quad \text{in } \mathbb{R}^3 \setminus \bar{B},$$

$$v(x) = w^i(x) + v^s(x),$$

$$\frac{\partial v}{\partial \nu} + i\lambda v = 0 \quad \text{on } \partial B,$$

$$\lim_{r \to \infty} r\left(\frac{\partial v^s}{\partial r} - ikv^s\right) = 0.$$

Hence, from (2.41), (10.35) and Theorem 2.13 we can conclude that

$$w^s(x) - v^s(x) = h_p^{(1)}(k|x|)\, Y_p^q(\hat{x}), \quad x \in \mathbb{R}^3 \setminus \bar{B},$$

i.e., w satisfies the boundary value problem

(10.37)
$$\Delta w + k^2 n(x) w = 0 \quad \text{in } B,$$

(10.38)
$$w(x) = w^i(x) + w^s(x),$$

(10.39)
$$\left(\frac{\partial}{\partial \nu} + i\lambda\right)(w - u_p^q)) = 0 \quad \text{on } \partial B,$$

(10.40)
$$\lim_{r \to \infty} r\left(\frac{\partial w^s}{\partial r} - ikw^s\right) = 0$$

uniformly for all directions. Again, as in Section 10.3, we have written

$$u_p^q(x) := h_p^{(1)}(k|x|)\, Y_p^q(\hat{x})$$

for the radiating spherical wave function.

The boundary value problem (10.37)–(10.40) can be understood as the problem of first solving the interior impedance problem (10.37), (10.39) and then decomposing the solution in the form (10.38) where w^i satisfies the Helmholtz equation in B and w^s is defined for all of $\mathbb{R}^3$, satisfies the Helmholtz equation in $\mathbb{R}^3 \setminus \bar{B}$ and the radiation condition (10.40). Note that for the identity (10.34) to be valid, w^i must be a Herglotz wave function with Herglotz kernel g_{pq}. In particular, from the above analysis we have the following theorem ([41, 42]).

Theorem 10.4 *Assume there exists a solution $w \in C^2(\mathbb{R}^3 \setminus \bar{B}) \cap C^1(\mathbb{R}^3 \setminus B)$ to the interior impedance problem (10.37), (10.39) such that w has the decomposition (10.38) with w^i and w^s as described in the paragraph above. Then there exists $g_{pq} \in L^2(\Omega)$ such that (10.34) is valid if and only if w^i is a Herglotz wave function with Herglotz kernel g_{pq}.*

We now turn our attention to when there exists a solution to (10.37), (10.39) having the decomposition (10.38). To establish the existence of a unique solution to (10.37), (10.39) we make use of potential theory. In particular, we recall from Section 3.1 the single-layer operator $S : C(\partial B) \to C^{0,\alpha}(\partial B)$ and the normal derivative operator $K' : C(\partial B) \to C(\partial B)$ and from Section 8.2 the volume potential $T_m : C(B) \to C(B)$. We can now prove the following theorem.

Theorem 10.5 *Suppose that $\lambda = 0$ if $\mathrm{Im}\, n(x) > 0$ for some $x \in B$ and $\lambda > 0$ if $\mathrm{Im}\, n(x) = 0$ for all $x \in B$. Then there exists a unique solution $w \in C^2(B) \cap C^1(\bar{B})$ to the impedance problem (10.37), (10.39) having the decomposition (10.38) where $w^i \in C^2(B) \cap C^1(\bar{B})$ is a solution of the Helmholtz equation in B and $w^s \in C^2(\mathbb{R}^3)$ satisfies the radiation condition (10.40).*

Proof. We first establish uniqueness for the boundary value problem (10.37), (10.39). Assume that w is a solution of (10.37) satisfying homogeneous impedance boundary data on ∂B. Then, by Green's theorem, we have that

$$(10.41) \qquad \lambda \int_{\partial B} |w|^2 ds = \mathrm{Im} \int_{\partial B} w \frac{\partial \bar{w}}{\partial \nu} \, ds = -k^2 \, \mathrm{Im} \int_B \bar{n} \, |w|^2 dx.$$

If $\mathrm{Im}\, n(x) > 0$ for some $x \in B$, then by assumption $\lambda = 0$ and we can conclude from (10.41) that $w(x) = 0$ for those points $x \in B$ where $\mathrm{Im}\, n(x) > 0$. Hence, by the Unique Continuation Theorem 8.6, we have $w(x) = 0$ for $x \in B$. On the other hand, if $\mathrm{Im}\, n(x) = 0$ for all $x \in B$ then $\lambda > 0$ and (10.41) implies that $w = 0$ on ∂B. Since w satisfies homogeneous impedance boundary data, we have that $\partial w / \partial \nu = 0$ on ∂B and Green's formula (2.4) now tells us that $w + k^2 T_m w = 0$ in B. Hence, we see from the invertibility of $I + k^2 T_m$ in $C(\bar{B})$ that $w(x) = 0$ for $x \in B$.

In order to establish existence for (10.37), (10.39), we look for a solution in the form

$$(10.42) \quad w(x) = \int_{\partial B} \Phi(x,y)\varphi(y) \, ds(y) - k^2 \int_B \Phi(x,y)m(y)\psi(y) \, dy, \quad x \in B,$$

where the densities $\varphi \in C(\partial B)$ and $\psi \in C(\bar{B})$ are assumed to satisfy the two integral equations

$$(10.43) \qquad\qquad \psi - \tilde{S}\varphi + k^2 T_m \psi = 0$$

and

$$(10.44) \qquad\qquad \varphi + (K' + i\lambda S)\varphi - 2k^2 T_{m,\lambda}\psi = 2f$$

with

$$f := \frac{\partial u_p^q}{\partial \nu} + i\lambda u_p^q \quad \text{on } \partial B.$$

Here we define $\tilde{S} : C(\partial B) \to C(\bar{B})$ by

$$(\tilde{S}\varphi)(x) := \int_{\partial B} \Phi(x,y)\varphi(y) \, ds(y), \quad x \in \bar{B},$$

and $T_{m,\lambda} : C(\bar{B}) \to C(\partial B)$ by

$$(T_{m,\lambda}\psi)(x) := \int_B \left\{ \frac{\partial \Phi(x,y)}{\partial \nu(x)} + i\lambda\Phi(x,y) \right\} m(y)\psi(y)\, dy, \quad x \in \partial B.$$

By the regularity of f, from Theorems 3.4 and 8.1 we have that for a continuous solution of (10.43), (10.44) we automatically have $\varphi \in C^{0,\alpha}(\partial B)$ and $\psi \in C^{0,\alpha}(B)$. Hence, defining w by (10.42), we see that $w \in C^2(B) \cap C^1(\bar{B})$, i.e., w has the required regularity. By Theorem 8.1, the integral equation (10.43) ensures that w solves the differential equation (10.37) and from the jump relations of Theorem 3.1 we see that the integral equation (10.44) implies that the boundary condition (10.39) is satisfied.

All integral operators in the system (10.43), (10.44) clearly have weakly singular kernels and therefore they are compact. Hence, by the Riesz–Fredholm theory, to show the existence of a unique solution of (10.43), (10.44) we must show that the only solution of the homogeneous problem is the trivial solution $\varphi = 0$, $\psi = 0$. If φ, ψ is a solution to (10.43), (10.44) with $f = 0$, then w defined by (10.42) is a solution to the homogeneous problem (10.37), (10.39) and therefore by uniqueness we have $w = 0$ in B. Applying $\Delta + k^2$ to both sides of (10.42) (with $w = 0$) now shows that $m\psi = 0$ and $\tilde{S}\varphi = 0$ in B. Hence, from (10.43), we have that $\psi = 0$ in B. Now, making use of the uniqueness for the exterior Dirichlet problem (Theorem 3.7) and the jump relations of Theorem 3.1, we can also conclude that $\varphi = 0$.

The decomposition (10.38) holds where $w^i = \tilde{S}\varphi \in C^2(B) \cap C^1(\bar{B})$ and $w^s = -k^2 T_m\psi \in C^2(\mathbb{R}^3)$. We have already shown above that the solution of (10.37), (10.39) is unique and all that remains is to show the uniqueness of the decomposition (10.38). But this follows from the fact that entire solutions of the Helmholtz equation satisfying the radiation condition must be identically zero (c.f. p. 19). □

Corollary 10.6 *In the decomposition (10.38), w^i can be approximated in $C^1(\bar{B})$ by a Herglotz wave function.*

Proof. From the regularity properties of surface and volume potentials we see that if $\psi \in C(\bar{B})$ and $\varphi \in C(\partial B)$ is a solution of (10.43), (10.44) then $\varphi \in C^{2,\alpha}(\partial B)$. The corollary now follows by approximating the surface density φ by a linear combination of spherical harmonics. □

With Theorem 10.5 and Corollary 10.6 at our disposal, we can now formulate an optimization scheme for solving the inverse scattering problem such that the minimum value of the cost functional is zero, use is made of an averaging process as in the dual space method of the previous section, and the problem of transmission eigenvalues is avoided. Indeed, using the same notation as in (10.22) and defining $F_k^\lambda : L^2(\Omega) \to L^2(\Omega)$ by

(10.45) $$(F_k^\lambda g)(d) := \int_\Omega h_\infty(\hat{x}; d, k)\, g(\hat{x})\, ds(\hat{x}), \quad d \in \Omega,$$

we can formulate the optimization problem

$$
\min_{\substack{g_{pq}\in W \\ w_{pq}\in U_1 \\ m\in U_2}} \left\{ \sum_{p=1}^{P}\sum_{q=-p}^{p}\sum_{r=1}^{R}\sum_{s=1}^{S} \left| ((F_{k_s} - F_{k_s}^{\lambda})g_{pq})(d_r) + \frac{i^{p+1}}{k_s} Y_p^q(d_r) \right|^2 \right.
$$

(10.46)
$$
+ \sum_{p=1}^{P}\sum_{q=-p}^{p}\sum_{s=1}^{S} \| w_{pq} + k_s^2 T_m w_{pq} - v_{pq} \|_{L^2(B)}^2
$$

$$
\left. + \sum_{p=1}^{P}\sum_{q=-p}^{p}\sum_{s=1}^{S} \left\| \left(\frac{\partial}{\partial r} + i\lambda \right)(v_{pq} - k_s^2 T_m w_{pq} + u_p^q) \right\|_{L^2(\partial B)}^2 \right\}.
$$

From Theorem 10.5 and Corollary 10.6 we see that if exact far field data is used the minimum value of the cost functional in (10.46) will be zero provided W, U_1 and U_2 are large enough. Numerical examples using this method for solving the inverse medium problem can be found in [41, 42] and Section 10.6 of this book.

10.5 The Inverse Medium Problem for Electromagnetic Waves

We recall from Chapter 9 that the direct scattering problem for electromagnetic waves can be formulated as that of determining the electric field E and magnetic field H such that

(10.47) $\operatorname{curl} E - ikH = 0, \quad \operatorname{curl} H + ikn(x)E = 0 \quad \text{in } \mathbb{R}^3,$

(10.48) $E(x) = \dfrac{i}{k} \operatorname{curl} \operatorname{curl} p \, e^{ik\,x\cdot d} + E^s(x), \quad H(x) = \operatorname{curl} p \, e^{ik\,x\cdot d} + H^s(x),$

(10.49) $\lim_{r\to\infty} (H^s \times x - rE^s) = 0$

uniformly for all directions where $k > 0$ is the wave number, $p \in \mathbb{R}^3$ is the polarization and $d \in \Omega$ the direction of the incident wave. The refractive index $n \in C^{1,\alpha}(\mathbb{R}^3)$ is of the form

$$
n(x) = \frac{1}{\varepsilon_0} \left\{ \varepsilon(x) + i\,\frac{\sigma(x)}{\omega} \right\}
$$

where $\varepsilon = \varepsilon(x)$ is the permittivity, $\sigma = \sigma(x)$ is the conductivity and ω is the frequency. We assume that $m := 1 - n$ is of compact support and, as usual, define $D := \{ x \in \mathbb{R}^3 : m(x) \neq 0 \}$. It is further assumed that D is connected with a connected C^2 boundary ∂D and D contains the origin. The existence of a unique solution to (10.47)–(10.49) was established in Chapter 9. It was also shown there that E^s has the asymptotic behavior

(10.50) $E^s(x; d, p) = \dfrac{e^{ik|x|}}{|x|} E_\infty(\hat{x}; d, p) + O\left(\dfrac{1}{|x|^2} \right), \quad |x| \to \infty,$

where E_∞ is the electric far field pattern. The *inverse medium problem* for electromagnetic waves is to determine n from $E_\infty(\hat{x}; d, p)$ for $\hat{x} \in \Omega$, $p \in \mathbb{R}^3$, a finite number of directions d and (possibly) different values of k. It can be shown that for k fixed, $\hat{x}, d \in \Omega$ and $p \in \mathbb{R}^3$, the electric far field pattern E_∞ uniquely determines n [47]. The proof of this fact is similar to the one for acoustic waves outlined after the proof of Theorem 10.2. The main difference is that we must now construct a solution E, H of (10.47) such that E has the form

$$E(x) = e^{i\zeta \cdot x}[\eta + R_\zeta(x)]$$

where $\zeta, \eta \in \mathbb{C}^3$, $\eta \cdot \zeta = 0$ and $\zeta \cdot \zeta = k^2$ and, in contrast to the case of acoustic waves, it is no longer true that R_ζ decays to zero as $|\zeta|$ tends to infinity. This makes the uniqueness proof for electromagnetic waves more complicated than the corresponding proof for acoustic waves and for details we refer to [47].

As with the case of acoustic waves, one approach for finding a solution to the inverse medium problem for electromagnetic waves is to use the integral equation (9.7). Since this can be done in precisely the same manner as in the case for acoustic waves (c.f. Section 10.1), we shall forego such an investigation and proceed directly to the derivation of dual space methods for solving the electromagnetic inverse medium problem that are analogous to these derived for acoustic waves in Sections 10.3 and 10.4. To this end, we recall the Hilbert space $T^2(\Omega)$ of L^2 tangential fields on the unit sphere Ω, let $\{d_n : n = 1, 2, 3, \ldots\}$ be a dense set of vectors on Ω and consider the set of electric far field patterns $\mathcal{F} := \{E_\infty(\cdot; d_n, e_j) : n = 1, 2, 3, \ldots, j = 1, 2, 3\}$ where e_1, e_2, e_3 are the unit coordinate vectors in $\mathbb{R}^3$. Following the proofs of Theorems 6.34 and 9.7, we can immediately deduce the following result.

Theorem 10.7 *For $q \in \mathbb{R}^3$, define the radiating solution E_q, H_q of the Maxwell equations by*

$$E_q(x) := \operatorname{curl} q\Phi(x, 0), \quad H_q(x) := \frac{1}{ik} \operatorname{curl} \operatorname{curl} q\Phi(x, 0), \quad x \in \mathbb{R}^3 \setminus \{0\}.$$

Then there exists $g \in T^2(\Omega)$ such that

$$(10.51) \qquad \int_\Omega E_\infty(\hat{x}; d, p) \cdot g(\hat{x}) \, dx(\hat{x}) = \frac{ik}{4\pi} p \cdot q \times d$$

for all $p \in \mathbb{R}^3$ and $d \in \Omega$ if and only if there exists a solution E_0, E_1, H_0, H_1 in $C^1(D) \cap C(\bar{D})$ of the electromagnetic interior transmission problem

$$(10.52) \qquad \operatorname{curl} E_1 - ikH_1 = 0, \quad \operatorname{curl} H_1 + ikn(x)E_1 = 0 \quad in \ D,$$

$$(10.53) \qquad \operatorname{curl} E_0 - ikH_0 = 0, \quad \operatorname{curl} H_0 + ikE_0 = 0 \quad in \ D,$$

$$(10.54) \quad \nu \times (E_1 - E_0) = \nu \times E_q, \quad \nu \times (H_1 - H_0) = \nu \times H_q \quad on \ \partial D,$$

such that E_0, H_0 is an electromagnetic Herglotz pair.

In order to make use of Theorem 10.7, we need to show that there exists a solution to the interior transmission problem (10.52)–(10.54) and that E_0, H_0 can be approximated by an electromagnetic Herglotz pair. Following the ideas of Colton and Päivärinta [45], we shall now proceed to do this for the special case when $\varepsilon(x) = \varepsilon_0$ for all $x \in \mathbb{R}^3$, i.e.,

$$(10.55) \qquad n(x) = 1 + i\,\frac{\sigma(x)}{\varepsilon_0\omega}$$

where $\sigma(x) > 0$ for $x \in D$.

We begin by introducing the Hilbert space $L^2_\sigma(D)$ defined by

$$L^2_\sigma(D) := \left\{ f : D \to \mathbb{C}^3 : f \text{ measurable}, \int_D \sigma |f|^2 < \infty \right\}$$

with scalar product

$$(f,g) := \int_D \sigma f \cdot \bar{g}\, dx.$$

Of special importance to us is the subspace $H \subset L^2_\sigma(D)$ defined by

$$H := \operatorname{span}\left\{ M_n^m, \operatorname{curl} M_n^m : n = 1, 2, \ldots, m = -n, \ldots, n \right\}$$

where, as in Section 6.5,

$$M_n^m(x) := \operatorname{curl}\{x j_n(k|x|)\, Y_n^m(\hat{x})\}.$$

Let $\bar{H}$ denote the closure of H in $L^2_\sigma(D)$. Instead of considering solutions of (10.52)–(10.54) in $C^1(D) \cap C(\bar{D})$, it is convenient for our purposes to consider a weak formulation based on Theorem 9.2.

Definition 10.8 *Let $\varepsilon(x) = \varepsilon_0$ for all $x \in \mathbb{R}^3$. Then the pair $E_0, E_1 \in L^2_\sigma(D)$ is said to be a weak solution of the interior transmission problem for electromagnetic waves with direction q if $E_0 \in \bar{H}$, $E_1 \in L^2_\sigma(D)$ satisfy the integral equation*

$$(10.56) \qquad E_1 = E_0 + T_\sigma E_1 \quad \text{in } D,$$

where

$$(T_\sigma E)(x) := i\mu_0\omega \int_D \Phi(x,y)\sigma(y)E(y)\,dy$$

$$(10.57)$$

$$+ \operatorname{grad} \int_D \frac{1}{n(y)}\, \operatorname{grad} n(y) \cdot E(y)\, \Phi(x,y)\,dy, \quad x \in \mathbb{R}^3,$$

for $n(x) = 1 + i\sigma(x)/\varepsilon_0\omega$ and

$$(10.58) \qquad T_\sigma E_1 = E_q \quad \text{in } \mathbb{R}^3 \setminus \bar{D}$$

where $E_q(x) := \operatorname{curl} q\Phi(x,0)$.

Before proceeding, we make some preliminary observations concerning Definition 10.8. To begin with, as in the acoustic case (see Section 8.6), we can view (10.58) as a generalized form of the boundary conditions (10.54). In order to ensure the existence of the second integral in (10.57), for the sake of simplicity, we assume that there exists a positive constant M such that

$$|\operatorname{grad}\sigma(x)|^2 \le M\sigma(x), \quad x \in D.$$

This also implies that we can write $T_\sigma(E) = \tilde{T}(\sqrt{\sigma}\,E)$ where $\tilde{T}$ has a weakly singular kernel. In particular, from this we can see that $T_\sigma : L^2_\sigma(D) \to L^2_\sigma(D)$ is compact. From the proof of Theorem 9.5 we recall that the inverse operator $(I - T_\sigma)^{-1} : C(\bar{D}) \to C(\bar{D})$ exists and is bounded. For our following analysis we also need to establish the boundedness of $(I - T_\sigma)^{-1} : L^2_\sigma(D) \to L^2_\sigma(D)$. To this end, we first note that since T_σ is an integral operator with weakly singular kernel it has an adjoint T^*_σ with respect to the L^2 bilinear form

$$\langle E, F \rangle := \int_D E \cdot F \, dx$$

which again is an integral operator with a weakly singular kernel and therefore compact from $C(\bar{D})$ into $C(\bar{D})$. Hence, by the Fredholm alternative, applied in the two dual systems $\langle C(\bar{D}), C(\bar{D})\rangle$ and $\langle L^2_\sigma(D), C(\bar{D})\rangle$ (see the proof of Theorem 3.20) the nullspaces of the operator $I - T_\sigma$ in $C(\bar{D})$ and $L^2_\sigma(D)$ coincide. Since from the proof of Theorem 9.5 we already know that the nullspace in $C(\bar{D})$ is trivial, by the Riesz–Fredholm theory, we have established existence and boundedness of $(I - T_\sigma)^{-1} : L^2_\sigma(D) \to L^2_\sigma(D)$.

If $E_0 \in H$, then given a solution $E_1 \in L^2_\sigma(D)$ of (10.56), we can use (10.56) and (10.57) to define E_1 also in $\mathbb{R}^3 \setminus \bar{D}$ such that (10.56) is satisfied in all of $\mathbb{R}^3$. From Theorem 9.2 we can then conclude that $E_0, H_0 = \operatorname{curl} E_0/ik$ and $E_1, H_1 = \operatorname{curl} E_1/ik$ satisfy (10.53) and (10.52) respectively. If $E_0 \in \bar{H}$, we choose a sequence $(E_{0,j})$ from H such that $E_{0,j} \to E_0$, $j \to \infty$, and define $E_{1,j}$ by $E_{1,j} := (I - T_\sigma)^{-1} E_{0,j}$. Then, by the boundeness of $(I - T_\sigma)^{-1} : L^2_\sigma(D) \to L^2_\sigma(D)$, we have $E_{1,j} \to E_1$, $j \to \infty$. From $E_{1,j} - E_{0,j} = T_\sigma E_{1,j}$ we see that $\operatorname{div} T_\sigma E_{1,j} = 0$ in $\mathbb{R}^3 \setminus \bar{D}$. Since from (10.57) we have that

$$(\operatorname{div} T_\sigma E_1)(x) = i\mu_0\omega \operatorname{div} \int_D \Phi(x,y)\sigma(y)E_1(y)\,dy$$

$$-k^2 \int_D \frac{1}{n(y)} \operatorname{grad} n(y) \cdot E_1(y)\,\Phi(x,y)\,dy, \quad x \in \mathbb{R}^3 \setminus \bar{D},$$

using Schwarz's inequality we can now deduce that $\operatorname{div} T_\sigma E_1 = 0$ in $\mathbb{R}^3 \setminus \bar{D}$. Hence, by Theorems 6.4 and 6.7 we conclude that $T_\sigma E_1$ is a solution to the Maxwell equations in $\mathbb{R}^3 \setminus \bar{D}$ satisfying the Silver–Müller radiation condition.

We now proceed to proving that there exists a unique weak solution to the interior transmission problem for electromagnetic waves.

Theorem 10.9 *Let $\varepsilon(x) = \varepsilon_0$ for all $x \in \mathbb{R}^3$. Then for any direction q and wave number $k > 0$ there exists at most one weak solution of the interior transmission problem for electromagnetic waves.*

Proof. It suffices to show that if

$$(10.59) \qquad E_1 = E_0 + T_\sigma E_1 \quad \text{in } D,$$

$$(10.60) \qquad T_\sigma E_1 = 0 \quad \text{in } \mathbb{R}^3 \setminus \bar{D},$$

where $E_0 \in \bar{H}$ then $E_1 = 0$. We first show that the only solution of (10.59) with $E_0 \in \bar{H}$ and $E_1 \in H^\perp$ is $E_0 = E_1 = 0$. To this end, let $(E_{0,j})$ be a sequence from H with $E_{0,j} \to E_0$, $j \to \infty$, and, as above, define $E_{1,j}$ by $E_{1,j} := (I - T_\sigma)^{-1} E_{0,j}$. Then $E_{1,j} - E_{0,j} = T_\sigma E_{1,j}$ and, with the aid of $\operatorname{curl} \operatorname{curl} E_{0,j} = k^2 E_{0,j}$ and $\operatorname{curl} \operatorname{curl} E_{1,j} = k^2 n E_{1,j}$, by Green's vector theorem (6.2) using $k^2 = \varepsilon_0 \mu_0 \omega^2$ we find that

$$
\int_{\partial B} \nu \cdot \overline{T_\sigma E_{1,j}} \times \operatorname{curl} T_\sigma E_{1,j} \, ds = \int_B |\operatorname{curl}(E_{0,j} - E_{1,j})|^2 dx
$$

$$(10.61)$$

$$
-k^2 \int_B |E_{0,j} - E_{1,j}|^2 dx - i\mu_0 \omega \int_B \sigma |E_{1,j}|^2 dx + i\mu_0 \omega \int_B \sigma E_{1,j} \cdot \overline{E_{0,j}} \, dx
$$

where B is an open ball with $\bar{D} \subset B$. Furthermore, from (10.57) we have

$$(10.62) \quad (\operatorname{curl} T_\sigma E_1)(x) = i\mu_0 \omega \operatorname{curl} \int_D \sigma(y) \Phi(x,y) E_1(y) \, dy, \quad x \in \mathbb{R}^3 \setminus \bar{D},$$

and from the Vector Addition Theorem 6.27 and the fact that $E_1 \in H^\perp$ we obtain that $\operatorname{curl} T_\sigma E_1 = 0$ in $\mathbb{R}^3 \setminus \bar{D}$. Therefore, using Schwarz's inequality in (10.57) and (10.62), we can now conclude that the integral on the left hand side of (10.61) tends to zero as $j \to \infty$. Hence, taking the imaginary part in (10.61) and letting $j \to \infty$, we see that

$$
\int_D \sigma |E_1|^2 dx = 0
$$

and hence $E_1 = E_0 = 0$.

We must now show that if E_0, E_1 is a solution of (10.59) and (10.60) with $E_0 \in \bar{H}$ then $E_1 \in H^\perp$. To this end, we note that from (10.60) we trivially have that $\operatorname{curl} T_\sigma E_1 = 0$ in $\mathbb{R}^3 \setminus \bar{D}$. From this, using (10.62) and the Vector Addition Theorem 6.27, we have by orthogonality that

$$
\int_D \sigma \, \overline{M_n^m} \cdot E_1 \, dy = 0 \quad \text{and} \quad \int_D \sigma \operatorname{curl} \overline{M_n^m} \cdot E_1 \, dy = 0
$$

for $n = 1, 2, \ldots$ and $m = -n, \ldots, n$, that is, $E_1 \in H^\perp$. $\qquad \square$

Having established the uniqueness of a weak solution to the interior transmission problem, we now want to show existence. Without loss of generality, we shall assume that $q = (0,0,1)$ and, in this case,

$$E_q(x) := \operatorname{curl} q\Phi(x,0) = \frac{ik^2}{\sqrt{12\pi}} N_1^0(x)$$

where, as in Section 6.5,

$$N_n^m(x) := \operatorname{curl}\{x h_n^{(1)}(k|x|) Y_n^m(\hat{x})\}.$$

Theorem 10.10 *Let $\varepsilon(x) = \varepsilon_0$ for all $x \in \mathbb{R}^3$. Then for any direction q and wave number $k > 0$ there exists a weak solution of the interior transmission problem for electromagnetic waves.*

Proof. We begin by defining the subspace H_0 by

$$H_0 := \{E \in \bar{H} : (E, M_1^0) = 0\}$$

and the associated orthogonal projection operator $P_0 : \bar{H} \to H_0$. We can then define the vector field $F \in \bar{H}$ by

$$F := M_1^0 - P_0 M_1^0$$

and note that $F \in H_0^\perp$ and $(F, M_1^0) \neq 0$ since $F \neq 0$ and, by orthogonality, $(F, M_1^0) = (F, F)$. Without loss of generality we can assume F is normalized such that

$$(F, M_1^0) = \frac{1}{i\sqrt{3\pi}} \sqrt{\frac{\varepsilon_0}{\mu_0}} .$$

We now want to construct a solution $E_0 \in \bar{H}$, $E_1 \in H_0^\perp$ of

$$E_1 = E_0 + T_\sigma E_1 \quad \text{in } D$$

such that

(10.63) $$(E_1, M_1^0) = \frac{1}{i\sqrt{3\pi}} \sqrt{\frac{\varepsilon_0}{\mu_0}} .$$

To this end, let $P : L_\sigma^2(D) \to H^\perp$ be the orthogonal projection operator. From the proof of Theorem 10.9, we see that $I - PT_\sigma$ has a trivial nullspace since $E_1 - PT_\sigma E_1 = 0$ implies that $E_1 - T_\sigma E_1 = E_0$ with $E_0 \in \bar{H}$ and $E_1 \in H^\perp$. Hence, by the Riesz–Fredholm theory, the equation

$$\tilde{E}_1 - PT_\sigma \tilde{E}_1 = PT_\sigma F$$

has a unique solution $\tilde{E}_1 \in H^\perp$. Setting

$$E_1 := \tilde{E}_1 + F$$

we have $E_1 - PT_\sigma E_1 = F$ and, since $T_\sigma E_1 = PT_\sigma E_1 + \tilde{E}_0$ with $\tilde{E}_0 \in \bar{H}$, we finally have $E_1 - T_\sigma E_1 = E_0$ with $E_0 = \tilde{E}_0 + F \in \bar{H}$. Since $\tilde{E}_1 \in H^\perp$, the condition (10.63) and $E_1 \in H_0^\perp$ are satisfied.

We will now show that $T_\sigma E_1 = E_q$ in $\mathbb{R}^3 \setminus \bar{D}$ which implies that (10.58) is satisfied, thus completing the proof of the theorem. To show this, we first note that from $E_1 \in H_0^\perp$ and (10.63) we have

(10.64)
$$\int_D \sigma \overline{M_n^m} \cdot E_1 \, dy = \delta_{n1} \delta_{m0} \frac{1}{i\sqrt{3\pi}} \sqrt{\frac{\varepsilon_0}{\mu_0}},$$

$$\int_D \sigma \operatorname{curl} \overline{M_n^m} \cdot E_1 \, dy = 0$$

for $n = 1, 2, \ldots$ and $m = -n, \ldots, n$ where δ_{nm} denotes the Kronecker delta symbol. From (10.62), (10.64) and the Vector Addition Theorem 6.27 we now see that for $|x|$ sufficiently large we have

$$(\operatorname{curl} T_\sigma E_1)(x) = \frac{ik^2}{\sqrt{12\pi}} \operatorname{curl} N_1^0(x) = \operatorname{curl} E_q(x).$$

By unique continuation, this holds for $x \in \mathbb{R}^3 \setminus \bar{D}$. Since from the analysis preceeding Theorem 10.9 we know that $T_\sigma E_1$ solves the Maxwell equations in $\mathbb{R}^3 \setminus \bar{D}$ this implies that $T_\sigma E_1 = E_q$ in $\mathbb{R}^3 \setminus \bar{D}$ and we are done. $\square$

Corollary 10.11 *Let $\varepsilon(x) = \varepsilon_0$ for all $x \in \mathbb{R}^3$ and let E_0, E_1 be a weak solution of the interior transmission problem for electromagnetic waves. Then E_0 can be approximated in $L_\sigma^2(D)$ by the electric field of an electromagnetic Herglotz pair.*

Proof. This follows from the facts that $E_0 \in \bar{H}$ and the elements of H are the electric fields of an electromagnetic Herglotz pair. $\square$

Using Theorem 10.7 and Definition 10.8, we can now set up an optimization scheme for solving the inverse scattering problem for electromagnetic waves in the special case when $\varepsilon(x) = \varepsilon_0$ for all $x \in \mathbb{R}^3$ that is analogous to the optimization scheme (10.22) for acoustic waves. We shall spare the reader the details of examining the optimization scheme more closely and instead now proceed to deriving the electromagnetic analogue of the scheme used to solve the acoustic inverse medium problem in Section 10.4.

The modified dual space method of Section 10.4 was based on two ingredients: the existence of a solution to the interior impedance problem (10.37), (10.39) having the decomposition (10.38) and a characterization of the Herglotz kernel satisfying the identity (10.34) in terms of this impedance problem. Following Colton and Kress [35], we shall now proceed to derive the electromagnetic analogue of these two ingredients from which the electromagnetic analogue of the modified dual space method for solving the acoustic inverse medium problem will follow immediately. Note that in the sequel we no longer assume that $\varepsilon(x) = \varepsilon_0$ for all $x \in \mathbb{R}^3$, i.e., we only assume that $n \in C^{2,\alpha}(\mathbb{R}^3)$ and $\operatorname{Im} n \geq 0$.

We first consider the following interior impedance problem for electromagnetic waves.

Interior Impedance Problem. *Let G be a bounded domain in $\mathbb{R}^3$ containing $D := \{x \in \mathbb{R}^3 : m(x) \neq 0\}$ with connected C^2 boundary ∂G, let c be a given Hölder continuous tangential field on ∂G and λ a complex constant. Find vector fields $E^i, H^i \in C^1(G) \cap C(\bar{G})$ and $E^s, H^s \in C^1(\mathbb{R}^3)$ satisfying*

$$(10.65) \qquad \operatorname{curl} E^i - ikH^i = 0, \quad \operatorname{curl} H^i + ikE^i = 0 \quad \text{in } G,$$

$$(10.66) \qquad \operatorname{curl} E^s - ikH^s = 0, \quad \operatorname{curl} H^s + ikE^s = 0 \quad \text{in } \mathbb{R}^3 \setminus \bar{G},$$

$$(10.67) \qquad \lim_{r \to \infty} (H^s \times x - rE^s) = 0,$$

such that $E = E^i + E^s$, $H = H^i + H^s$ satisfies

$$(10.68) \qquad \operatorname{curl} E - ikH = 0, \quad \operatorname{curl} H + ikn(x)E = 0 \quad \text{in } G,$$

$$(10.69) \qquad \nu \times \operatorname{curl} E - i\lambda (\nu \times E) \times \nu = c \quad \text{on } \partial G,$$

where, as usual, ν is the unit outward normal to ∂G and the radiation condition (10.67) is assumed to hold uniformly for all directions.

We note that the interior impedance problem can be viewed as the problem of first solving the impedance boundary value problem (10.68), (10.69) and then decomposing the solution such that (10.65)–(10.67) hold.

Theorem 10.12 *Assume $\lambda < 0$. Then the interior impedance problem has at most one solution.*

Proof. Let E, H denote the difference between two solutions. Then from (9.17) and the homogeneous form of the boundary condition (10.69) we see that

$$i\lambda \int_{\partial G} |\nu \times E|^2 ds = -k^2 \int_G (\bar{n} |E|^2 - |H|^2) \, dx$$

and hence, taking the imaginary part,

$$\lambda \int_{\partial G} |\nu \times E|^2 ds = k^2 \int_G \operatorname{Im} n |E|^2 dx \geq 0.$$

Since $\lambda < 0$ it follows that $\nu \times E = 0$ on ∂G and hence, from the boundary condition, $\nu \times H = 0$ on ∂G. Applying Theorem 6.2 to E and H in the domain $G \setminus \operatorname{supp} m$, we can conclude that E, H can be extended to all of $\mathbb{R}^3$ as a solution to (10.68) satisfying the radiation condition. Hence, by Theorem 9.4, we can conclude that $E = H = 0$ in G.

To show uniqueness for the decomposition (10.65)–(10.67), assume that $E^i + E^s = 0$ and $H^i + H^s = 0$ such that (10.65)–(10.67) is valid. Then E^i, H^i

can be extended to all of $\mathbb{R}^3$ as an entire solution of the Maxwell equations (10.65) satisfying the radiation condition whence $E^i = H^i = 0$ in $\mathbb{R}^3$ follows (c.f. p. 156). This in turn implies $E^s = H^s = 0$. □

Motivated by the methods of Chapter 9, we now seek a solution of the interior impedance problem by solving the integral equation

$$E(x) = \text{curl} \int_{\partial G} \Phi(x,y)a(y)\,ds(y) - k^2 \int_G \Phi(x,y)m(y)E(y)\,dy$$

(10.70)

$$+ \text{grad} \int_G \frac{1}{n(y)} \text{grad}\,n(y) \cdot E(y)\,\Phi(x,y)\,dy, \quad x \in \bar{G},$$

where the surface density $a \in T_d^{0,\alpha}(\partial G)$ is determined from the boundary condition and, having found a and E, we define H by

$$H(x) := \frac{1}{ik}\,\text{curl}\,E(x), \quad x \in G.$$

After recalling the operators M, N and R from our investigation of the exterior impedance problem in Section 9.5, the electromagnetic operator T_e from (9.18) in the proof of Theorem 9.5 (with D replaced by G) and introducing the two additional operators $W : T_d^{0,\alpha}(\partial G) \rightarrow C(\bar{G})$ and $T_{e,\lambda} : C(\bar{G}) \rightarrow T^{0,\alpha}(\partial G)$ by

$$(Wa)(x) := \text{curl} \int_{\partial G} \Phi(x,y)a(y)\,ds(y), \quad x \in \bar{G},$$

$$(T_{e,\lambda}E)(x) := \nu(x) \times (\text{curl}\,T_e E)(x) - i\lambda\,(\nu(x) \times (T_e E)(x)) \times \nu(x), \quad x \in \partial G,$$

we consider the system of integral equations

$$NRa - i\lambda RMa + i\lambda Ra + T_{e,\lambda}E = 2c$$

(10.71)

$$E - Wa - T_e E = 0.$$

Analogous to the proof of Theorem 9.13 (c.f. (9.49)), the first integral equation in (10.71) ensures that the impedance boundary condition (10.69) is satisfied. Proceeding as in the proof of Theorem 9.2, it can be seen that the second integral equation guarantues that E and $H = \text{curl}\,E/ik$ satisfy the differential equations (10.68). The decomposition $E = E^i + E^s$, $H = H^i + H^s$, follows in an obvious way from (10.70). Hence, to show the existence of a solution of the interior impedance problem we must show the existence of a solution to the system of integral equations (10.71).

Theorem 10.13 *Assume $\lambda < 0$. Then there exists a solution to the interior impedance problem.*

Proof. We need to show the existence of a solution to (10.71). To this end, we have by Theorem 3.3 that the operator W is bounded from $\bar{T}_d^{0,\alpha}(\partial G)$ into

$C^{0,\alpha}(\bar{G})$ and hence $W : T_d^{0,\alpha}(\partial G) \to C(\bar{G})$ is compact by Theorem 3.2. Similarly, using Theorem 8.1, we see that $T_{e,\lambda} : C(\bar{G}) \to T^{0,\alpha}(\partial G)$ is bounded.

Now choose a real wave number $\tilde{k}$ which is not a Maxwell eigenvalue for G and denote the operators corresponding to M and N by $\tilde{M}$ and $\tilde{N}$. Then, since $\lambda < 0$, we have from Theorem 9.13 that

$$\tilde{N}R + i\lambda R(I + \tilde{M}) : T_d^{0,\alpha}(\partial G) \to T^{0,\alpha}(\partial G)$$

has a bounded inverse

$$B := (\tilde{N}R + i\lambda RI + i\lambda R\tilde{M})^{-1} : T^{0,\alpha}(\partial G) \to T_d^{0,\alpha}(\partial G).$$

Therefore, by setting $b := B^{-1}a$ we can equivalently transform the system (10.71) into the form

$$\begin{pmatrix} I & T_{e,\lambda}B \\ 0 & I \end{pmatrix} \begin{pmatrix} b \\ E \end{pmatrix} - \begin{pmatrix} \tilde{B} & 0 \\ W & T_e \end{pmatrix} \begin{pmatrix} b \\ E \end{pmatrix} = \begin{pmatrix} 2c \\ 0 \end{pmatrix},$$

where $\tilde{B} := (\tilde{N} - N)RB + i\lambda R(M + \tilde{M})B$. In this system, the first matrix operator has a bounded inverse and the second is compact from $T^{0,\alpha}(\partial G) \times C(\bar{G})$ into itself since the operator $(N - \tilde{N})R : T_d^{0,\alpha}(\partial G) \to T^{0,\alpha}(\partial G)$ is compact by Theorem 2.23 of [32]. Hence, the Riesz-Fredholm theory can be applied.

Suppose a and E is a solution of the homogeneous form of (10.71). Then E and $H := \operatorname{curl} E/ik$ solve the homogeneous interior impedance problem and hence by Theorem 10.12 we have that $E = 0$ in D. This implies from (10.70) that the field $\tilde{E}$ defined by

$$\tilde{E} := \operatorname{curl} \int_{\partial D} \Phi(x,y)a(y)\,ds(y), \quad x \in \mathbb{R}^3 \setminus \partial D,$$

vanishes in D. By the jump relations of Theorem 6.11 we see that the exterior field $\operatorname{curl}\tilde{E}$ in $\mathbb{R}^3 \setminus \bar{G}$ satifies $\nu \times \operatorname{curl}\tilde{E} = 0$ on ∂D whence $\operatorname{curl}\tilde{E} = 0$ in $\mathbb{R}^3 \setminus \bar{D}$ follows by Theorem 6.18. This implies $\tilde{E} = 0$ in $\mathbb{R}^3 \setminus \bar{D}$ since $\operatorname{curl}\operatorname{curl}\tilde{E} - k^2\tilde{E} = 0$ holds. Hence, by Theorem 6.11 we can conclude that $a = 0$. The proof of the theorem is now complete. $\qquad\square$

We now turn our attention to the second ingredient that is needed in order to extend the modified dual space method of Section 10.4 to the case of electromagnetic waves, i.e., the generalization of the identity (10.34). To this end, let E_∞ be the electric far field pattern corresponding to the scattering problem (10.47)–(10.49) and E_∞^λ the electric far field pattern for the exterior impedance problem (9.42)–(9.44) with $c = -\nu \times \operatorname{curl} E^i + i\lambda(\nu \times E^i) \times \nu$ and with H^i and E^i given by (9.19). (We note that by the Analytic Fredholm Theorem 8.19 applied to the integral equation (9.46) we have that there exists a solution of the exterior impedance problem not only for $\lambda > 0$ but in fact for all $\lambda \in \mathbb{C}$ with the exception of a countable set of values of λ accumulating only at zero and infinity.) Our aim is to find a vector field $g \in T^2(\Omega)$ such that, given $q \in \mathbb{R}^3$, we have

$$(10.72) \qquad \int_\Omega [E_\infty(\hat{x}; d, p) - E_\infty^\lambda(\hat{x}; d, p)] \cdot g(\hat{x})\, ds(\hat{x}) = \frac{ik}{4\pi}\, p \cdot q \times d$$

for all $d \in \Omega$, $p \in \mathbb{R}^3$. To this end, we have the following theorem.

Theorem 10.14 *Let $q \in \mathbb{R}^3$ and define E_q and H_q by*

$$E_q(x) := \operatorname{curl} q\Phi(x,0), \quad H_q(x) := \frac{1}{ik} \operatorname{curl} \operatorname{curl} q\Phi(x,0), \quad x \in \mathbb{R}^3 \setminus \{0\}.$$

Suppose $\lambda < 0$ is such that there exists a solution to the exterior impedance problem. Then there exists $g \in T^2(\Omega)$ such that the integral equation (10.72) is satisfied for all $d \in \Omega$, $p \in \mathbb{R}^3$, if and only if the solution of the interior impedance problem for $c = \nu \times \operatorname{curl} E_q - i\lambda(\nu \times E_q) \times \nu$ is such that E^i, H^i is an electromagnetic Herglotz pair.

Proof. First assume that there exists $g \in T^2(\Omega)$ such that (10.72) is true. Then, by the reciprocity relations given in Theorems 9.6 and 9.13 we have that

$$p \cdot \int_\Omega [E_\infty(-d; -\hat{x}, g(\hat{x})) - E_\infty^\lambda(-d; -\hat{x}, g(\hat{x}))]\, ds(\hat{x}) = \frac{ik}{4\pi}\, p \cdot q \times d$$

for every $p \in \mathbb{R}^3$ and hence

$$\int_\Omega [E_\infty(-d; -\hat{x}, g(\hat{x})) - E_\infty^\lambda(-d; -\hat{x}, g(\hat{x}))]\, ds(\hat{x}) = \frac{ik}{4\pi}\, q \times d,$$

or

$$(10.73) \qquad \int_\Omega [E_\infty(\hat{x}; -d, h(d)) - E_\infty^\lambda(\hat{x}; d, h(d))]\, ds(d) = \frac{ik}{4\pi}\, \hat{x} \times q$$

where $h(d) = g(-d)$. For this h, we define the electromagnetic Herglotz pair E_0^i, H_0^i by

$$(10.74) \quad H_0^i(x) = \operatorname{curl} \int_\Omega h(d)\, e^{ik\, x \cdot d}\, ds(d), \quad E_0^i(x) = -\frac{1}{ik} \operatorname{curl} H_0^i(x),$$

denote by E_0^s, H_0^s the radiating field of (9.3)–(9.6) with incident field given by (10.74) and let $E_{\lambda 0}^s$ be the radiating field of the exterior impedance problem with E^i replaced by E_0^i in the definition of c above. Then the left hand side of (10.73) represents the electric far field pattern of the scattered field $E_0^s - E_{\lambda 0}^s$. Since E_q, H_q is a radiating solution of the Maxwell equations with electric far field given by the right hand side of (10.73), we can conclude from Theorem 6.9 that

$$(10.75) \qquad E_0^s - E_{\lambda 0}^s = E_q \quad \text{in } \mathbb{R}^3 \setminus \bar{D}.$$

It can now be verified that $E_0 = E_0^s + E_0^i$, $H_0 = H_0^s + H_0^i$, satisfies the interior impedance problem with c as given in the theorem.

Now suppose there exists a solution to the interior impedance problem with c given as in the theorem such that E^i, H^i is an electromagnetic Herglotz pair. Relabel E^i, H^i by E_0^i, H_0^i and let E_0^s, H_0^s be the relabeled scattered fields. Then, defining $E_{\lambda 0}^s$ as above, we see that (10.75) is valid. Retracing our steps, we see that (10.72) is true, and the proof of the theorem is complete.   □

With Theorems 10.12, 10.13 and 10.14 at our disposal, we can now follow Section 10.4 for the case of acoustic waves and formulate an optimization scheme for solving the inverse medium problem for electromagnetic waves. In particular, from the proof of Theorem 10.13, we see that if G is a ball then the solution of the interior impedance problem can be approximated in $C(\bar{G})$ by a continuously differentiable solution such that E^i, H^i is an electromagnetic Herglotz pair (Approximate the surface density by a finite sum of the surface gradients of spherical harmonics and the rotation of these functions by ninety degrees; see Theorem 6.23). This implies that (10.72) can be approximated in $C(\Omega)$. We can now reformulate the inverse medium problem for electromagnetic waves as a problem in constrained optimization in precisely the same manner as in the case of acoustic waves in Section 10.4. By the above remarks, the cost functional of this optimization scheme has infimum equal to zero provided the constraint set is sufficiently large. Since this approach for solving the electromagnetic inverse medium problem is completely analogous to the method for solving the acoustic inverse medium problem given in Section 10.4, we shall omit giving further details.

10.6 Numerical Examples

We shall now proceed to give some simple numerical examples of the methods discussed in this chapter for solving the acoustic inverse medium problem. We shall refer to the dual space method discussed in Section 10.3 as Method A and the modified dual space method of Section 10.4 as Method B. For the sake of simplicity we shall restrict ourself to the case of a spherically stratified medium, i.e., it is known a priori that $m(x) = m(r)$ and $m(r) = 0$ for $r > a$. Numerical examples for the case of non-spherically stratified media in $\mathbb{R}^2$ can be found in [44].

We first consider the direct problem. When $m(x) = m(r)$, the total field u in (10.1)–(10.3) can be written as

$$(10.76) \qquad u(x) = \sum_{p=0}^{\infty} u_p(r) P_p(\cos \theta),$$

where P_p is Legendre's polynomial, $r = |x|$ and

$$\cos \theta = \frac{x \cdot d}{r}.$$

Without loss of generality we can assume that $d = (0,0,1)$. Using (10.76) in (10.5), it is seen that u_p satisfies

(10.77) $$u_p(r) = i^p(2p+1)j_p(kr) - ik^3 \int_0^a K_p(r,\rho)m(\rho)u_p(\rho)\rho^2\, d\rho$$

where

$$K_p(r,\rho) := \begin{cases} j_p(kr)\, h_p^{(1)}(k\rho), & \rho > r, \\[2mm] j_p(k\rho)\, h_p^{(1)}(kr), & \rho \leq r, \end{cases}$$

and j_p and $h_p^{(1)}$ are, respectively, the p–th order spherical Bessel and first kind Hankel functions. Then, from Theorem 2.15 and (10.77), we obtain

(10.78) $$u_\infty(\hat{x}; d) = \sum_{p=0}^{\infty} f_p(k)\, P_p(\cos\theta)$$

where

(10.79) $$f_p(k) = (-i)^{p+2}k^2 \int_0^a j_p(k\rho)m(\rho)u_p(\rho)\rho^2\, d\rho.$$

The Fourier coefficients $f_p(k)$ of the far field pattern can thus be found by solving (10.77) and using (10.79).

Turning now to the inverse problem, we discuss Method A first and, in the case that m is real, assume that k is not a transmission eigenvalue. From (10.15) and (10.78) we have

(10.80) $$g_{pq} = i^{p-1}\frac{(2p+1)Y_p^q}{4\pi k f_p(k)}.$$

Under the above assumptions, we have that $f_p(k) \neq 0$. As to be expected, g_{pq} depends on q in a way that is independent of u_∞ and hence independent of m. Thus, we only consider the case $q = 0$. Having computed g_{p0}, we can compute v from (10.18) and (10.80) as

(10.81) $$v_p(x) = -\frac{i(2p+1)j_p(kr)}{k f_p(k)}\sqrt{\frac{2p+1}{4\pi}}\, P_p(\cos\theta).$$

From (10.16) and (10.17) we now have that w is given by

$$w(x) = w_p(r)\sqrt{\frac{2p+1}{4\pi}}\, P_p(\cos\theta)$$

where w_p satisfies

(10.82) $$w_p(r) = -\frac{i(2p+1)j_p(kr)}{k f_p(k)} - ik^3 \int_0^a K_p(r,\rho)m(\rho)w_p(\rho)\rho^2\, d\rho,$$

(10.83) $$w_p(b) + \frac{i(2p+1)j_p(kb)}{k f_p(k)} = h_p^{(1)}(kb)$$

and

$$(10.84) \qquad \frac{\partial}{\partial r}\left(w_p(r) + \frac{i(2p+1)j_p(kr)}{kf_p(k)}\right)\Bigg|_{r=b} = \left(\frac{\partial}{\partial r}h_p^{(1)}(kr)\right)\Bigg|_{r=b}$$

where $b > a$. Note that (10.83) and (10.84) each imply the other since, if we replace b by r, from (10.82) we see that both sides of (10.83) are radiating solutions to the Helmholtz equation and hence are equal for $r \geq a$. To summarize, Method A consists of finding m and w_p for $p = 0, 1, 2, \ldots$ such that (10.82) and either (10.83) or (10.84) hold for each p and for k in some interval.

A similar derivation to that given above can be carried out for Method B and we only summarize the results here. Let

$$\gamma_p(k) := \frac{i(2p+1)}{k} \frac{\dfrac{\partial}{\partial r}j_p(kr) + i\lambda j_p(kr)}{\dfrac{\partial}{\partial r}h_p^{(1)}(kr) + i\lambda h_p^{(1)}(kr)}\Bigg|_{r=b}, \qquad p = 0, 1, 2, \ldots,$$

i.e., the $\gamma_p(k)$ are the Fourier coefficients of h_∞ where h_∞ is the far field pattern associated with (10.29)–(10.32). Then w_p satisfies the integral equation

$$(10.85) \qquad w_p(r) = -\frac{i(2p+1)j_p(kr)}{k[f_p(k) - \gamma_p(k)]} - ik^3 \int_0^a K_p(r, \rho)m(\rho)w_p(\rho)\rho^2\, d\rho$$

and the impedance condition (10.39) becomes

$$(10.86) \qquad \left(\frac{\partial}{\partial r} + i\lambda\right)\left(w_p(r) - h_p^{(1)}(kr)\right)\Bigg|_{r=b} = 0.$$

Thus, for Method B, we must find m and w_p for $p = 0, 1, 2, \ldots$, such that (10.85) and (10.86) are satisfied for an interval of k values.

To construct our synthetic far field data, we use a N_f point trapezoidal Nyström method to approximate u_p as a solution of (10.77). Then we compute the far field pattern for

$$k = k_j = k_{\min} + (k_{\max} - k_{\min})\frac{j-1}{N_k - 1}, \qquad j = 1, 2, \ldots, N_k,$$

by using the trapezoidal rule with N_f points to discretize (10.79). Thus the data for the inverse solver is an approximation to $f_p(k_j)$ for $p = 0, 1, \ldots, P$ and $j = 1, 2, \ldots, N_k$.

We now present some numerical results for Methods A and B applied to the inverse problem using the synthetic far field data obtained above. To discretize m we use a cubic spline basis with N_m equally spaced knots in $[0, a]$ under the constraints that $m(a) = m'(0) = 0$. This implies that the expansion for m has N_m free parameters that must be computed via an appropriate inverse algorithm. To implement Methods A and B, we approximate, respectively, (10.82) and (10.85) using the trapezoidal Nyström method with N_i equally spaced quadrature points on $[0, a]$. The approximate solution to (10.82) or (10.85) can be computed away from the Nyström points by using the N_i

point trapezoidal rule to approximate the integral in (10.82) or (10.85). Let $w_p^a(r, \tilde{m}, k)$ represent the Nyström solution of either (10.82) or (10.85) with m replaced by an arbitrary function $\tilde{m}$.

For Method A, we choose to work with (10.82) and (10.84). For this case the inverse algorithm consists of finding $m^\star$ such that the sum of the squares of

$$(10.87) \quad F_{p,j}^A(\tilde{m}) := \frac{\partial}{\partial r}\left(w_p^a(r, \tilde{m}, k_j) + i\,\frac{(2p+1)j_p(k_j r)}{k_j f_p(k_j)} - h_p^{(1)}(k_j r)\right)\Bigg|_{r=b}$$

is as small as possible when $\tilde{m} = m^\star$. Similarly, from (10.86), Method B consists of finding $m^\star$ such that the sum of the squares of

$$(10.88) \quad F_{p,j}^B(\tilde{m}) := \left(\frac{\partial}{\partial r} + i\lambda\right)\left(w_p^a(r, \tilde{m}, k_j) - h_p^{(1)}(k_j r)\right)\Bigg|_{r=b}$$

is as small as possible when $\tilde{m} = m^\star$. Since the inverse problem is ill-posed, we use a Tikhonov regularization technique to minimize (10.87) or (10.88). Let

$$(10.89) \quad J_\alpha(\tilde{m}) := \frac{1}{N_k(P+1)} \sum_{p=0}^{P}\sum_{j=1}^{N_k} |F_{p,j}(\tilde{m})|^2 + \alpha^2\|\tilde{m}'\|_{L^2(0,a)}^2$$

where $F_{p,j}$ is given by either (10.87) or (10.88) and $\alpha > 0$ is a regularization parameter. The approximate solution $m^\star$ for either Method A or B is obtained by minimizing $J_\alpha(\tilde{m})$ over the spline space for $\tilde{m}$ with $F_{p,j}(\tilde{m})$ replaced by (10.87) for Method A or (10.88) for Method B. Since $\tilde{m}$ is given by a cubic spline, the minimization of (10.89) is a finite dimensional optimization problem and we use a Levenberg–Marquardt method to implement the optimization scheme.

Before presenting some numerical examples, some comments are in order regarding the design of numerical tests for our inverse algorithms. Care must be taken that interactions between the inverse and forward solvers do not result in excessively optimistic predictions regarding the stability or accuracy of an inverse algorithm. For example, if (10.4) and (10.5) are discretized and used to generate far field data for a given profile, and the same discretization is used to solve the inverse problem, it is possible that the essential ill-posedness of the inverse problem may not be evident. This is avoided in our example, since different discretizations are used in the forward and the inverse algorithms. A similar problem can occur if the subspace containing the discrete coefficient $\tilde{m}$ contains or conforms with the exact solution. This problem is particularly acute if a piecewise constant approximation is used to approximate a discontinuous coefficient. If the grid lines for the discrete coefficient correspond to actual discontinuities in the exact solution, the inverse solver may again show spurious accuracy. In particular, one can not choose the mesh consistent with the coefficient to be reconstructed. In our examples, we use a cubic spline basis for approximating m, and choose coefficients m that are not contained in the cubic spline space.

We now discuss a few numerical examples. In what follows, we write the refractive index as in (8.7), i.e.,

$$n(x) = n_1(r) + i\,\frac{n_2(r)}{k}\,,$$

and identify n_2 as the absorption coefficient of the inhomogeneous medium.

Example 1. We take

(10.90)
$$n_1(r) = 1 + \frac{1}{2}\,\cos\frac{9}{2}\,\pi r,$$
$$0 \le r \le 1 = a.$$
$$n_2(r) = \frac{1}{2}\,(1-r)^2(1+2r),$$

The absorption n_2 is a cubic spline, but n_1 is not. We choose $k_{\min} = 1$, $k_{\max} = 7$, $N_f = 513$, $N_i = 150$, $N_m = 15$ and $\lambda = 0$. The solution is computed for the regularization parameter $\alpha = 0.001$ by first computing m^* for $\alpha = 0.1$, taking as initial guess for the coefficients of m the value -0.3. The solution for $\alpha = 0.1$ is used as an initial guess for $\alpha = 0.01$ and this solution is used as initial guess for $\alpha = 0.001$. We report only results for $\alpha = 0.001$. In Table 10.1 we report the relative L^2 error in the reconstructions defined by

$$\left[\frac{\|\operatorname{Re}(n^* - n_e)\|_{L^2(0,a)}^2 + k^2\|\operatorname{Im}(n^* - n_e)\|_{L^2(0,a)}^2}{\|\operatorname{Re}(n_e)\|_{L^2(0,a)}^2 + k^2\|\operatorname{Im}(n_e)\|_{L^2(0,a)}^2}\right]^{1/2}$$

expressed as a percentage. Here, n_e is the exact solution (in this case given by (10.90) and (10.91)) and n^* is the approximate reconstruction. Table 10.1 shows that both methods work comparably well on this problem except in the case $P = 1$, when Method A is markedly superior to Method B.

Table 10.1. Numerical reconstructions for (10.90)

P	Method A $N_k = 9$	Method A $N_k = 15$	Method B $N_k = 9$	Method B $N_k = 15$
0	26.50	26.20	27.2	27.1
1	13.20	13.30	41.0	40.3
2	9.59	9.66	12.0	11.7
3	9.35	8.66	11.7	11.8

Example 2. Our next example is a discontinuous coefficient given by

(10.91)
$$n_1(r) = \begin{cases} 3.5, & 0 \le r < 0.5, \\ 1, & 0.5 \le r \le 1 = a, \end{cases}$$

$$n_2(r) = 0.3\,(1 + \cos 3\pi r), \quad 0 \le r \le 1 = a.$$

In this case, neither n_1 nor n_2 are cubic spline functions. Table 10.2 shows results for reconstructing this coefficient (the parameters are the same as in Example 1). When both methods yield satisfactory results, they possess similar errors. However, Method A is somewhat less robust than Method B and fails to work in two cases. This failure may be due to insufficient absorption to stabilize Method A. The effect of absorption on the reconstruction is investigated further in the next example.

Table 10.2. Numerical reconstructions for (10.91)

P	Method A $N_k = 9$	Method A $N_k = 15$	Method B $N_k = 9$	Method B $N_k = 15$
0	11.2	10.50	11.10	10.80
1	10.0	9.85	10.50	11.20
2	111.0	9.51	10.10	9.88
3	131.0	9.38	9.67	9.49

Example 3. In this example we allow the maximum value of the absorption to be variable.

$$n_1(r) = 1 + \frac{1}{2} \cos \frac{5}{2} \pi r,$$

(10.92)
$$0 \leq r \leq 1 = a.$$

$$n_2(r) = \frac{\gamma}{2}(1 - r)^2(1 + 2r),$$

To reconstruct (10.92), we take $N_f = 129$, $N_k = 15$, $N_i = 50$ and $N_m = 15$. Independent of γ, we take $\lambda = k$, since we want to vary only a single parameter in our numerical experiments.

Table 10.3. Reconstruction of (10.92) as γ varies

γ	Method A $P = 0$	$P = 1$	$P = 2$	$P = 3$	Method B $P = 0$	$P = 1$	$P = 2$	$P = 3$
0.0	107.00	87.40	53.50	52.00	5.89	2.44	2.59	2.26
0.01	109.00	88.60	2.72	25.90	–	–	–	–
0.025	105.00	88.90	3.88	2.53	–	–	–	–
0.05	213.00	96.50	2.73	2.56	–	–	–	–
0.1	124.00	165.00	2.76	2.62	5.87	2.44	2.59	2.26
0.15	6.24	320.00	2.78	2.68	–	–	–	–
0.2	6.24	2.60	2.80	2.74	5.81	2.44	2.58	2.26
0.3	6.24	2.65	2.83	2.83	5.73	2.44	2.58	2.24
0.4	6.21	2.70	2.86	2.92	5.62	2.43	2.57	2.25
0.5	6.15	2.76	2.90	3.00	5.50	2.43	2.56	2.25

First we investigate changing the amount of absorption in the problem for $k_{max} = 7$, $k_{min} = 1$. Table 10.3 shows that results of varying γ between 0 and 0.5. As to be expected, Method B is insensitive to γ (although if γ is made large enough, we would expect that the quality of reconstruction would deteriorate). Method A does not work for $\gamma = 0$. It is believed that this failure is due to transmission eigenvalues in $[k_{min}, k_{max}]$. The value of γ below which Method A is unstable depends on P. For example, when $P = 0$ Method A works satisfactorily when $\gamma = 0.15$ but not when $\gamma = 0.1$, whereas when $P = 3$ the method works satisfactorily when $\gamma = 0.025$ but not when $\gamma = 0.01$. These results are the principle reason for preferring Method B over Method A in the case of low absorption.

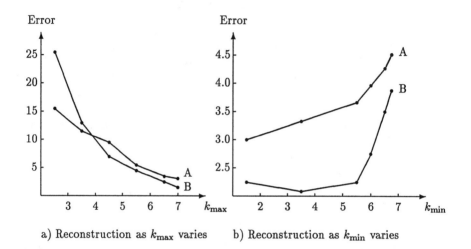

a) Reconstruction as k_{max} varies   b) Reconstruction as k_{min} varies

Fig. 10.1. A graph of the percentage relative L^2 error in reconstructing (10.92)

Finally, we investigate the dependence on k_{max} and k_{min} of the reconstruction error. Figure 10.1a shows the relative L^2 error in the reconstruction of (10.92) when $\gamma = 0.5$, $k_{min} = 1$ and k_{max} varies. Clearly, in this example, increasing k_{max} greatly improves the reconstruction. Figure 10.1b shows results in the reconstruction of (10.92) with $\gamma = 0.5$, $k_{max} = 7$ and k_{min} variable. For this example, it is clear that lower wave numbers contribute little information to the reconstruction. However, in the presence of large absorption, the lower wave number may well be important.

References

[1] Adams, R.A.: *Sobolev Spaces.* Academic Press, New York 1975.

[2] Angell, T.S., Colton, D., and Kirsch, A.: The three dimensional inverse scattering problem for acoustic waves. J. Diff. Equations **46**, 46–58 (1982).

[3] Angell, T.S., Colton, D., and Kress, R.: Far field patterns and inverse scattering problems for imperfectly conducting obstacles. Math. Proc. Camb. Soc. **106**, 553–569 (1989).

[4] Angell, T.S., and Kirsch, A.: The conductive boundary condition for Maxwell's equations. SIAM J. Appl. Math. (to appear).

[5] Angell, T.S., Kleinman, R.E., and Hettlich, F.: The resistive and conductive problems for the exterior Helmholtz equation. SIAM J. Appl. Math. **6**, 1607–1622 (1990).

[6] Angell, T.S., Kleinman, R.E., Kok, B., and Roach, G.F.: A constructive method for identification of an impenetrable scatterer. Wave Motion **11**, 185–200 (1989).

[7] Angell, T.S., Kleinman, R.E., Kok, B., and Roach, G.F.: Target reconstruction from scattered far field data. Ann. des Télécommunications **44**, 456–463 (1989).

[8] Angell, T.S., Kleinman, R.E., and Roach, G.F.: An inverse transmission problem for the Helmholtz equation. Inverse Problems **3**, 149–180 (1987).

[9] Ari, N., and Firth, J.R.: Acoustic inverse scattering problems of polygonal shape reconstruction. Inverse Problems **6**, 299–310 (1990).

[10] Atkinson, K.E.: The numerical solution of Laplace's equation in three dimensions. SIAM J. Numer. Anal. **19**, 263–274 (1982).

[11] Baumeister, J.: *Stable Solution of Inverse Problems.* Vieweg, Braunschweig 1986.

[12] Bers, L., John, F., and Schechter, M.: *Partial Differential Equations.* John Wiley, New York 1964.

[13] Bleistein, N.: *Mathematical Methods for Wave Phenomena.* Academic Press, Orlando 1984.

[14] Blöhbaum, J.: Optimisation methods for an inverse problem with time-harmonic electromagnetic waves: an inverse problem in electromagnetic scattering. Inverse Problems **5**, 463–482 (1989).

[15] Bojarski, N.N.: Three dimensional electromagnetic short pulse inverse scattering. Spec. Proj. Lab. Rep. Syracuse Univ. Res. Corp., Syracuse 1967.

[16] Bojarski, N.N.: A survey of the physical optics inverse scattering identity. IEEE Trans. Ant. Prop. **AP-20**, 980–989 (1982).

[17] Brakhage, H., and Werner, P.: Über das Dirichletsche Aussenraumproblem für die Helmholtzsche Schwingungsgleichung. Arch. Math. **16**, 325–329 (1965).

[18] Brebbia, C.A., Telles, J.C.F., and Wrobel, L.C.: *Boundary Element Techniques.* Springer-Verlag, Berlin Heidelberg New York 1984.

[19] Calderón, A.P.: The multipole expansions of radiation fields. J. Rat. Mech. Anal. **3**, 523–537 (1954).

[20] Calderón, A.P.: On an inverse boundary value problem. In: *Seminar on Numerical Analysis and its Applications to Continuum Mechanics.* Soc. Brasileira de Matemática, Rio de Janerio, 65–73 (1980).

[21] Chadan, K., and Sabatier, P. C.: *Inverse Problems in Quantum Scattering Theory.* Springer-Verlag, Berlin Heidelberg New York 1989.

[22] Chew, W: *Waves and Fields in Inhomogeneous Media.* Van Nostrand Reinhold, New York 1990.

[23] Colonius, F., and Kunisch, K.: Stability for parameter estimation in two point boundary value problems. J. Reine Angew. Math. **370**, 1–29 (1986).

[24] Colonius, F., and Kunisch, K.: Output least squares stability in elliptic systems. Appl. Math. Opt. **19**, 33–63 (1989).

[25] Colton, D.: *Partial Differential Equations.* Random House, New York 1988.

[26] Colton, D., and Kirsch, A.: Dense sets and far field patterns in acoustic wave propagation. SIAM J. Math. Anal. **15**, 996–1006 (1984).

[27] Colton, D., and Kirsch, A.: Karp's theorem in acoustic scattering theory. Proc. Amer. Math. Soc. **103**, 783–788 (1988).

[28] Colton, D., and Kirsch, A.: An approximation problem in inverse scattering theory. Applicable Analysis **41**, 23–32 (1991).

[29] Colton, D., and Kirsch, A.: The use of polarization effects in electromagnetic inverse scattering problems. Math. Meth. in the Appl. Sci. **15**, 1–10 (1992).

[30] Colton, D., Kirsch, A., and Päivärinta, L.: Far field patterns for acoustic waves in an inhomogeneous medium. SIAM J. Math. Anal. **20**, 1472–1483 (1989).

[31] Colton, D., and Kress, R.: The impedance boundary value problem for the time harmonic Maxwell equations. Math. Meth. in the Appl. Sci. **3**, 475–487 (1981).

[32] Colton, D., and Kress, R.: *Integral Equation Methods in Scattering Theory.* Wiley-Interscience Publication, New York 1983.

[33] Colton, D., and Kress, R.: Dense sets and far field patterns in electromagnetic wave propagation. SIAM J. Math. Anal. **16**, 1049–1060 (1985).

[34] Colton, D., and Kress, R.: Karp's theorem in electromagnetic scattering theory. Proc. Amer. Math. Soc. **104**, 764–769 (1988).

[35] Colton, D., and Kress, R.: Time harmonic electromagnetic waves in an inhomogeneous medium. Proc. Royal Soc. Edinburgh **116 A**, 279–293 (1990).

[36] Colton, D., and Monk, P.: A novel method for solving the inverse scattering problem for time-harmonic acoustic waves in the resonance region. SIAM J. Appl. Math. **45**, 1039–1053 (1985).

[37] Colton, D., and Monk, P.: A novel method for solving the inverse scattering problem for time-harmonic acoustic waves in the resonance region II. SIAM J. Appl. Math. **46**, 506–523 (1986).

[38] Colton, D., and Monk, P.: The numerical solution of the three dimensional inverse scattering problem for time-harmonic acoustic waves. SIAM J. Sci. Stat. Comp. **8**, 278–291 (1987).

[39] Colton, D., and Monk, P: The inverse scattering problem for time harmonic acoustic waves in a penetrable medium. Quart. J. Mech. Appl. Math. **40**, 189–212 (1987).

[40] Colton, D., and Monk, P: The inverse scattering problem for acoustic waves in an inhomogeneous medium. Quart. J. Mech. Appl. Math. **41**, 97–125 (1988).

[41] Colton, D., and Monk, P: A new method for solving the inverse scattering problem for acoustic waves in an inhomogeneous medium. Inverse Problems **5**, 1013–1026 (1989).

[42] Colton, D., and Monk, P: A new method for solving the inverse scattering problem for acoustic waves in an inhomogeneous medium II. Inverse Problems **6**, 935–947 (1990).

[43] Colton, D., and Monk, P: A comparison of two methods for solving the inverse scattering problem for acoustic waves in an inhomogeneous medium. J. Comp. Appl. Math. (to appear).

[44] Colton, D., and Monk, P: The numerical solution of an inverse scattering problem for acoustic waves. IMA J. Appl. Math. (to appear).

[45] Colton, D., and Päivärinta, L.: Far field patterns and the inverse scattering problem for electromagnetic waves in an inhomogeneous medium. Math. Proc. Camb. Phil. Soc. **103**, 561–575 (1988).

[46] Colton, D., and L. Päivärinta, L.: Far-field patterns for electromagnetic waves in an inhomogeneous medium. SIAM J. Math. Anal. **21**, 1537–1549 (1990).

[47] Colton, D. and Päivärinta, L.: The uniqueness of a solution to an inverse scattering problem for electromagnetic waves. Arch. Rational Mech. Anal. (to appear).

[48] Colton, D., and Sleeman, B.D.: Uniqueness theorems for the inverse problem of acoustic scattering. IMA J. Appl. Math. **31**, 253–259 (1983).

[49] Davis, P.J.: *Interpolation and Approximation*. Blaisdell Publishing Company, Waltham 1963.

[50] Davis, P.J., and Rabinowitz, P.: *Methods of Numerical Integration*. Academic Press, New York 1975.

[51] Devaney, A.J.: Acoustic tomography. In: *Inverse Problems of Acoustic and Elastic Waves* (Santosa et al, eds). SIAM, Philadelphia, 250–273 (1984).

[52] Dolph, C. L.: The integral equation method in scattering theory. In: *Problems in Analysis* (Gunning, ed). Princeton University Press, Princeton, 201–227 (1970).

[53] Engl, H.W., Kunisch, K., and Neubauer, A.: Tikhonov regularisation of nonlinear ill-posed problems. Inverse Problems **5**, 523–540 (1989).

[54] Erdélyi, A.: *Asymptotic Expansions*. Dover Publications, New York 1956.

[55] Gieseke, B.: Zum Dirichletschen Prinzip für selbstadjungierte elliptische Differentialoperatoren. Math. Z. **68**, 54–62 (1964).

[56] Gilbarg, D., and Trudinger, N.S.: *Elliptic Partial Differential Equations of Second Order*. Springer-Verlag, Berlin Heidelberg New York 1977.

[57] Gilbert, R.P., and Xu, Yongzhi: Dense sets and the projection theorem for acoustic harmonic waves in a homogeneous finite depth ocean. Math. Meth. in the Appl. Sci. **12**, 69–76 (1990).

[58] Gilbert, R.P., and Xu, Yongzhi: The propagation problem and far field patterns in a stratified finite depth ocean. Math. Meth. in the Appl. Sci. **12**, 199–208 (1990).

[59] Goldstein, C.I.: The finite element method with non-uniform mesh sizes applied to the exterior Helmholtz problem. Numer. Math. **38**, 61–82 (1981).

[60] Grisvard, P.: *Elliptic Problems in Nonsmooth Domains*. Pitman, Boston 1985.

[61] Groetsch, C.W.: *The Theory of Tikhonov Regularization for Fredholm Equations of the First Kind*. Pitman, Boston 1984.

[62] Hackbusch, W.: *Multi-grid Methods and Applications.* Springer-Verlag, Berlin Heidelberg New York 1985.

[63] Hadamard, J.: *Lectures on Cauchy's Problem in Linear Partial Differential Equations.* Yale University Press, New Haven 1923.

[64] Hähner, P.: Abbildungseigenschaften der Randwertoperatoren bei Randwertaufgaben für die Maxwellschen Gleichungen und die vektorielle Helmholtzgleichung in Hölder- und L^2-Räumen mit einer Anwendung auf vollständige Flächenfeldsysteme. Diplomarbeit, Göttingen 1987.

[65] Hähner, P.: An exterior boundary-value problem for the Maxwell equations with boundary data in a Sobolev space. Proc. Roy. Soc. Edinburgh **109A**, 213–224 (1988).

[66] Hähner, P.: Eindeutigkeits- und Regularitätssätze für Randwertprobleme bei der skalaren und vektoriellen Helmholtzgleichung. Dissertation, Göttingen 1990.

[67] Hähner, P.: A uniqueness theorem for the Maxwell equations with L^2 Dirichlet boundary conditions. Meth. Verf. Math. Phys. **37**, 85–96 (1991).

[68] Hartman, P., and Wilcox, C.: On solutions of the Helmholtz equation in exterior domains. Math. Z. **75**, 228–255 (1961).

[69] Hettlich, F.: Die Integralgleichungsmethode bei Streuung an Körpern mit einer dünnen Schicht. Diplomarbeit, Göttingen 1989.

[70] Hsiao, G.C.: The coupling of boundary element and finite element methods. Z. Angew. Math. Mech. **70**, T493–T503 (1990).

[71] Imbriale, W.A., and Mittra, R.: The two-dimensional inverse scattering problem. IEEE Trans. Ant. Prop. **AP-18**, 633–642 (1970).

[72] Isakov, V.: On uniqueness in the inverse transmission scattering problem. Comm. Part. Diff. Equa. **15**, 1565–1587 (1990).

[73] Isakov, V.: Stability estimates for obstacles in inverse scattering. J. Comp. Appl. Math. (to appear).

[74] Ivanov, K.V.: Integral equations of the first kind and an approximate solution for the inverse problem of potential. Soviet Math. Doklady **3**, 210–212 (1962) (English translation).

[75] Ivanov, K.V.: On linear problems which are not well-posed. Soviet Math. Doklady **3**, 981–983 (1962) (English translation).

[76] Jörgens, K.: *Lineare Integraloperatoren.* Teubner–Verlag, Stuttgart 1970.

[77] Johnson, S.A. and Tracy, M.L.: Inverse scattering solutions by a sinc basis, multiple source, moment method - Part I: theory. Ultrasonic Imaging **5**, 361–375 (1983).

[78] Jones, D.S.: *Methods in Electromagnetic Wave Propagation.* Clarendon Press, Oxford 1979.

[79] Jones, D.S.: *Acoustic and Electromagnetic Waves.* Clarendon Press, Oxford 1986.

[80] Jones, D.S. and Mao, X.Q.: The inverse problem in hard acoustic scattering. Inverse Problems **5**, 731–748 (1989).

[81] Jones, D.S. and Mao, X.Q.: A method for solving the inverse problem in soft acoustic scattering. IMA Jour. Appl. Math. **44**, 127–143 (1990).

[82] Karp, S.N.: Far field amplitudes and inverse diffraction theory. In: *Electromagnetic Waves* (Langer, ed). Univ. of Wisconsin Press, Madison, 291-300 (1962).

[83] Kedzierawski, A.: Inverse Scattering Problems for Acoustic Waves in an Inhomogeneous Medium. Ph.D. Thesis, University of Delaware 1990.

[84] Kellogg, O.D.: *Foundations of Potential Theory*. Springer-Verlag, Berlin Heidelberg New York 1929.

[85] Kersten, H.: Grenz- und Sprungrelationen für Potentiale mit quadratsummierbarer Dichte. Resultate d. Math. **3**, 17–24 (1980).

[86] Kirsch, A.: Generalized Boundary Value and Control Problems for the Helmholtz Equation. Habilitationsschrift, Göttingen 1984.

[87] Kirsch, A.: The denseness of the far field patterns for the transmission problem. IMA J. Appl. Math. **37**, 213–225 (1986).

[88] Kirsch, A.: Properties of far field operators in acoustic scattering. Math. Meth. in the Appl. Sci. **11**, 773–787 (1989).

[89] Kirsch, A.: Surface gradients and continuity properties for some integral operators in classical scattering theory. Math. Meth. in the Appl. Sci. **11**, 789–804 (1989).

[90] Kirsch, A.: Remarks on some notions of weak solutions for the Helmholtz equation. Applicable Analysis (to appear).

[91] Kirsch, A.: The domain derivative and an application in inverse scattering (to appear).

[92] Kirsch, A., and Kress, R.: On an integral equation of the first kind in inverse acoustic scattering. In: *Inverse Problems* (Cannon and Hornung, eds). ISNM **77**, 93–102 (1986).

[93] Kirsch, A., and Kress, R.: A numerical method for an inverse scattering problem. In: *Inverse Problems* (Engl and Groetsch, eds). Academic Press, Orlando, 279–290 (1987).

[94] Kirsch, A., and Kress, R.: An optimization method in inverse acoustic scattering. In: *Boundary elements IX, Vol 3. Fluid Flow and Potential Applications* (Brebbia, Wendland and Kuhn, eds). Springer-Verlag, Berlin Heidelberg New York, 3–18 (1987).

[95] Kirsch, A., and Kress, R.: Uniqueness in inverse obstacle scattering. (to appear).

[96] Kirsch, A., Kress, R., Monk, P., and Zinn, A.: Two methods for solving the inverse acoustic scattering problem. Inverse Problems 4, 749–770 (1988).

[97] Kirsch, A., and Monk, P.: An analysis of the coupling of finite element and Nyström methods in acoustic scattering. (to appear).

[98] Kleinman, R., and van den Berg, P.: A hybrid method for two dimensional problems in tomography. J. Comp. Appl. Math. (to appear).

[99] Knauff, W., and Kress, R.: On the exterior boundary value problem for the time-harmonic Maxwell equations. J. Math. Anal. Appl. **72**, 215–235 (1979).

[100] Kravaris, C., and Seinfeld, J.H.: Identification of parameters in distributed parameter systems by regularization. SIAM J. Control Opt. **23**, 217–241 (1985).

[101] Kress, R.: Ein ableitungsfreies Restglied für die trigonometrische Interpolation periodischer analytischer Funktionen. Numer. Math. **16**, 389–396 (1971).

[102] Kress, R.: On boundary integral methods in stationary electromagnetic reflection. In: *Ordinary and Partial Differential Equations* (Everitt and Sleeman, eds). Springer-Verlag Lecture Notes in Mathematics **846**, Berlin Heidelberg New York, 210–226 (1980).

[103] Kress, R.: Minimizing the condition number of boundary integral operators in acoustic and electromagnetic scattering. Q. Jl. Mech. appl. Math. **38**, 323–341 (1985).

[104] Kress, R.: On the boundary operator in electromagnetic scattering. Proc. Royal Soc. Edinburgh **103A**, 91–98 (1986).

[105] Kress, R.: On the low wave number asymptotics for the two-dimensional exterior Dirichlet problem for the reduced wave equation. Math. Meth. in the Appl. Sci. **9**, 335–341 (1987).

[106] Kress, R.: *Linear Integral Equations.* Springer-Verlag, Berlin Heidelberg New York 1989.

[107] Kress, R.: A Nyström method for boundary integral equations in domains with corners. Numer. Math. **58**, 145–161 (1990).

[108] Kress, R.: Boundary integral equations in time-harmonic acoustic scattering. Mathl. Comput. Modelling **15**, 229–243 (1991).

[109] Kress, R., and Zinn, A.: Three dimensional reconstructions in inverse obstacle scattering. In: *Mathematical Methods in Tomography* (Hermans, Louis and Natterer, eds). Springer-Verlag Lecture Notes in Mathematics **1497**, Berlin Heidelberg New York, 125–138 (1991).

[110] Kress, R., and Zinn, A.: Three dimensional reconstructions from near-field data in obstacle scattering. In: *Inverse Problems in Engineering Sciences* (Yamaguti et al, ed). ICM-90 Satellite Conference Proceedings, Springer-Verlag, Tokyo Berlin Heidelberg, 43–51 (1991).

[111] Kress, R., and Zinn, A.: On the numerical solution of the three dimensional inverse obstacle scattering problem. J. Comp. Appl. Math. (to appear).

[112] Kristensson, G., and Vogel, C.R.: Inverse problems for acoustic waves using the penalised likelihood method. Inverse Problems **2**, 461–479 (1986).

[113] Kussmaul, R.: Ein numerisches Verfahren zur Lösung des Neumannschen Aussenraumproblems für die Helmholtzsche Schwingungsgleichung. Computing **4**, 246–273 (1969).

[114] Langenberg, K.J.: Applied inverse problems for acoustic, electromagnetic and elastic wave scattering. In: *Basic Methods of Tomography and Inverse Problems* (Sabatier, ed). Adam Hilger, Bristol and Philadelphia, 127–467 (1987).

[115] Lax, P.D.: Symmetrizable linear transformations. Comm. Pure Appl. Math. **7**, 633–647 (1954).

[116] Lax, P.D., and Phillips, R.S.: *Scattering Theory.* Academic Press, New York 1967.

[117] Lebedev, N.N.: *Special Functions and Their Applications.* Prentice-Hall, Englewood Cliffs 1965.

[118] Leis, R.: Zur Dirichletschen Randwertaufgabe des Aussenraums der Schwingungsgleichung. Math. Z. **90**, 205–211 (1965)

[119] Leis, R.: *Initial Boundary Value Problems in Mathematical Physics.* John Wiley, New York 1986.

[120] Levine, L.M.: A uniqueness theorem for the reduced wave equation. Comm. Pure Appl. Math. **17**, 147–176 (1964).

[121] Lin, T.C.: The numerical solution of Helmholtz's equation for the exterior Dirichlet problem in three dimensions. SIAM J. Numer. Anal. **22**, 670–686 (1985).

[122] Louis, A.K.: *Inverse und schlecht gestellte Probleme.* Teubner, Stuttgart 1989.

[123] Magnus, W.: Fragen der Eindeutigkeit und des Verhaltens im Unendlichen für Lösungen von $\Delta u + k^2 u = 0$. Abh. Math. Sem. Hamburg **16**, 77–94 (1949).

[124] Maponi, P., Misici, L., and Zirilli, F.: An inverse problem for the three dimensional vector Helmholtz equation for a perfectly conducting obstacle. Computers Math. Applic. **22**, 137–146 (1991).

[125] Martensen, E.: Über eine Methode zum räumlichen Neumannschen Problem mit einer Anwendung für torusartige Berandungen. Acta Math. **109**, 75–135 (1963).

[126] Martensen, E.: *Potentialtheorie*. Teubner-Verlag, Stuttgart 1968.

[127] Mautz, J.R., and Harrington, R.F.: A combinded-source solution for radiating and scattering from a perfectly conducting body. IEEE Trans. Ant. and Prop. **AP-27**, 445–454 (1979).

[128] McLaughlin, J., and Polyakov, P.: On the uniqueness of a spherically symmetric speed of sound from transmission eigenvalues. J. Diff. Equations (to appear).

[129] Mikhlin, S.G.: *Mathematical Physics, an Advanced Course*. North-Holland, Amsterdam 1970.

[130] Misici, L. and Zirilli, F.: An inverse problem for the three dimensional Helmholtz equation with Neumann or mixed boundary conditions. In: *Mathematical and Numerical Aspects of Wave Propagation Phenomena* (Cohen et al, eds). SIAM, Philadelphia, 497–508 (1991).

[131] Moré, J.J.: The Levenberg–Marquardt algorithm, implementatiion and theory. In: *Numerical analysis* (Watson, ed). Springer-Verlag Lecture Notes in Mathematics **630**, Berlin Heidelberg New York, 105–116 (1977).

[132] Morozov, V.A.: On the solution of functional equations by the method of regularization. Soviet Math. Doklady **7**, 414–417 (1966) (English translation).

[133] Morozov, V.A.: Choice of parameter for the solution of functional equations by the regularization method. Soviet Math. Doklady **8**, 1000–1003 (1967) (English translation).

[134] Morozov, V.A.: *Methods for Solving Incorrectly Posed Problems*. Springer-Verlag, Berlin Heidelberg New York 1984.

[135] Morse, P.M., and Ingard, K.U.: Linear acoustic theory. In: *Encyclopedia of Physics* (Flügge, ed). Springer-Verlag, Berlin Heidelberg New York, 1–128 (1961).

[136] Müller, C.: Zur mathematischen Theorie elektromagnetischer Schwingungen. Abh. deutsch. Akad. Wiss. Berlin **3**, 5–56 (1945/46).

[137] Müller, C.: Über die Beugung elektromagnetischer Schwingungen an endlichen homogenen Körpern. Math. Ann. **123**, 345–378 (1951)

[138] Müller, C.: Über die ganzen Lösungen der Wellengleichung. Math. Annalen **124**, 235–264 (1952).

[139] Müller, C.: Zur Methode der Strahlungskapazität von H. Weyl. Math. Z. **56**, 80–83 (1952).

[140] Müller, C.: Randwertprobleme der Theorie elektromagnetischer Schwingungen. Math. Z. **56**, 261–270 (1952).

[141] Müller, C: On the behavior of solutions of the differential equation $\Delta u = F(x, u)$ in the neighborhood of a point. Comm. Pure Appl. Math. **7**, 505–515 (1954).

[142] Müller, C.: *Foundations of the Mathematical Theory of Electromagnetic Waves*. Springer-Verlag, Berlin Heidelberg New York 1969.

[143] Murch, R.D., Tan, D.G.H., and Wall, D.J.N.: Newton–Kantorovich method applied to two-dimensional inverse scattering for an exterior Helmholtz problem. Inverse Problems 4, 1117–1128 (1988).

[144] Nachman, A.: Reconstructions from boundary measurements. Annals of Math. **128**, 531–576 (1988).

[145] Natterer, F.: *The Mathematics of Computerized Tomography*. Teubner, Stuttgart and Wiley, New York 1986.

[146] Newton, R.G.: *Scattering Theory of Waves and Particles*. Springer-Verlag, Berlin Heidelberg New York 1982.

[147] Newton, R.G.: *Inverse Schrödinger Scattering in Three Dimensions*. Springer-Verlag, Berlin Heidelberg New York 1989.

[148] Novikov, R.: Multidimensional inverse spectral problems for the equation $- \Delta \psi + (v(x) - E u(x)) \psi = 0$. Translations in Func. Anal. and its Appl. **22**, 263–272 (1988).

[149] Ochs, R.L.: The limited aperture problem of inverse acoustic scattering: Dirichlet boundary conditions. SIAM J. Appl. Math. **47**, 1320–1341 (1987).

[150] Olver, F.W.J: *Asymptotics and Special Functions*. Academic Press, New York 1974.

[151] Onishi, K., Ohura, Y., and Kobayashi, K.: Inverse scattering in 2D for shape identification by BEM. In: *Boundary elements XII, Vol 2. Applications in Fluid Mechanics and Field Problems* (Tanaka, Brebbia and Honma, eds). Springer-Verlag, Berlin Heidelberg New York, 435–446 (1990).

[152] Panich, O.I.: On the question of the solvability of the exterior boundary-value problems for the wave equation and Maxwell's equations. Usp. Mat. Nauk **20A**, 221–226 (1965) (in Russian).

[153] Pironneau, O.: *Optimal Shape Design for Elliptic Systems*. Springer-Verlag, Berlin Heidelberg New York 1984.

[154] Protter, M.H.: Unique continuation for elliptic equations. Trans. Amer. Math. Soc. **95**, 81–90, (1960).

[155] Protter, M.H., and Weinberger, H.F.: *Maximum Principles in Differential Equations*. Prentice-Hall, Englewood Cliffs 1967.

[156] Ramm, A.G.: *Scattering by Obstacles*. D. Reidel Publishing Company, Dordrecht 1986.

[157] Ramm, A.G.: On completeness of the products of harmonic functions. Proc. Amer. Math. Soc. **98**, 253–256 (1986).

[158] Ramm, A.G.: Recovery of the potential from fixed energy scattering data. Inverse Problems 4, 877–886 (1988).

[159] Ramm, A.G.: Symmetry properties of scattering amplitudes and applications to inverse problems. J. Math. Anal. Appl. **156**, 333–340 (1991).

[160] Reed, M., and Simon, B.: *Scattering Theory*. Academic Press, New York 1979.

[161] Rellich, F.: Über das asymptotische Verhalten der Lösungen von $\Delta u + \lambda u = 0$ in unendlichen Gebieten. Jber. Deutsch. Math. Verein. **53**, 57–65 (1943).

[162] Roger, A.: Newton Kantorovich algorithm applied to an electromagnetic inverse problem. IEEE Trans. Ant. Prop. **AP-29**, 232–238 (1981).

[163] Ruland, C.: Ein Verfahren zur Lösung von $(\Delta + k^2)u = 0$ in Aussengebieten mit Ecken. Applicable Analysis **7**, 69–79 (1978).

[164] Rynne, B.P., and Sleeman, B.D.: The interior transmission problem and inverse scattering from inhomogeneous media. SIAM J. Math. Anal. **22**, 1755–1762 (1991).

[165] Schatz, A.H.: An observation concerning Ritz–Galerkin methods with indefinite bilinear forms. Math. Comp. **28**, 959–962 (1974).

[166] Schechter, M.: *Principles of Functional Analysis*. Academic Press, New York 1971.

[167] Seidman, T.I., and Vogel, C.R.: Well-posedness and convergence of some regularization methods for nonlinear ill-posed problems. Inverse Problems **5**, 277–238 (1989).

[168] Silver, S.: *Microwave Antenna Theory and Design*. M.I.T. Radiation Laboratory Series Vol. 12, McGraw-Hill, New York 1949.

[169] Sommerfeld, A.: Die Greensche Funktion der Schwingungsgleichung. Jber. Deutsch. Math. Verein. **21**, 309–353 (1912).

[170] Stratton, J.A., and Chu, L.J.: Diffraction theory of electromagnetic waves. Phys. Rev. **56**, 99–107 (1939).

[171] Tabbara, W., Duchêne, B., Pichot, Ch., Lesselier, D., Chommeloux, L., and Joachimowicz, N.: Diffraction tomography: contribution to the analysis of some applications in microwaves and ultrasonics. Inverse Problems **4**, 305–331 (1988).

[172] Tikhonov, A.N.: On the solution of incorrectly formulated problems and the regularization method. Soviet Math. Doklady **4**, 1035–1038 (1963) (English translation).

[173] Tikhonov, A.N.: Regularization of incorrectly posed problems. Soviet Math. Doklady **4**, 1624–1627 (1963) (English translation).

[174] Tikhonov, A.N., and Arsenin, V.Y.: *Solutions of Ill-posed Problems*. Winston and Sons, Washington 1977.

[175] Tobocman, W.: Inverse acoustic wave scattering in two dimensions from impenetrable targets. Inverse Problems **5**, 1131–1144 (1989).

[176] Tracy, M.L., and Johnson, S.A.: Inverse scattering solutions by a sinc basis, multiple source, moment method - part II: numerical evaluations. Ultrasonic Imaging **5**, 376–392 (1983).

[177] Treves, F.: *Basic Linear Partial Differential Equations*. Academic Press, New York 1975.

[178] van Bladel, J.: *Electromagnetic Fields*. Hemisphere Publishing Company, Washington 1985.

[179] Vekua, I.N.: Metaharmonic functions. Trudy Tbilisskogo matematichesgo Instituta **12**, 105–174 (1943).

[180] Wang, S.L., and Chen, Y.M.: An efficient numerical method for exterior and interior inverse problems of Helmholtz equation. Wave Motion **13**, 387–399 (1991).

[181] Wang, Y.M., and Chew, W.C.: An iterative solution of two dimensional electromagnetic inverse scattering problems. Inter. Jour. Imaging Systems and Technology **1**, 100–108 (1989).

[182] Weck, N.: Klassische Lösungen sind auch schwache Lösungen. Arch. Math. **20**, 628–637 (1969).

[183] Werner, P.: Zur mathematischen Theorie akustischer Wellenfelder. Arch. Rational Mech. Anal. **6**, 231–260 (1961).

[184] Werner, P.: Randwertprobleme der mathematischen Akustik. Arch. Rational Mech. Anal. **10**, 29–66 (1962).

[185] Werner, P. : On the exterior boundary value problem of perfect reflection for stationary electromagnetic wave fields. J. Math. Anal. Appl. **7**, 348–396 (1963).

[186] Werner, P.: Low frequency asymptotics for the reduced wave equation in two-dimensional exterior spaces. Math. Meth. in the Appl. Sc. **8**, 134–156 (1986).

[187] Weston, V.H.: Multifrequency inverse problem for the reduced wave equation with sparse data. J. Math. Physics **25**, 1382–1390 (1984).

[188] Weston, V.H.: Multifrequency inverse problem for the reduced wave equation: resolution cell and stability. J. Math. Physics **25**, 3483–3488 (1984).

[189] Weston, V.H., and Boerner, W.M.: An inverse scattering technique for electromagnetic bistatic scattering. Canadian J. Physics **47**, 1177–1184 (1969).

[190] Weyl, H.: Kapazität von Strahlungsfeldern. Math. Z. **55**, 187–198 (1952).

[191] Weyl, H.: Die natürlichen Randwertaufgaben im Aussenraum für Strahlungsfelder beliebiger Dimensionen und beliebigen Ranges. Math. Z. **56**, 105–119 (1952).

[192] Wienert, L.: Die numerische Approximation von Randintegraloperatoren für die Helmholtzgleichung im $\mathbb{R}^3$. Dissertation, Göttingen 1990.

[193] Wilcox, C.H.: A generalization of theorems of Rellich and Atkinson. Proc. Amer. Math. Soc. **7**, 271–276 (1956).

[194] Wilcox, C.H.: An expansion theorem for electromagnetic fields. Comm. Pure Appl. Math. **9**, 115–134 (1956).

[195] Wilcox, C.H.: *Scattering Theory for the d'Alembert Equation in Exterior Domains*. Springer-Verlag Lecture Notes in Mathematics **442**, Berlin Heidelberg New York, 1975.

[196] Wloka, J.: *Partial Differential Equations*. University Press, Cambridge 1987.

[197] Zinn, A.: On an optimisation method for the full- and limited-aperture problem in inverse acoustic scattering for a sound-soft obstacle. Inverse Problems **5**, 239–253 (1989).

[198] Zinn, A.: Ein Rekonstruktionsverfahren für ein inverses Streuproblem bei der zeitharmonischen Wellengleichung. Dissertation, Göttingen 1990.

[199] Zinn, A.: The numerical solution of an inverse scattering problem for time-harmonic acoustic waves. In: *Inverse Problems and Imaging* (Roach, ed). Pitman Research Notes in Mathematics Series **245**. Longman, London, 242–263 (1991).

Index

Applied Mathematical Sciences

cont. from page ii